NOTES OF A POTATO WATCHER

NUMBER FOUR
Texas A&M University Agriculture Series
C. Allan Jones, General Editor

Notes of a Potato Watcher

JAMES LANG

Foreword by Hubert G. Zandstra

TEXAS A&M UNIVERSITY PRESS
COLLEGE STATION

Library of Congress Cataloging-in-Publication Data

Lang, James, 1944–
      Notes of a potato watcher / James Lang ; Foreword by
   Hubert G. Zandstra.—1st ed.
            p. cm. — (Texas A&M University agriculture series ; no. 4)
         Includes bibliographical references and index (p.   ).
         ISBN 1-58544-138-4 (cloth : alk. paper)—
         ISBN 1-58544-154-6 (paper : alk. paper)
         1. Potatoes.   2. Potatoes—History. ·
      I. Title.   II. Series.
      SB211.P8 L36   2001
      635'.21'09—dc21   2001000830

To Cecilia,

"She walks in beauty like the night."

—Lord Byron

CONTENTS

ILLUSTRATIONS

TABLES

FOREWORD

The potato is a crop that inspires enthusiasm in many. In Jim Lang, it has a devotee par excellence. The chapters that follow are a fascinating travelogue through the world of potatoes, seen through the lens of a keen observer of agriculture in its broadest cultural significance.

The potato is more efficient, more nutritious, and more profitable than any other staple crop. It is unparalleled as a producer of food, jobs, and cash, and is ideally suited to places where land is limited and labor is abundant—conditions that characterize much of the developing world. This argument is backed up by numbers. A plant's "harvest index"—the ratio of the weight of usable food to the weight of the entire plant—provides a good gauge of a crop's efficiency. Potatoes typically have a harvest index of 75–85 percent. That means that less than one-fourth of the plant material produced by sunlight, water, nutrients, labor, and other inputs is wasted.

Compared with other crops, that is an astounding figure. In simple terms, it translates into a huge potential for solving some of the most pressing problems of farmers and consumers in the developing world.

Over the next two decades, the world population is expected to grow by an average of more than 100 million a year. More than 95 percent of the increase will occur in developing countries, where pressure on land, water, and other resources is already intense. At the International Potato Center, we believe that the potato will make a significant contribution to dealing with these challenges efficiently and equitably. It is high in energy and rich in important nutrients. It commands a competitive price nearly everywhere it is sold. With existing technology alone— such as eliminating diseased seed or planting improved varieties—farmers could increase potato yields by proportions that would be unthinkable in grains. In China, for example, researchers estimate that a 30 percent jump is within reach. The impact of that would be staggering. And science continues to explore new and innovative ways of unleashing the full potential of this crop. Since the early 1960s, the land area planted to potatoes in developing countries has grown faster than that planted to any other major food crop. During the first two decades of this century, production of potatoes in the world's poorest countries is expected to continue increasing at a higher rate than production of rice, wheat, or corn. An unprecedented explosion in demand will accompany this increase in production. By 2020, the developing world's appetite for potatoes will be more than double what it was in 1993.

Over the past decade, Jim Lang has tracked CIP's global activities in potato research. In *Notes of a Potato Watcher,* he has joined his own wealth of knowledge on themes ranging from art to archaeology with the information he has gained in the course of these travels. We thank him for his vote of confidence in CIP, and for putting down on paper a story that is rarely told—the story of the commitment of numerous men and women who, on a daily basis, use their scientific know-how to construct better lives. By realizing the potato's potential, they are helping to satisfy the most basic of human requirements—nutritious food, better health, secure livelihoods, peace, and a sound environment—for the people most in need.

—Hubert G. Zandstra
Director General
International Potato Center
Lima, Peru

ACKNOWLEDGMENTS

I began my career with *Conquest and Commerce,* an analysis of Spanish and English colonization in the New World. Later, as a fellow at the Woodrow Wilson Center for Scholars, I finished *Portuguese Brazil: The King's Plantation,* which studies the impact the colonial sugar trade had on Brazil's early development. There has been many a detour since. Eric R. Wolf, Charles Tilly, and Charles Gibson encouraged me to follow my own path, come what may. I have taken their advice.

As a Kellogg Fellow (1981–84), I abandoned colonial sugar exports for community health and rural development projects. During my tour of duty, I visited thirty-four projects in six countries: Costa Rica (where I visited six projects), Brazil (sixteen), the Dominican Republic (two), Colombia (seven), Paraguay (one), and Argentina (two). Of these, fourteen were rural development projects and twenty were health projects. When I wrote my report, *Inside Development,* I drew on a sample I knew firsthand. At the Kellogg Foundation, I owe a special debt to Mario Chavez, who kept me on the right track.

The first agrarian project I covered took me to the Colombian Andes for a rendezvous with the potato. The agronomist at the site was César Villamizar, who kept a tattered handbook on potato diseases in his back pocket. That handbook was my first contact with the International Potato Center (CIP); prior to that, I did not realize such research centers existed. Subsequently, with support from the Rockefeller Foundation (1988–91), I visited six such centers: CIP in Peru, CIAT (the International Center for Tropical Agriculture) in Colombia, CIMMYT (the International Maize and Wheat Improvement Center) in Mexico, IRRI (the International Rice Research Institute) in the Philippines, ICRISAT (the International Crops Research Institute for the Semi-Arid Tropics) in India, and AVRDC (the Asian Vegetable Research and Development Center) in Taiwan.

Originally, I had promised the Rockefeller Foundation a book "that helps people see development in a new way, a book about solutions." I am grateful to Robert W. Herdt for his confidence in this project. I am also grateful to William A. Christian, who wrote to the Rockefeller Foundation on my behalf, assuring them that I could make spittle bugs interesting. The result was *Feeding a Hungry Planet,* a book about rice and research in Asia and Latin America. I considered it a good book, but it was not the book I had promised the Rockefeller Foundation. *Notes of a Potato Watcher* is.

I almost gave up on writing this book. The rice book was in press, and I had a semester of leave ahead. My plan was to go back to the high road of historical analysis. Before I did that, however, I wrote to Ed Sulzberger at CIP. I told him that if CIP helped organize the fieldwork, I would write a book featuring its mandated research crops, the potato and the sweet potato. Ed wrote back and suggested I start in Indonesia. Thank you, Ed. I did not look back.

Between 1995 and 1998, I visited CIP offices in Asia, Latin America, and Africa. After Ed Sulzberger left CIP, Steve Kearl took up my cause. Using my fieldwork as a basis, I wrote articles for CIP's annual reports, including short versions of what would later become chapters 19 ("Natural Enemies"), 21 ("Mother Plants"), 24 ("True Potato Seed"), 27 ("Food Security"), and 29 ("Feeding China"). I am grateful to CIP for permission to draw on its published illustrations, drawings, and photographs. John Stares helped me get the necessary approvals, had slides sent to me, and kept track of my requests. A special thanks to Peter Schmiediche, André Devaux, Aziz Lagnaoui, Ramzy El-Bedewy, Sarath Ilangantileke, Charles Crissman, and Peter Ewell, who helped plan my itinerary at research sites. CIP's research staff, national program scientists, local extension workers, and farmers shared their expertise about production problems, crops, and solutions. The originality is theirs. The mistakes are mine.

CIP did not impose its views on my work; it certainly is not responsible for my conclusions. CIP paid me for the articles I wrote for its annual reports. Between 1996 and 1998, their support came to $7,500. Most of the funds for research, travel, and fieldwork came out of my own pocket. CIP did not pay for this book.

For over twenty years, Vanderbilt's Center for Latin American and Iberian Studies has been a mainstay of my research and teaching; from 1996 to 2000, it was my privilege to serve as the center's director. When I was a young assistant professor, the center sponsored a research trip to Brazil that included intensive training in Portuguese. Since then, small travel grants from Vanderbilt's Research Council have contributed greatly to my work. I owe a special debt to Russell Hamilton, Dean of the Graduate School and advocate of research abroad.

I am grateful to Vanderbilt's College of Arts and Science for its many years of support. A special thanks to Deans Jacque Voegeli, Madeleine Goodman, and Ettore Infante.

In the Department of Sociology, Gary Jensen approved the leave that got my fieldwork off to a good start, and Dan Cornfield approved the leave that brought it to a conclusion.

The potato's true home is in the Andes Mountains of South America. The potato's story cannot be told apart from its many travels, trials, and tribulations. So as fate would have it, *Notes of a Potato Watcher* brought me back to the historical path I started out on years before.

NOTES OF A POTATO WATCHER

Introduction

"You're writing about potatoes?" The tone of voice suggested I had some explaining to do. Modern academia is without imagination, not to mention a sense of humor. Why should I do the explaining? I rather think the shoe is on the other foot.

History used to be the story of famous people, great battles, palace intrigue, and tragic crimes of passion. Plutarch's exemplary *Lives of the Noble Grecians and Romans* featured such highbrow subject matter.[1] Now we are not so sure what history's subject should be. Fernand Braudel's *Civilization and Capitalism* dotes on diets, fads, and fashion.[2] When everyday life is in the driver's seat, it's the kings who hitch the ride. The discovery of America featured great navigators and conquistadors, noble savages and Puritan divines. But in the end, how important is one conquistador more or less, or one sermon more or less? Not very. That cannot be said of the crops that took root in the Americas. Sugar and tobacco created plantation agriculture, slavery, and a landed gentry. The corn and potatoes were left to small farmers. Change the crops, and history is changed.

The potato ended up on the lowbrow side of the crop ledger. That's because of the kind of people who ate it. The Irish were thought to be lazy and irresponsible, and they ate potatoes. That made the spud undesirable company. But that's fine with me. One can learn a lot in unexpected places.

Notes of a Potato Watcher owes a great debt to *Dirt*, by William Bryant Logan.[3] *Dirt* is divided into eight parts and forty-four chapters, the longest of which is ten pages. Under this rubric, Logan takes on an enormous range of topics, from star dust and dung beetles to Virgil's *Georgics* and John Adams's manure piles. Even before I started writing *Notes*, I decided that I wanted the same kind of flexibility. That meant short chapters. For the title, I must thank Lewis Thomas and his *The*

Lives of a Cell, which is subtitled *Notes of a Biology Watcher.*[4] While watching biology, Thomas also took the time to notice music, language, and culture. In calling my book *Notes of a Potato Watcher,* I hope to notice other things as well.

Notes begins at the beginning: with art and agriculture, with domestication. We owe much more to ancient people than we admit; their understanding of the stars, of time, and of cycles is an achievement on par with agriculture. The potato was domesticated by Andean farmers, probably in the Lake Titicaca basin. But the potato was not alone. The ancient people of the Americas domesticated more crops than in all of Asia; they were the greatest farmers ever. They began farming almost as early as people in the Near East did. Archaeologists now think South America was settled many thousands of years prior to the last ice age.

The potato, then, marks but one episode in a larger story about Andean agriculture and the many crops that originated in the New World. These crops included four species of beans, plus an assortment of tomatoes, peppers, corn, and squashes, to name but a few. Andean markets sell an array of tuber and root crops not found in most U.S. grocery stores. The potato is one of the few Andean crops that traveled internationally; most stayed behind, as the so-called lost crops of the Incas. In small plots near their potato fields, Andean farmers still plant oca, ulluco, and mashua. At lower altitudes, where the weather is warmer and wetter, come the sweet potatoes, ahipa, arracacha, achira, and yacón. The diversity of Andean crops underscores the region's rich agricultural heritage. Nonetheless, such crops play but minor parts in the story told here. In *Notes,* the potato is the star, and the sweet potato plays a supporting role.

While the meeting of the Old World with the New greatly enhanced the world's food supply, it also led to an exchange of pathogens that decimated the Aztec and Inca Empires. Hundreds of years later, the New World had its revenge. The potato famine laid waste to Ireland much as smallpox had debilitated Andean cultures.

To many a critic, the Irish famine shows what happens when population growth outstrips the food supply. The famine, however, struck with equal ferocity in the Netherlands, France, and Scotland. Nonetheless, impoverished folk in those countries did not suffer the fate of Ireland's cottager class. In the end, Ireland's plight owed as much to the country's social structure as it did to the potato. Thomas Malthus worried that the hare of population growth would outdistance the tortoise of food production. This concern led to the notion that adding to subsistence was itself the culprit. More food, in other words, meant more people. If that were the case, then helping populous countries grow more potatoes is the last thing an international organization like CIP ought to do. *Notes* sorts out the demographics. The shift from large to small families started in Europe and has now taken hold in Asia and Latin America. The drop is not a consequence of food shortages.

Quite the contrary. The best-fed and best-educated countries have the lowest fertility rates.

Notes presents the facts about the potato. It is nutritious and has no fat. I hope readers will never shun the spud again to lose weight. After a long decline in the United States, the microwave oven and the frozen french fry have given the potato a new lease on life. Even so, the potato's redoubt today is in developing countries. China is the world's largest potato producer; India is not far behind. To tell the potato's story, I did fieldwork in South America, Asia, and Africa. Stories sometimes overlap, but they are never identical. When it comes to freezing a french fry, to see one factory is to see them all. Not so with a potato patch. The production problems faced, the technology applied, the varieties planted, and the nature of the seed used can be so different that separate stories must be told. For Bolivia and Ecuador, the story is about protected beds, weevils, and sprout multiplication; in Tunisia and Egypt, it's about the tuber moth; Uganda features mother plants and the "flush out" system; in India, it's about true potato seed (TPS). For the sweet potato, the stories focus on food security in Uganda, Indonesia's farmer field schools, and China's crop utilization.

The Flemish master, Pieter Bruegel, found his inspiration in hay fields, in people skating on a winter's day or reveling in a village at Mardi Gras. He depicted everyday life with almost scientific realism.[5] Hidden within Bruegel's mundane paintings, however, great events unfold: the nativity, the conversion of Saint Paul, the fall of Icarus. The more I looked at the potato of farm and field, the more I saw a deeper truth. The way projects set priorities, the cooperation that CIP's work required, and the way ordinary farmers reaped the benefits—all of these hold lessons beyond the potato.

These lessons reach back to the village world, where agriculture began thousands of years ago. Village life teaches the value of small things. In West Bengal's Midnapore district, most farm families have less than half a hectare of land. The village is reality. Revered at every shrine and temple, the goddess Kali holds that world together. Of course, we all know that villagers do not see the big picture, that they do not know what the modern world is. The reality sponsored by a village may be noble, we think, but it is inferior; it falls short technologically. Villages are like endangered species. We would like to save them, but they are losers.

Such a view is fundamentally flawed. Potato projects teach a simple lesson: how to address basic problems with practical solutions. That seems like a small truth for modern times. The greatest truths, however, can spring forth from the tiniest of seeds. Village culture recognizes that solutions must be tailored to fit the circumstances. Whether the problem is seed production, controlling crop pests, genetic improvement, or storage, the key is to take the diversity imposed by place, by farming traditions, and by ecology as a starting point.

To solve the problems that matter most, we must learn how to see things as villagers would. Whether it is producing enough food, educating our children, providing employment, or switching to new energy sources, the place to start is closer to home than we think. The big picture is the result of many small slices of reality. We must learn to value each slice, to see that the whole depends on the part. This is how villagers see the world. Far from being technologically inferior, villagers see reality more like a computer does: every bit and byte makes a difference.

PART I
Beginnings

CHAPTER 1

Art and Agriculture

Ancient people thirty thousand years ago etched and painted complex lines, glyphs, and majestic animals on rocks and inside caves. They did so prodigiously, they did so everywhere, and they did so over many thousands of years.[1] Did such people live together peacefully or did they kill each other? Did they have matriarchal or patriarchal societies? Did they eat mostly meat, or were they vegetarians? We do not know. That they liked to draw and paint is one of the few things we know for sure. In 1994, exploring the gorges along the Ardèche River in France, investigators discovered a magnificent new cave site now called Chauvet. Its brilliant, bold sketches and paintings rival those of Lascaux. And it is much older. Lascaux is dated to about seventeen thousand years ago; carbon-14 dating suggests that the artwork at Chauvet is at least thirty thousand years old.[2]

Chauvet Cave supposedly has the oldest known paintings in the world. That does not make them the first paintings. The distinction is important. Chauvet's artists already had a mastery of palettes and pigment, of technique, of how to incorporate the rough surfaces of the cave into the painting itself. They knew how to extract, mix, and apply iron-oxide reds, the natural yellow of ocher, and the deep black-blues of manganese dioxide. They knew how to find the kind of caves they wanted and how to keep them lit while they worked. Chauvet Cave embodies technical knowledge remembered and passed on across many generations. How many? At sites in France, archaeologists have discovered caches of natural pigment dating back at least fifty thousand years. Long before people bothered painting rocks, they probably painted each other. Experts on prehistoric art typically posit a period of protoart prior to the great cave paintings. So if Chauvet Cave is thirty thousand years old, then less refined, more tentative, less durable artistic expression preceded the achievement.[3]

Chauvet Cave is the creation of modern humans, that is, of *Homo sapiens sapiens*. Modern humans have been around for at least one hundred thousand years.[4] What does it mean to be a modern human? In what sense can such ancient people be considered as modern as we are? Go back thirty thousand years to Ardèche and kidnap a newborn Stone Age baby. If that child were raised by an American middle-class family, what would happen? Chances are the kid would play video games, surf the internet, and wear T-shirts. After high school, proms, and sports, the kid might end up an anthropology major. That is because, as members of the same species, we share the same genome. They are us. We are them.[5]

The people who lived thirty thousand years ago were at least as smart as we are, maybe smarter. What are the greatest technological achievements of all time? Writing, the combustion engine, and the computer quickly come to mind. Hardly anyone would think of agriculture. Yet, the knowledge contained therein is cumulative, extensive, and revolutionary. Agriculture is the great achievement of ancient people; it may be the greatest achievement of humankind ever. Prehistoric farmers worked out the principles of husbandry for virtually every crop produced today. They acquired this knowledge in bits and spurts, they remembered it, and they passed it on accurately across generations. They did this without artificial forms of intelligence. Today, that knowledge is the common heritage of humankind; it is far older than we generally suppose it to be.

Stone Age people were hunters and gatherers. Because they did not make pottery does not mean they lacked comparable artifacts. Because ancient people did not have agriculture does not mean they knew nothing about plants. Remnants of bowls and food-processing tools predate Chauvet Cave.[6]

Nonagricultural societies can accumulate botanical knowledge of great sophistication.[7] Australia's aborigines did not domesticate a single plant. Yet they did all the things that agricultural societies do; "they altered vegetation, sowed seeds, planted tubers, and protected plants." They knew the habits, cycles, and uses, both harmful and beneficial, of hundreds, if not thousands, of plants. They practiced "delayed-return strategies." When they dug up yams *(Dioscorea)*, for example, they replaced part of the root to insure a future crop. They diverted streams to irrigate forestland during the dry season and burned over the land to clear the way for desirable plants.[8] Like Australia's aborigines, ancient food collectors modified ecosystems to their advantage. Such deliberate manipulation can push back the proto-origins of agriculture by one hundred thousand years.[9]

Most gathering societies, including those of ancient times, had an expert knowledge of narcotics and medicines, of poisons, dyes, and paints. They knew the barks from which drugs could be extracted and which tree species and reeds made the best household utensils, traps, and ceremonial objects. Early ethnologists recorded that Australia's aborigines gathered and used over four hundred species of plants belonging to over 250 genera.[10] Overall, ancient gathering societies collected the progenitors of virtually every crop subsequently domesticated, from wild rices and grains to roots and tubers, melons, legumes, oil seeds, fruits, and nuts. The line between collection and domestication, between gathering and agriculture, is easily crossed. Perhaps it was in the primitive garden that the metamorphosis took place.[11]

How big is the step from replanting a piece of a yam root in the ground from whence it came to taking that yam to a different place, to a protected spot? It is a small step, probably taken many thousands of times. So figuring out precisely how agriculture originated is virtually impossible to do. One thing, however, is certain. The view that agriculture is something discovered or invented is flawed. We do not imagine that anyone discovered or invented cave painting. It must have been a collective cultural enterprise of long standing. So too with gathering. How much knowledge is implied in collecting the seeds, fruits, and leaves from hundreds of plants, finding out which can be eaten fresh, which should be cooked, which can be dried, and which must be detoxified? What makes for a poisonous plant or for a hallucinogenic one? What is each plant's life cycle? Where is each plant found, what conditions favor it, and which collective interventions best help preserve it?

Such knowledge is built up over time; it must be learned and applied, or it will be lost. Such knowledge is experimental; it is based on testing and replication. Knowledge of the botanical world is essentially scientific; assumptions can be verified, tests are run, and results are observed and evaluated. Modern humans are scientific by nature and have been so for many thousands of years. The evidence for this is everywhere: in tools, in the technology that cave painting requires, in the kind of knowledge protoagriculture implies.[12]

The difference between protoagriculture and full-fledged agriculture is domestication, that is, bringing a plant home. In the lowland tropics, that home place was most likely a garden. Perhaps gardens were started to nurture sacred plants; maybe they were started for convenience or for food security. Different things happened in different places for different reasons. Forms of protoagriculture probably coexisted with gardening, hunting, and fishing in different proportions for millennia before true dependence on agriculture set in.[13]

Domestication means that people select the seed they want to plant; this is a clear mark of agriculture as opposed to gathering. With domestication, the degree of genetic engineering—the role people play in the lives of plants—goes up dramatically. When gatherers burn over a stand of forest, divert water to a field, or scatter the seeds of a favorite plant, they intervene in the course of nature, favoring some wild plants over others. More intrusive still is when gardener-collectors select seed only from plants with traits they consider desirable. The plant species so favored no longer have to compete for survival in the wild; they can depend on farmers to help them survive. They become domesticated. Such selection can engender rapid genetic change.

Consider corn. Compared to its ancient progenitors, domesticated corn is almost a different plant. Ancient farmers selected seed from cobs with tightly packed kernels, which makes for more grain. Such crowding together, however, is a poor survival strategy in the wild. When such an ear is left on its own to germinate, all the kernels sprout at once; the competition is such that few, if any, survive. Domesticated corn would not have survived without the careful husbandry of ancient farmers.[14]

Which crops were domesticated first, and where? Archaeologists used to take a grain-focused approach to agricultural origins. Today, revisions are in order. Just a few starchy taro roots or yam tubers make a good meal. And they are easy to cook; no complicated processing is required first. By contrast, getting a meal from the grassy *Gramineae* grain family is a lot more work. Whether wild marsh rice or primitive wheats, lots of stalks must be cut. The grains cannot be eaten directly; they must first be softened, perhaps by soaking them in water. The versatility we associate with grain depends on threshing, drying, and milling, not to mention culinary innovations from porridges to breads. The great popularity of the grain family could be a late development—except, perhaps, for its use in alcohol. Mead is probably the first of all fermented drinks, antedating even protoagriculture. Honey left in water produces an alcoholic fermentation. Whole grains left in warm water will also ferment. Such simple technologies could have preceded domestication by thousands of years.[15]

The dynamics of domestication—why ancient people favored certain crops—depend on the kind of technology people had to process and cook what they collected or what they produced selectively in their gardens. Roots and tubers can be roasted over hot coals or baked in ovens of hot stones; no special container is needed. Boiling water, by contrast, requires some kind of receptacle, whether fashioned from skins, hollowed out from charred wood, or made from baked clay. For rendering a hard grain like rice palatable, boiling is a key innovation. Almost all grains, including wheat, started out in porridges. So the steps from protoagriculture to gardens to seed-based agriculture were not just a question of botanical knowledge; sudden leaps and innovations were possible, propelled by advances in food processing, culinary skills, and utensils.[16]

Contrary to the claims of grain-school orthodoxy, roots and tubers were among the first crops cultivated, and the lowland tropics are a cradle of agriculture. Yams, taro root, manioc (cassava), yautia (cocoyam), arrowroot, and sweet potato reproduce vegetatively from vines, stem cuttings, or pieces of root or tuber. This greatly simplifies domestication. Progeny can be selected for their taste, productivity, or size; the offspring turn out the same as the parent plant.[17] In the Americas, yautia, which thrives in soggy soil, was first domesticated in the lowland tropics. So too were sweet potatoes and manioc; only later did their cultivation spread into the highlands.[18] In the Pacific Islands, irrigation and terracing systems are linked to growing taro root, a crop that flourishes in the monsoon rains and probably predates rice.[19] In Africa, ancient people domesticated yams in the tropical forests and savannas prior to grains like millet, sorghum, and glabberima rice.[20]

How old is agriculture? This is where the grain-theory advocates weigh in with the evidence. And they are right. The earliest crop remains are typically grains, found at archaeological sites in dry zones. Hard, dried grains can simply be stored better and longer than roots and tubers. Climatically, the arid Near East has an archaeological selective advantage over the humid tropics. Samples of primitive einkorn wheat date back eleven thousand to twelve thousand years; the region's cultivated barley and nonshattering emmer wheats go back nine thousand to ten thousand years.[21]

Finding crop remains, however, is a bit like finding Chauvet Cave. In neither case are the remains found the first of their kind. Behind Chauvet is an artistic heritage thousands of years old. Behind the primitive wheats and barleys of the Near East stand perhaps a thousand generations of gathering, testing, learning, and dissemination. But what if domesticating grains is a late development and agriculture actually got its start in the tropics with roots and tubers? Then agriculture is older still. Regardless of how and where agriculture started, however, its impact was revolutionary everywhere. Crops, handicrafts, technology, and culture evolved together. The ancient city and the cultivation of cereal crops are intertwined. With the domestication of crops, humankind started down the path to the city. We are about to end that journey decisively and forever. The global reach of the modern city is uprooting the last great village cultures of Asia. Humankind will never be the same.[22]

CHAPTER 2

Cycles

A date is how a culture locates itself relative to an arbitrary starting point. As a place to start counting, the birth of Christ is as valid as any. A civilization's chronology, however, is not immortal. By the Aztec calendar, Cortés invaded Mexico in the year 1-Reed.[1] Who bothers with Aztec dates today? Today's textbooks convert the dates used by Ramses' Egyptians, Caesar's Romans, and Wu Ti's Chinese into conventional Gregorian ones. Such transformations are imaginary constructs, less real than papyrus scrolls or old coins. The year 2000, and the millennial fervor that went with it, was more cultural fiction than fact. The year 2000 will not be 2000 forever.

Time itself is a culturally shaped concept. In the West, we package it into decades, centuries, and millennia. This seems obvious, but it is primarily a reflection of how ingrained our decimal-based counting system is. In ancient Babylonia and China, sixty was the basic unit of the number system. Even today, the Chinese use a sixty-year zodiac cycle.[2] One Mayan calendar, the Tzolkin, pairs cycles of twenty sacred days and thirteen numbers, which yields 260 unique permutations.[3] Regardless of how ancient cultures organized counting, they looked to the heavens to keep track of time. The moon, sun, planets, and constellations journey across the celestial stage in predictable cycles. Their movements can be used to mark off seasonal cycles of hunting and fishing, of planting, harvesting, and resting. The discovery of time—how to count it and locate one's self in its passage—is one of the great achievements of ancient people, on par with hunting and protoagriculture.[4]

Neolithic Stonehenge, whose circular embankments were laid out some five thousand years ago, is considered a great calendar-clock that ticks off the movement of the sun from winter to summer solstice.[5] Were Stonehenge and the sacred knowledge it embodied needed to fix ancient Britain's agricultural calendar? Anyone watching the daily sunrise from a fixed point can see that the sun journeys south in the winter (as the earth tilts north and the southern hemisphere gets direct sunlight) and north in the summer (as the earth tilts south, warming the northern hemisphere). A priestly caste is not a prerequisite for such recognition. The knowledge of cycles, built up through observation, must have preceded recording that knowledge in stone by thousands of years. Astronomically aligned sites are not so much calendars as they are ancient cathedrals, sacred places where people worshipped the cycles they perceived in nature, cycles whose presence and importance daily experience reinforced.[6] Priestly specialists may have organized rituals, composed liturgies, and defined orthodoxies, but it is unlikely farmers needed them to notice it was spring and time to plant.

More ancient than the solar calendar is tallying time's passage by the moon. Our cave-painting ancestors used vertical lines and crescents to notch the lunar sequence on their artifacts. Such engraved bits of bone and stone date back twenty thousand years or more.[7] Ancient people also counted a year by watching constellations rise, journey across the sky, disappear, and then return to start a new cycle.[8] The Inca, for example, marked the start of the agricultural cycle by the return of the constellation Pleiades (the Seven Sisters) to a position in the eastern horizon. Mayan astronomers calculated the cycles of the planet Venus.[9] Whatever celestial tracking device was used, the conclusions reached were similar. The cycle of seasons and the movements observed in the heavens were interrelated. The lives of plants, animals, and man could be marked off and predicted according to the stars.[10]

In *The Story of B*, fiction writer Daniel Quinn argues that human consciousness and the close observation of nature evolved together.[11] By looking deeply into the biological world that surrounds us, we have built up our mental capacities and created ourselves. Our brains evolved through interaction with nature, in tandem with watching its flora and fauna, even its insects.[12] At the same time, however, we looked up into the stars. Finding signs in the vast orb above us is as old as art and much older than agriculture.

In the nineteenth century, Alfred Russel Wallace argued that the human brain was vastly overdesigned for what "primitive" societies required.[13] That prejudice has endured. The truth may be the opposite. Constant observation of and interaction with the complex biota of habitats—the building up and remembering of the knowledge acquired through continuous engagement—may have used far more brain capacity than we use in modern cultures, where so much is done for us. It is in prehistory that everything of importance happened to humankind, when we stole fire from the gods and stoked the flames of intelligence.

Western culture is not well disposed to the lessons of prehistory. This reflects its religiously based predispositions. Old Testament calculations concluded that creation itself could be no more than about 6,000 years old.[14] Linear time and authentic history begin with the birth of Christ. In such a temporal universe, a date of 30,000 years ago is hard to accept. Ancient cultures, by contrast, discerned immense cycles of time. They drew such knowledge from the heavens, in particular, from the precession of equinoctial and solstitial points through the zodiac. Due to the wobbly movement of the earth's axis, the constellations slide forward.[15] Egypt's ancient astrologers recognized this phenomena over 8,000 years ago. Eventually, they worked out the mathematics. They calculated that at the spring equinox, the sun rises in a new segment of the zodiac every 2,160 years, ushering in a new world age. The sun's spring equinoctial point is currently moving from Pisces into Aquarius; hence, we have the dawning of the Aquarian age. Multiplying the twelve zodiacal signs in which the sun rises by 2,160 yields what ancient astronomers considered a "great year," composed of 25,920 years. The compilation of sacred cycles spread throughout the Near East, typically using multiples of 2,160. In ancient Sumer, an eon was calculated to last 432,000 years (2,160 x 50); in Hinduism, the world's oldest religion, a great yuga is a cycle of 4,320,000 years (2,160 x 2000).[16] In Mesoamerica, the Maya constructed their cycles differently, but they still encompassed great spans of time. The Mayan Long Count multiples a 360-day year by twenty to constitute a *katun;* twenty *katuns* equal a *baktun* of 400 years. The passage of thirteen *baktuns* makes up a Great Cycle of 5,200 years (400 x 13). The current Great Cycle, which is calculated to end in the Gregorian year 2012, is itself part of thirteen *piktuns.* Each *piktun* lasts for twenty *baktuns,* or 104,000 years.[17]

A cave painting 32,000 years old or a piece of charred potato 12,000 years old would not have seemed anomalous to ancient people. They imagined vast cycles, golden ages in the past where great things had been accomplished. The eons posited by ancient cultures parallel modern views of the solar system and the earth's geological history. Evidence from the Hubble space telescope puts the age of the universe at 13.5 billion years.[18] Single-cell organisms came a couple of billion years ago, after the earth's surface had cooled and the oceans had formed. Then came the green plants. Vertebrates showed up about 600 million years ago, dinosaurs 200 million years ago, and mammals 100 million years ago, including the tree shrew, alleged ancestor of the primates. "Lucy," the oldest known humanoid fossil, is dated to 3.7 million years ago. Handy man *(Homo habilis),* the first tool user and fire builder, is perhaps 2 million years old; upright man *(Homo erectus)* started out in Africa between 1.7 and 1.5 million years ago and walked as far as China and Java. Smart man *(Homo sapiens sapiens),* so-called because of an enlarged cranial capacity, has been around for at least 100,000 years and maybe for as long as 300,000 years. It all depends on when one presumes our sapiens ancestors started talking, that is, invented language.[19]

Was language worked out in different places by different groups of archaic *Homo sapiens*—like the Neanderthal in Europe or Asian forms of *Homo erectus*—such that they all ended up smart, evolving towards and converging in what is now the *Homo sapiens sapiens* category? Such a view fits the evidence on tools and fire, which show up in disparate locations.[20] The same is true of rock painting, time keeping, and agriculture. Radically different cultures worked them out independently. So why not language?

Regardless of how we learned to talk, the accomplishments of prehistory are extraordinary: tools and fire; language; the knowledge of plants, of cycles, and of seasons; and the reckoning of time. Ancient people worked out the domestication of plants, which led to agriculture. To judge from prehistoric art, aesthetic sensibility, social ritual, and a sense of common purpose are likewise ancient achievements.

In *Spiritwalker,* anthropologist Hank Wesselman imagines a future hundreds of years hence in which civilization has collapsed under ecological stress.[21] The cities are gone, reclaimed by jungle. The climate has changed, and with it, North America's flora and fauna. The old landscape is unrecognizable. We are Stone Age hunters and gatherers once again. On the outside, the way people live is completely different from the rhythm of our times. On the inside, however, these people are very much like us. Perhaps, in a similar way, the great accomplishments of prehistory are also on the inside. In our emotional makeup, we are the same people, today and thirty thousand years ago. The foundation upon which all humanity stands is not just technological but aesthetic, emotional, and spiritual, capacities enriched and strengthened at the dawn of time.

CHAPTER 3

Columbian Exchange

We are on an immense journey. From astronomy to earth history and archaeology, the universe, the planet, our species, and our cultures are much older than previously assumed. The same is true for crops. And it is true for the settlement of the Americas.

The old theory is that big-game hunters from Asia pursued their quarry into North America across the Bering Strait landbridge. Eventually, their descendants made it all the way to Tierra del Fuego at the tip of South America. How long ago did this happen? The textbook answer is ten thousand to twelve thousand years ago. New archaeological evidence suggests that revisions are in order. The earliest evidence comes from the southern hemisphere, precisely where the landbridge theory says it should not be.[1]

At Monte Verde in Chile, a campsite with footprints, foundation holes, animal skins, and plant remains has been radiocarbon dated to about thirteen thousand years ago; tools and other debris are estimated to be some thirty-three thousand years old.[2] At the rock shelter of Pedra Furada in the Piauí region of Brazil, archaeologists uncovered charcoal from hearths, plus tools, pebble flakes, and painted sandstone fragments dated between seventeen thousand and thirty-two thousand years ago.[3] Even where preservation is difficult, as in the humid Amazon basin, weathered cave and rock paintings are reputedly eleven thousand years old.[4] In Patagonia's Cave of the Hands, an ancient people stenciled their prints in enormous, brightly colored clusters; the cave is dated to over nine thousand years ago.[5] A series of Peruvian cave sites likewise demonstrates early human occupation: stone tools, bones, and plant remains are dated between eleven thousand and about twenty-two thousand years ago.[6]

Biological and linguistic data likewise point to a long history of human occupation of the Americas.[7] DNA analysis of American aboriginal groups suggests a first wave of migrants about forty-two thousand years ago. Considering the linguistic evidence, if aboriginal languages diverged from a single mother tongue, then the first migrations occurred fifty thousand to sixty thousand years ago; if successive waves of migrants with different languages is posited, a date of forty thousand years ago is likely.[8]

How did such early migrants, or waves of migrants, get to the American continents? Did they cross over a landbridge between Siberia and Alaska, or did they reach the Americas by sea? During an interglacial age sixty thousand to twenty-five thousand years ago, temperatures were warmer and sea levels lower. This is when people from Southeast Asia crossed the sea to the Australian land mass, which then included Tasmania and New Guinea. Their seafaring descendants colonized the Pacific Islands. Perhaps they got as far as South America. Of course, this is just a guess; rising seas would have hidden much of the evidence. We know for sure that Polynesians made great voyages across the tropical Pacific, sailing against the prevailing winds and ocean currents. They got as far as Easter Island, off the coast of Chile, thousands of miles from Polynesia.[9] Regardless of how the first migrants reached the Americas, they did so much earlier than scholars used to think. Nonetheless, caution is in order. A record of archaic human settlement dating back at least a million years can be found in Africa, the Near East, and Asia, but not, so far, in South America.[10] Even if settlement in South America started forty thousand years ago, that is still comparatively recent. On the other hand, if settlement started just twelve thousand years ago, as older theories propose, that did not put the region behind in its agriculture. The dates for crop remains in South America are comparable to those found in the Old World, including the Near East. The difference is in the crops found.[11]

So far, the first wild einkorn wheats, ryes, and barleys unearthed in the Near East date to about eleven thousand years ago.[12] Domesticated versions of these crops, which produced larger grains that shattered less when harvested, turned up about two thousand years later. So too did plant remains from peas and lentils and evidence for the domestication of wild sheep, goats, pigs, and cattle.[13]

Taken together, these crops and animals constitute a unique, Near-Eastern food-production complex. Over time, its elements migrated and intermixed with pre-existing patterns across Eurasia and into Africa. The result was a mutual exchange that enriched the agriculture of the Old World. Sorghum, various types of millet, cowpeas, okra, and coffee originated in Africa; from Asia came rice, soybeans, and eggplants, plus assorted yams. Bananas are from the Malay Peninsula, while sugarcane is originally from New Guinea and Polynesia. The exchange of such domesticated plants across the vast Eurasian-African complex began thousands of years ago. The ancient Indian subcontinent was a crossroads. It got wheat and barley from the Near East, tea and ginger from Southeast Asia, and sorghum and cow-peas from Africa. From its own local flora came cucumbers, rice, eggplants, and pigeon peas.[14]

In ancient America, an exchange of flora and fauna with the Old World never took place. Long before the time of the Maya, Aztecs, or Incas, the region's neolithic farmers drew forth from their habitat a unique set of domesticated plants, many of which were not duplicated in the Old World. (See table 1.) Along the spine of the Andes, microclimates are staggered up and down the mountainsides. In a day, a traveler can hike down from the temperate valleys up to the altiplano or down further to the rainy subtropics. At the time of the Spanish Conquest, Andean farmers cultivated almost as many different species of plants as in all of Asia.[15] They had been farming for thousands of years.

Archaeological remains from south-central Chile show that wild potatoes were collected in the region thirteen thousand years ago.[16] The oldest remains of a cultivated potato come from Peru and are estimated to be ten thousand years old.[17] The cultivation of beans, peppers, and squash likewise goes back nearly ten thousand years.[18] For domesticated corn, a date of at least seven thousand years ago is well established.[19] Potatoes, squash, corn, and peppers are all uniquely American crops.

China's farmers domesticated the soybean, the Near East is home to the fava bean, and Africa's farmers can claim the cowpea. But for sheer variety, it is hard to surpass the *Phaseolus* genus from which the New World's four bean species (pulses) are derived. *Phaseolus* beans are especially rich in proteins, oils, and carbohydrates. This is fortunate. The diversity of animal protein in Mesoamerica and the Andes was more restricted than in the Old World. Farmers kept turkeys, guinea pigs, rabbits, and other small animals for food, but they produced no dairy products or red meat.[20] Beans were thus an indispensable source of protein. They included scarlet runners; tepary beans; the large, moon-shaped limas (so-called because when first imported into the United States in the 1840s they came from Lima, Peru); and the common bean, *Phaseolus vulgaris*. The "vulgar" type is the most diverse. Green beans, pole beans, snap beans, string beans, and yellow wax beans, all of which are typically cooked fresh, are beans of the common type. So too are the kidney, pinto, navy, butter, red, and black bean types, which are dried before cooking.[21] To this day, beans are a staple throughout the Americas. Venezuela is famous for its bean soups. Mexico has its refried beans. Brazil's national dish, *feijoada,* is made from black beans and pork jowl. In the United States, Tex-Mex chili con carne is partial to the kidney bean, the Old South has its pinto beans, and New England stakes its pride on Boston's baked navy beans.

New World squashes include the pumpkins, the acorns, and the hubbards, not to mention the zucchinis and the yellow squashes, whether crookneck or straight.[22]

Table 1. Useful Plants Domesticated in the Americas, by Category

Roots	Peppers	Nuts and Seeds	Cereals
achira	hot chili	sunflower	corn
arracacha	sweet bell	Brazil nut	wild rice
manioc (cassava)		cacao (chocolate)	kañiwa
maca		pecan	kiwicha
yautia (cocoyam)		walnut	(amaranth)
mauka		vanilla	quinoa
sweet potato			
yacón			
ahipa			

Tubers	Squashes	Berries	Stimulants
oca	crookneck	blackberry	tobacco
potato	pumpkin	blueberry	coca leaf
ulluco	winter	cranberry	maté
mashua	hubbard	gooseberry	peyote
	acorn	raspberry	maguey
	zucchini	strawberry	guaraná

Legumes	Fruits	Fibers
common beans	capuli cherry	cotton[a]
green, pole	tomato	henequin
snap, string	tree tomato	sisal
yellow wax	guanábana	
kidney, pinto	naranjilla (lulo)	
navy, butter	papaya	
red, black	guava	
lima beans	passion fruit	
scarlet runner beans	pineapple	
tepary beans	avocado	
tarwi		
peanut		

Source: Compiled from the appendix in Foster and Cordell, eds., *Chilies to Chocolate*, pp. 163–67; and from Harlan, *Crops and Man*, pp. 76-79.

Note: This list is not complete.

[a] Also domesticated in the Old World.

The transformation of the bitter, unpromising gourd, the common ancestor of all squash, is a great feat, but it cannot compare to the transformation of corn, perhaps the greatest breeding achievement ever. From Argentina to Quebec, corn was the staple grain of America's indigenous cultures. The harder flints and dents originated in Mesoamerica; the soft-grained or "floury" types are from the Andes. Corn started out as teosinte, a wild progenitor with a tiny cob that held perhaps fifteen kernels. By 1520, at the time of the Spanish Conquest of Mexico, "corn" had changed radically; cobs averaged thirteen centimeters in length (about six inches) and had hundreds of tightly packed kernels.[23]

In Mesoamerica, beans, squash, and corn made up the core diet. For the Andean world, potatoes must be added. These four crops alone make for an impressive list, but it is just a start. The ancient farmers of the Americas also domesticated tomatoes, peppers, and peanuts, plus root staples of the wet tropics such as yautia, sweet potatoes, and manioc. The cocoa bean, for chocolate, as well as vanilla are American crops; so too is the pineapple.[24] Such crops are well traveled and respected, but there are still many lesser-known crops left out. Andean farmers domesticated more tuber and root crops than any other people on earth. The tubers included ulluco, oca, and mashua, plus eight species of potato; the roots included arracacha, ahipa, mauka, maca, and yacón. There were grains besides corn too, such as kañiwa (amaranth) and, for high altitudes, quinoa.[25]

America's ancient farmers were good biochemists. Many of their roots and tubers must be detoxified, leached, soaked, cured in the sun, or freeze-dried before they are edible. Bitter varieties of manioc, for example, must be purged of cyanide before cooking, a feat accomplished by grating the roots and squeezing out the juice from the pulp. Indian cultures also synthesized an array of psychoactive drugs. They identified such substances in some 130 plant species, including 10 species of tobacco; by contrast, Old World farmers identified just 20 psychoactive species. The Aztecs understood the chemistry of hundreds of plants that healed wounds and cured diseases. In the Amazon, Indian cultures had extensive knowledge of poisons, which they used to stun fish and game. By comparison, in 1492, "the knowledge Europeans had of efficacious plants was pathetic."[26]

Much of the knowledge of crops, drugs, and medicines accumulated in the Americas is now lost. Even so, it is hard to imagine today's world without American crops—Italy without tomatoes, Ireland without potatoes, or India without chili peppers for its curries and chutneys. Corn now competes with rice and wheat for the title of the world's most important crop. In 1997–99, corn production worldwide averaged 596 million metric tons, compared to 578 million tons for rice and 593 million tons for wheat.[27] Initially, however, most American crops faced a hostile reception, the pepper excepted.

The hot chili pepper was the first American crop to get established abroad, but in Asia, not Europe. Part of the explanation for this is geopolitical. In the sixteenth century, Portugal was the dominant European power in the spice trade with Asia. It also had an American connection via its colonies in Brazil. Moreover, Spain and Portugal were neighbors, both in the Americas and on the Iberian Peninsula; whatever novelty ended up in one empire easily made it to the other. So there are many ways peppers could have made the voyage to Asia: from Mexico to Spain and then on to Portugal and Asia; from Brazil to Lisbon and from there to Asian ports of call; or directly from Brazil. Whatever the route, farmers were growing peppers in India in 1542; from India, Portuguese traders took the crop to Indonesia, China, and Japan. Why peppers instead of, say, the noble potato? Peppers simply make better travelers. In their dried form, they could survive the transatlantic voyage to Portugal and then go on to Asia without spoiling. And they fit well with the spicy eating habits of Asians, who are partial to curries, soy sauces, and black pepper. Even today, Asians consume some 60 percent of the world's seventeen million metric tons of peppers. In Iberia, by contrast, the pepper remained an exotic curiosity too hot to handle. It was a century before the pepper made a successful debut in Europe, through the backdoor. Arab traders brought peppers from India across the Levant and then into the Balkans and Central Europe. Is goulash Hungarian without peppers? Well into modern times, botanists thought peppers had originated in India.[28]

The American food crops that ultimately had the greatest impact in Europe were corn, tomatoes, and potatoes. Corn took hold in part because it was a grain, and grains were familiar crops. But what a grain! The plant itself, its cobs, and its kernels are giants compared to the stalks, rachises, and grains of a wheat plant. Fortunately, corn could be planted, harvested, and processed much like wheat. So despite its novelty, farmers could grow corn without major changes in production technology. In fact, growing corn was much less work. In the Americas, farmers had no draft animals. The llama aside, there were no beasts of burden, no carts or wheels. To plant corn, Andean farmers dug up the ground with a wooden spade.[29] In Europe, farmers could cultivate a plot of corn without the draft animals and plows required for wheat. Corn has another advantage: a few rows yield an impressive quantity of food. By comparison, a farmer cannot do much with just a few rows of wheat. When harvested fresh, corn can be eaten immediately, without threshing or milling. A little grating and straining separates out the corn's milky starch from the kernel's soft shell. The starch can then be cooked into porridges and polentas; for puddings, add sugar.

Corn fares poorly in the semiarid climate that favors wheat. The new crop's earliest success in Europe was probably in the northeast of the Iberian Peninsula, a zone of abundant rainfall. Evidently, this was the case in Portugal, where corn was introduced from Spain in about 1520; it soon transformed the region's agriculture and eventually even rivaled wheat in production.[30] *Broa,* the national bread of Portugal, is made from a mixture of corn and wheat flour. Corn also quickly spread to Italy's Po Valley, in the country's rainy north. To be successful in the drier climates of Spanish Valencia and Andalucia, the crop has to be irrigated. Corn crossed the Mediterranean to the Islamic world, from whence it circled back into Europe from the Ottoman Empire. In 1650, cornmeal mush was reportedly a stable of the Balkan peasantry. From the Balkans it went north as far as the Danube plains.[31] Corn, however, did not take to northern Europe, where summers are too cool and too short for the crop to ripen fully. Even today, farmers in Germany harvest their corn green and then grind it up into silage for feed.[32]

The tomato's fame in Europe is of recent vintage. In the 1890s, English country folk still considered it a novelty. The tomato's problem was that it looked like a fruit but did not taste like one. Working itself into culinary traditions was an up-hill battle. The tomato reached Spain from Mesoamerica; tomatoes were not cultivated in the Andes. In the 1590s, Spanish authors reported that tomatoes were used in Mexican sauces. Even so, it took another century for a tomato-sauce recipe to show up in an Italian cookbook; it was published in Naples, then part of the Spanish Empire. Subsequently, the tomato found a home in Italy, where it mixed in with garlic, basil, and oregano. The tomato's mass-market appeal, however, was a result of canning and the Industrial Revolution. In their canned form, tomatoes do not spoil. As a canned sauce, the tomato conquered Europe's lower classes. The same was true in the United States, where the tomato disguised itself as ketchup. In 1878, the Heinz Company introduced the product, which brought "gustatory brightness and excitement" to the country's "greasy, under-flavored, and overcooked food." With production at ninety million metric tons a year, the multifaceted tomato is now listed among the world's top ten crops. But it took some four hundred years to get there.[33] In 1997–99, Italy and Spain grew over nine million metric tons of tomatoes, half of Europe's total production.[34]

Like its saucy solanaceous cousin, the potato also had acceptance problems. Aboveground, its foliage was easily confused with Europe's nightshade (belladonna), a plant notorious for its narcotic properties. Guilty by association, the first potatoes were rejected as poisonous, leprous, and flatulent. Worse yet, the potato was not mentioned in the Bible.[35] Prejudice aside, the potato's problems were as much biological as ideological. The first potatoes to reach Spain were andigena types *(Solanum tuberosum andigena)* from the Andes, adapted to the short days of equatorial latitudes.[36] Mesoamerican farmers did not domesticate the potato, nor did they produce the crop in 1520. In Europe, andigena types did not set tubers until the fall equinox in September, when shorter days prevailed; hence, the first potatoes were unsuited to northern Europe and its fall frosts, but they did well enough in the south of Spain. The first evidence of potato production comes from records at the Hospital de la Sangre in Seville, which purchased large quantities of them in the 1570s. The potato harvest took place from November into December; stored potatoes were purchased into March. To figure out how best to grow a potato, plus how to cook and eat one, Spanish farmers must have experimented with the crop before it showed up in the hospital records.[37]

The potato's great success story came in Ireland, where it was introduced in the 1580s. A cool but frost-free fall gave the crop enough time to mature. In 1640, it was already a mainstay of Irish tenant farmers, or cottagers (cottiers). A century later, it was indispensable. By then, farmer selection had bred an andigena variant that set tubers earlier.[38] Arthur Young, who toured Ireland in 1776–79, calculated that a cottager family of five went though a barrel weighing 127 kilos in just a week; consumption for an adult was estimated at 5.5 kilos per person per day.[39]

Both the climate and the social structure favored the potato in Ireland. The potato likes a moist, cool climate and a deep, loose soil. Ireland had both. The island's blustery weather checked diseases spread by insect vectors, especially viruses; the bogs, once spaded, were particularly suitable to the crop. Cottagers rented the subsistence plots they tilled; as rents increased, they had to produce as much food as possible on as little land as possible. No crop produced more food per acre, demanded less cultivation, and stored as easily as the potato.[40]

From the British Isles, the potato spread across northern Europe, west to east. In the 1650s, farmers were growing potatoes in the Low Countries; the crop was established in Germany, Prussia, and Poland by 1770 and in Russia by the 1840s.[41] What made the potato so welcome on the mainland was its versatility. The spud could be left in the ground for weeks, to be dug up as needed. Such piecemeal harvesting hid the crop from tax collectors and protected the peasant's food supply in war time. Marauding soldiers laid waste to field crops and raided grain stores. They rarely stopped to dig up an acre of spuds.

The potato found a secure niche across the plains of northern Europe as far as Russia, where rye was the dominant cereal crop. To keep weeds in check, farmers followed two successive rye crops with a fallow. When the weeds came up in the spring, they plowed them under before they set seed. Then they left the field fallow for the rest of the season. A potato crop, however, could substitute for a fallow, as it competed with the kind of weeds that infested rye fields. Potatoes put out a dense canopy; they took different nutrients from the soil and attracted different insects than rye did. Thus, potatoes made an excellent rotation crop with rye, adding to the food supply without reducing the grain harvest. The prolific potato produced four times the calories per acre that rye did.[42] Today, a recipe for German, Swedish, or Russian rye bread typically specifies a hefty glob of mashed potato. Finally, when the starch is squeezed out and fermented, potatoes make a potent alcoholic beverage; vodka is a mainstay from Poland to Russia.[43]

Despite such merits, on Europe's mainland the spud never displaced grains or mixed farming, as it did in Ireland. There, the peculiar intersection of the potato with population growth, land rents, and the late-blight fungus unleashed the Great Famine of 1845–49.[44] Ireland's plight is well known. Less appreciated is how the meeting between the Old World and the New unleashed a demographic disaster of even greater magnitude in the Americas.

In 1500, the Aztec and Inca Empires were densely populated. Tenochtitlán, the Aztec capital, had three hundred thousand inhabitants and covered over twenty-five square kilometers, making it larger than any city in Europe.[45] Conservative estimates put the population of central Mexico at eleven to twelve million; for the Inca Empire, the figure is put at about six million.[46] Such densities are comparable to those found in Europe. In 1500, England's population was about three million, Spain's eight million, and France's fifteen million.[47] The New World's population, however, may have been much larger than the figures above suggest. Contemporaries thought that at the time of its discovery the island of Hispaniola had a population of more than a million.[48] Meticulous studies by historians Sherburne Cook and Woodrow Borah calculated that in 1518 central Mexico had as many as twenty-five million inhabitants.[49] For the Andes, the pre-Conquest population just in Peru is put at nine million.[50] Archaeologists now believe that in 1500 the Amazon basin had at least five million people; the density along the Amazon River's flood plain is put at twenty-eight inhabitants per square kilometer.[51]

Why is the size of the contact population so difficult to determine? Because diseases from the Old World decimated the Americas so quickly and so decisively. Smallpox, measles, typhoid, and influenza proved terrible, swift killers, far beyond anything that European conquerors intentionally inflicted. In the Old World, people had coexisted with such maladies for centuries. The result of such contagion was high infant mortality. But the children who passed through the gauntlet ended up resistant to some of the worst pathogens. Smallpox is an example. An attack confers immunity; survivors can carry the virus, but a repeat exposure is no longer life threatening.[52]

Prior to 1492, the New World was not a disease-free paradise, but contagious diseases were certainly less common—no smallpox, measles, whooping cough, bubonic plague, diphtheria, typhus, cholera, yellow fever, or malaria.[53] Against such silent invaders, the population of the Americas, including adults, had built up no immunity. The result was a devastation without precedent. No disease proved more deadly than smallpox. An airborne disease, it is extremely communicable, infecting its host through the respiratory tract. The first recorded epidemic was on the island of Hispaniola in 1518–19; the disease spread "like a fire through dry brush."[54] The next year, it hit the Aztec Empire. The very day that Aztec warriors drove Cortés and his allies from Tenochtitlán, smallpox struck. It raged through the city for sixty days, providing a chance for the Spanish to regroup and counterattack. Had there been no epidemic, it is hard to believe that Cortés could have vanquished the Aztecs. The siege of Tenochtitlán lasted eighty-five days, but the greatest Aztec losses came from the unchecked pestilence that gripped the city.[55]

Wherever the Spanish went, smallpox prepared the way. A great epidemic swept through the Inca Empire in the early 1520s, more than a decade before Francisco Pizarro and the Spanish entered Peru. By then, the disease had killed the Inca emperor, Huayna Capac; his designated successor died the same day. The result was an empire without a ruler and a civil war over succession.[56] Even so, Inca resistance was not quelled until the last Inca ruler was captured and executed in 1572.[57] Pestilence was likewise an ally of the Pilgrims in English North America. In 1616, more than half the coastal Indians died in a terrible epidemic.[58]

It is almost impossible today to imagine a mortality rate so high. We do know, however, that the Arawaks of the Bahamas and the Antilles died out completely, victims of forced labor and contagious diseases.[59] This occurred across the Caribbean, from Cuba and Hispaniola to tiny Montserrat. For the Aztec and Inca Empires, mortality rates are estimated at between a third and a half.[60] Such figures still understate the impact. Smallpox did not respect royalty or social class. It cut through the ranks, undermining leadership and morale at a critical moment. And it destroyed invaluable social knowledge in architecture, agriculture, medicine, and cosmology.

The demoralized, impoverished Indian culture of 1620 was but a pale reflection of the Inca Empire conquered a century before. Jesuit scholar Bernabé Cobo, whose *History of the New World* was completed in 1653, considered Peru's Indians to be "inconstant, docile, and unreliable"; he disliked their "foul smell, squalor, and filth."[61] In Cobo's taciturn, ignorant *indio,* subsequent generations could not see the imperial administrators, the astronomers, or the engineers who built the floating gardens and the irrigation systems. Perhaps we need to look closer. For in their botanical knowledge, their cropping systems, and the engineering of their fields, the ancient cultures of the Americas had no peers.[62]

In their exchange with the Old World, the Indian cultures of the Americas lost much more than they gained. It could have been otherwise. Diseases from the New World could have devastated the Old. In fact, syphilis probably originated in the Americas.[63] But compared to New World crops, the disease export list is short indeed. Historians credit American crops, especially corn and potatoes, with helping to make the Old World's demographic expansion possible.[64] Today, of course, there is but one world. We are linked together in our crops and in our pathologies come what may. The protection that we imagine place provides is an illusion. No place now is far from home.

In 1997, fifty-four million U.S. residents made trips to foreign countries and forty-seven million foreigners entered the United States, twice the numbers of 1985. Together, their 1997 travel expenses totaled $164 billion.[65] How many contagious diseases did these travelers exchange? How many fruits and fresh vegetables made the trip illegally, hidden inside flight bags and carry-on luggage? Which insects, bacteria, and lethal viruses hitched a ride? When people travel, more than money gets exchanged. The health status of countries far away may be more relevant than most of us think.[66] Can we be sure the exchange will always be in our favor?

PART II
The Potato

CHAPTER 4

Potato Facts

Specialists argue over which of the world's top crops is the best. Historian Earl J. Hamilton praises the potato but is critical of rice. Polished rice, he says, "is notoriously deficient in vitamins"; moreover, a hectare of rice yields only half as much food as a hectare of potatoes. A potato patch is also easier to plant than a rice field; all a farmer needs is a spade. By contrast, the work required to puddle and transplant a paddy makes rice "one of the most labor-intensive of all food crops." The potato can be cooked in more ways than rice and converts into more versatile products. Add in a bit of milk and you can "live indefinitely on potatoes, a claim neither rice nor wheat can match." Of course, the spud also tastes better—at least to the potato minded.[1]

If the potato is so superior, why don't rice farmers switch crops? Because each crop has its own special niche. In Asia, farmers plant rice in flooded paddies during the summer monsoons. The potato cannot take the water, heat, and humidity. For potatoes in Asia, farmers must wait for winter weather. Wheat and corn cannot substitute for rice either. Wheat prefers a semiarid climate and is not a monsoon crop. Corn likes rain and can take humidity, but it will rot in a flooded field. Much of Asia's rice belt is under water in the summer. So for monsoon-ridden Asia, rice is the best alternative. How productive, flexible, nutritious, and labor saving the potato might be is irrelevant.

Nevertheless, Asia is where potato output is growing the fastest. Between 1961–63 and 1997–99, harvests increased almost threefold, from an average of 24 million to 91 million metric tons. For the same period, Asia's share of global potato output rose from 9 to 31 percent. China is now the world's largest potato producer; its 1997–99 output averaged almost 48 million metric tons. (See table 2.) The potato thrives in the cool, sandy soils of China's northern provinces. Asia's

Table 2. Distribution of Potato Production for Selected Regions and Countries, by Average Total Production and Percentage of World Production: 1961–63 and 1997–99

Region	Production (000MT)		Percentage	
	1961–63	1997–99	1961–63	1997–99
World	265,114	291,116	100.0	100.0
Western Europe	84,981	47,931	32.0	16.4
Eastern Europe	56,271	34,476	21.2	11.8
Poland	42,629	24,241	16.1	8.3
Former Soviet Union	75,274	65,627	28.3	22.5
Russian Federation	43,174	33,292	16.3	11.4
Ukraine	17,640	15,835	6.7	5.4
Belarus	9,273	8,980	3.5	3.1
North America[a]	14,625	25,606	5.5	8.8
United States	12,543	21,365	4.7	7.3
Canada	2,082	4,241	< 1	1.5
Asia	23,932	91,254	9.0	31.3
China	12,908	48,413	4.9	16.6
India	2,844	21,588	1.1	7.4
Africa	2,181	8,876	< 1	3.0
Latin America and the Caribbean	6,959	15,503	2.6	5.3

Sources: For 1961–63, see CIP, *Potatoes in the 1990s*, p. 29. The 1997–99 data is compiled from FAO, *FAO Quarterly Bulletin of Statistics* 1999, pp. 38–39. For comparability, production in former Soviet Republics is included in the total for the Former Soviet Union.
[a]Mexico and Central America were added to the Latin American total.

potato story, however, is not just about China. Even tropical India is a big producer; in 1997–99, its farmers harvested over 21 million metric tons a year.[2]

Growing potatoes in Asia's lowlands, which are besieged by monsoons, is a recent innovation. The change came on the heels of the green revolution, which reduced the time it took to grow rice and wheat by six weeks or more. A potato surge followed as farmers added spud rotations between rice crops or between wheat and rice. In this system, the potatoes go in during the fall, when temperatures start to drop; they get harvested in the winter. Asia's irrigated potato grows quickly. In the Ganges valley, farmers get a crop in just seventy-five days; if they are impatient and don't mind their spuds small, they can harvest even earlier. To keep potato stocks through the monsoonal summer, however, farmers must put them in cold storage. The potato's popularity in South Asia reflects the region's rapid urbanization and its affluent middle class. Bored with rice-based diets, consumers have turned to the flexible potato.[3]

In 1997–99, worldwide potato harvests came to 291 million metric tons a year, a modest increase over the 1961–63 average of 265 million.[4] By comparison, the world's rice harvests more than doubled, rising from 230 million metric tons to 578 million tons.[5] Why didn't the potato keep up with rice? Because what the spud gained in Asia it lost in Europe. Between 1961–63 and 1997–99, western Europe's average production fell over 40 percent, from about 85 million metric tons to just 48 million tons.[6] In eastern Europe, even devoted potato eaters in Poland changed allegiance; harvests dropped from approximately 43 million metric tons to 24 million tons. Poland still claims the world's potato-eating title, but at a much reduced rate. Per-capita intake of potatoes dropped from 220 kilos in 1961–63 to 125 kilos for 1997–99. The Russian Federation plus the Ukraine and Belarus still make up the world's largest potato block; together, production averages some 58 million metric tons. Even so, the block's per-capita consumption has declined from 132 kilos in 1961–63 to 95 kilos in 1997–99.[7]

The big losses in Europe follow shifts in eating habits and crop utilization. Affluent consumers cut back on the spuds to add more grains, fruits, and vegetables. In western Europe, the potato's use as an animal feed declined from 35 percent of the crop in 1961–63 to 22 percent in 1991–92; in eastern Europe, the drop was from 45 percent to 39 percent. Outside Europe, the potato's use as feed stock is minimal. In most developing countries, China excepted, 75 to 80 percent of the crop goes to direct human consumption; the rest is processed or held back to seed the next crop.[8]

Compared to Europe, where production dropped, the potato gained ground in the United States. Between 1961–63 and 1997–99, production rose from 12.5 million metric tons a year to 21.4 million tons, an increase of 68 percent. Per-capita consumption went up as well, from 49 kilos to 65 kilos; virtually no potatoes in the United States are used as feed. Unlike developing countries, only a third of the U.S. potato crop is consumed fresh. The rest is processed into frozen french fries, chips, dried flakes, and other potato products.[9] If Europe follows the United States down the processing trail, potato output could recover.

Fresh potatoes are too bulky, perishable, and virus ridden to trade internationally. Only about 3 percent of global potato output crosses a border, and most of that is within the European Community. When processed products like frozen french fries and starch are included under potato trade, then the total goes up a bit to about 4 percent.[10]

For 1997–99, rice, wheat, and corn harvests totaled 1.8 billion metric tons, accounting for half the calories consumed by the world's 5.9 billion people. The potato comes next, but compared to the grain triumvirate's output, its 291 million tons does not seem like much. In the potato's defense, at least it outranks all the vegetables. Metric tonnage for the runner-up, the tomato, is a mere 90 million tons. The potato also outstrips the beans, including Asia's soybeans, from whence comes tofu. The world's 1997–99 soybean harvest weighed in at 143 million metric tons a year.[11]

Tonnage, however, is but one measure of a crop's stature. How much food does a crop yield per hectare, and what is the nutritional value of the food so produced? In both productivity and nutritional terms, the potato can hold its own against the grains, even against rice. Some 85 percent of the global rice crop goes to direct human consumption, compared to just 60 percent for wheat and 40 percent for corn.[12] For 1997–99, the world's rice harvests averaged 578 million metric tons grown on 152 million hectares, with yields of 3.8 metric tons per hectare. With potatoes, 18 million hectares sufficed to produce a 1997–99 average of 291 million tons, for an overall yield of 16 metric tons.[13] That looks like a four-to-one productivity advantage for the potato. But appearances are deceptive. A kilo of uncooked rice translates into a lot more food than a kilo of uncooked potatoes. The difference is in the dry-matter content of each. A fresh potato is roughly 23 percent dry matter and 78 percent water; uncooked rough rice is about 86 percent dry matter and 14 percent water.[14] So precisely how much edible food each crop yields gets a bit complicated. Still, in rough terms, the potato still comes out ahead of rice. A hectare of potatoes produces 3.7 metric tons of dry matter (16 metric tons per hectare x 23 percent dry matter), compared to 3.2 metric tons for rice (3.8 metric tons per hectare x 84 percent dry matter). Of course, the calculations are based on world averages. Nonetheless, as yields for both crops go up, the advantage shifts even more decisively to the potato. In the United States, potato yields weigh in at 39 metric tons per hectare and rice yields at 6.6 metric tons. The result is 9 metric tons of dry matter from potatoes (39 metric tons per hectare x 23 percent dry matter) versus 5.5 metric tons for rice (6.6 metric tons per hectare x 84 percent dry matter).[15] The same calculations can be made for wheat and corn, with comparisons equally favorable to the spud.[16]

"The problem with the potato," I've been told, "is that it makes you fat." What a lie! The potato has so little fat it should be labelled "fat free." The fat content of a fresh tuber is just two-tenths of 1 percent.[17] By comparison, wheat and rice seem like fatty foods; even a carrot has more fat than a potato.[18] As for calories, a medium-sized potato has approximately 100 calories, which is comparable to a pear.[19]

I refer to the baked, microwaved, or boiled spud. As a french fry, the calories add up fast: 274 in 100 grams; as for the traditional potato chip, don't even think about it. And don't add any butter, sour cream, or melted cheese. With these provisos, the potato passes the fat test. It belongs in the arsenal of any serious weight watcher. The potato is filling, virtually fat free, low in calories, and low in cholesterol. It does not have much sugar either; about 1 percent of the dry matter in a potato is sucrose.[20]

The potato is nutritious. (See table 3.) Just 100 grams of a freshly harvested tuber contains on average twenty milligrams of ascorbic acid. That is half the minimum daily vitamin C requirement of forty milligrams.[21] Grains like rice and wheat, by contrast, have no vitamin C.[22] In times past, sailors ate raw potato as an antidote to scurvy; just ten milligrams a day of ascorbic acid prevents the malady.[23] The vitamin C in two fresh potatoes is equivalent to that in a tomato, an orange, a grapefruit, or three apples.[24] In fact, early references to the Irish potato dubbed it "earth apple," no doubt influenced by the French *pomme de terre.*[25] Potatoes are

Table 3. Percentage of Daily Requirements Met Eating Half a Kilo of Potatoes, by Cooking Method

Minerals and Vitamins	Peeled and Boiled	Unpeeled, Boiled, Eaten in Skin
Calcium	5.6	9.4
Iron	27.0	41.6
Phosphorus	17.5	25.0
Potassium	Over 100	Over 100
Vitamin C	Over 100	Over 100
Vitamin A	1.5	1.5
Vitamin B1	30.0	40.0
Riboflavin	2.5	5.0
Protein		
Nitrogen	15.5	16.5
Nicotinic acid	15.0	15.0
Fat	Almost None	Almost None
Calories[a]	11.5	11.5

Source: Burton, *The Potato,* pp. 173, 174, 182. The percentage of the calcium, iron, and phosphorus requirements met are calculated from the milligrams in 100 grams of potatoes for each preparation (from Burton's table 35) multiplied by 5 (for 500 grams) and divided by the daily requirement (from Burton's table 38, note 1). For vitamins and proteins, divide the percentages in Burton's table 38 by 2; for fat and potassium, see Burton's table 34.

Note: Half a kilo of potatoes equals about a pound.

[a]Or 345 calories, which is 11.5 percent of 3,000 calories, the approximate adult daily requirement for men, or 15.7 percent of 2,200 calories, the approximate adult daily requirement for women.

also rich in the B complex vitamins. Eaten regularly, they have enough niacin (B5) to prevent pellagra and skin disorders; they are a good source of vitamin B6, thiamine (B1), and riboflavin (B2).[26] Eaten with their skins on, potatoes provide dietary fiber; in fact, they have more fiber than rice. Admittedly, the potato falls behind the grains in protein: just 2.1 grams in 100 grams of fresh, uncooked tubers.[27] Still, potato protein is of high quality, being rich in amino acids and lysine; eating a kilo a day supplies half the adult requirement.[28] Potatoes contain essential minerals, notably calcium, iron, phosphorus, and potassium.[29] Low in sodium, but rich in potash and alkaline salts, potatoes qualify for salt-free diets and help correct for acidity.[30]

The plain truth, however, is that 70 percent of the dry matter in a potato is starch.[31] In a culture fixated on protein, starch is almost a dirty word. This dualistic thinking is both unbalanced and erroneous. We need protein for essential amino acids and nitrogen. We do not need much of it; 50 grams a day is more than enough. For fine tuning, protein is just what we want, but to fill up the tank, we need starch. As a healthy source of energy, starch is far better than a diet skewed to sugars and fats. The starch in a cooked potato gets high marks for digestibility, as its starch grains are but loosely combined with fiber and cellulose. The body readily assimilates 95 percent of the carbohydrates in a potato, and it does so slowly; the result is a steady energy supply.[32]

Critics claim that when a potato is cooked its luster diminishes. Of course, cooking almost anything reduces its nutritional value, hence the good reputation of fresh vegetables and the ill repute of highly processed fast food. How much of the potato's value is lost in the kitchen, however, depends on its preparation.

Peeling wastes a full quarter of the potato. Household economy, however, is not the main issue. Much of a tuber's insoluble protein is near the surface; peeling largely removes it. As to soluble components, some 20 percent dissolve in the water. If potatoes must be boiled, spare them the paring knife. If they must be peeled, then steam them first, which reduces leaching. Do not add salt to the pot, as this draws out vitamins and minerals. It is best not to cut potatoes into pieces if you value the vitamin C; oxidation of ascorbic acid is much greater in cut tubers.[33]

The best way to prepare potatoes is unpeeled; then they can be steamed, baked, or roasted. Still, a boiled potato, even if peeled, is not without its nutritional merits. For regardless of what happens in the kitchen, a potato still delivers. Two kilos, even of the boiled variety, will meet the minimum requirement for thiamin (B1) and essential amino acids. The body's utilization of thiamin from potatoes is more efficient than from brown rice. When steamed, potatoes provide significant quantities of niacin and riboflavin. As to vitamin C, the key is a fresh tuber cooked rapidly and eaten immediately. When steamed and unpeeled, vitamin C losses are 10 to 15 percent; when peeled and boiled, 15 to 25 percent; when fried, 20 to 45 percent. Even so, one hundred grams of potato boiled fresh in its skin still contains twelve to sixteen milligrams of vitamin C. When boiled and peeled, 15 percent of the iron is lost, 10 percent of the phosphorus, and about 12 percent of the calcium; what remains, if one eats three to four medium-sized spuds, adds up to 27, 17.5, and 5.6 percent of the adult daily requirement of each mineral.[34]

Take a close look at a potato. Try Red Pontiac, a rose-colored, round variety with rather deep eyes. It has a well-defined front end with buds from which new stems emerge. And it has a back end where it was once attached underground to the mother plant—look for a tiny piece of dried stem. Now hold the potato with the stem end down and look at the eyes. They all point up. Look closer. Each eye has a moon-shaped brow underneath, and inside each eye are tiny buds. When planted, the main stems sprout from the apical, or front end, with lateral shoots developing from the sides. A potato, in fact, is a very large bud broken off from the plant's underground stems; that is why it is classified botanically as a tuber rather than as a root.[35]

Potatoes do well in cool climates. With daytime temperatures as low as ten degrees centigrade (fifty degrees Fahrenheit), growth still takes place, albeit slowly. The optimum daytime temperature is about twenty degrees centigrade (sixty-eight degrees Fahrenheit); it needs to be even cooler at night. Relatively low night temperatures and short days speed up tuber formation; high temperatures and long days, by contrast, promote growth aboveground and draw out tuberization. Consequently, how long it takes a potato to mature depends on location, the season, the amount of rainfall, prevailing temperatures, day length, and the variety in question. Potatoes are ready to harvest at full size when the foliage dies back. When dug out earlier, tubers are smaller, with thin skins that damage easily. Such early harvesting results in what we call "new" potatoes; it has nothing to do with the variety. In the Andes of Ecuador, where farmers harvest potatoes only at full maturity, the crop takes at least 150 days.[36]

In Tennessee, our potatoes go in by mid-March; we harvest them piecemeal from the end of June into August. For seed tubers, we take whatever the local feed store happens to stock, typically some combination of Irish Cobbler, Red Pontiac, and Kennebec. This so-called seed consists of tubers and is not true sexually produced seed at all. Each tuber has eyes with buds from which stems spring forth. Underground, these stems produce clones of the original mother tuber. In Tennessee, potato plants usually bloom, putting out delicate, star-shaped flowers, white to lavender in color. In the Andes, the potato's true home, the flowers set seed balls the size of large cherry tomatoes. This rarely happens in Tennessee, where the crop matures as spring days grow longer. To set true seed, the potato plant prefers shorter days; for Tennessee, that means the fall.

We plant our seed tubers whole. Cutting them up exposes the surface directly to soil-borne diseases; if spring rains are heavy and the soil is poorly drained, they tend to rot. A whole tuber, by contrast, is better protected by its thick skin. It also puts out many stems and sets more potatoes per plant. When cut up, each piece tends to generate a single, dominant stem from which fewer tubers sprout. Our potatoes go into boxes or raised beds planted the "Irish" way, that is, with spade, sand, and lime. We add a little fertilizer to the soil and cover the box with straw. Then we wait. If there is a late spring frost, which happens occasionally, we cut back the damaged foliage; in Tennessee, there is plenty of time for the plant to recover. We have few potato beetles and no late-blight spores, so we apply no pesticides. We do have woodchucks. Fortunately, the potato's solanaceous foliage is much less appetizing than the tender flowers of the squashes and melons.

The potato is not a fussy plant. Most gardeners have their own technique. My neighbor plants them on top of old newspapers covered up with compost. As the plants grow, she piles more compost around them. Planting in barrels, trash cans, and old tires are simply variations on the mulch mound. Planting potatoes in trenched, raised beds is very popular with small farmers. The tubers are set just below ground level in loose soil; as they sprout, farmers hill up around the stems. Whichever way you plant them, a light, loamy soil neither too acidic nor too alkaline (with a pH of 5.0 to 6.8) is best.[37] Old leaves left to decompose over the winter make a good potato compost. The best crop we ever got had piles of old leaves worked into the soil.

The yield potential in potatoes far surpasses the grains. The rule of thumb in the *Rodale* guide is between one and two kilos per plant.[38] The range is remarkable. The world average is almost sixteen metric tons per hectare. Depending on seed size and spacing, it takes between two and three metric tons of seed tubers to plant a hectare. So on average, the multiplication rate worldwide is at best eight to one.[39] In Central Asia, yields drop below five metric tons. Potato yields in the United States and most of western Europe are in the forty-ton range, with a multiplication rate of twenty to one.[40] Western Australia, which combines ideal soils and temperatures with few pests and diseases, holds the record: an incredible one hundred metric tons per hectare, with irrigation.[41] With traditional cereal crops, the architecture of the plant limits yields. To boost the yield ceiling for rice and wheat, green-revolution scientists had to redesign the plant.[42] With the potato, the path to higher yields is blocked by pests, diseases, and poor quality seed tubers; no one seems worried about the plant's design.

We tend to treat all potatoes equally. We should be more discriminating. Russett-Burbank is on the starchy, flaky side of the dry-matter ledger, which makes it good for baking. Red Pontiac, by comparison, is watery; it likes to be boiled and made into potato salad. That said, the potato is a most accommodating vegetable. It can be steamed, fried, mashed, roasted, baked, or microwaved. Cut it up into wedges;

slice it across for chips, or laterally for fries; grate it up for hash browns and potato pancakes. A boiled potato can be refried for breakfast or diced up as the mainstay in hash. Mashed potatoes can be recycled into bread. In the Andes, potatoes go into almost every soup, including Ecuador's justly famous *locros*. In Italy, mashed potato ends up in pasta as gnocchi. The Portuguese and the Brazilians make a salted codfish stew, *bacalhoada*, which relies on the potato to counter the sodium. For au gratin, mix cheese in with potatoes scalloped for the baking dish. For Peru's *huancaina papas*, add cottage cheese, onions, and lemon juice; then chill. From Germany to Russia, the potato keeps the kraut and cabbage company. In the United States, potato salads are served cold with globs of mayonnaise and hard-boiled eggs or hot with bacon and a heavy dose of vinegar. As a thickener and binder the potato has few rivals. Many a watery soup, sauce, and casserole has been saved by the mashed potato's starch, not to mention the potato's place on top of every shepherd's pie. The potato is multicultural, accepting diverse flavor principles equally. It gets along with the Mediterranean's olive oil, basil, and oregano and takes to the rosemary, sage, and thyme of southern France. The potato accepts India's curries, Mexico's chilies, and America's ketchup; it is not adverse to China's soy sauce and ginger or to West Africa's peanuts, peppers, and tomatoes. Wherever it goes, the flexible potato is never a troublesome guest.[43]

I wish I could say that for all the potato's relatives. Unfortunately, the potato's botanical family *(Solanaceae)* includes the notorious *Nicotiana tabacum*—in a word, tobacco. Native American farmers domesticated some ten different species of the crop, becoming experts in its cultivation and curing. First they planted the tiny tobacco seeds, some three hundred thousand to an ounce (thirty grams), in a nursery bed. After germination and thinning of the seedlings, they transplanted them to larger plots. To channel the tobacco plant's energy into leaf making, they topped it off to prevent flowering. Finally, the leaves had to be carefully selected, dried, and cured, usually for several years. In Virginia, Native Americans grew *Nicotiana rustica*, a tobacco so strong its use left smokers in a trance, even unconscious. Today, this rustic type of tobacco is used in nicotine-based insecticides. What saved the struggling Virginia colony's economy was a much milder tobacco from the Caribbean. Colonists quickly took to the new type. In 1615, Virginia settlers exported about 1,000 kilos of tobacco; by 1629, exports reached some 680,000 kilos.[44]

To Native Americans, tobacco was a divine plant with magical properties and ceremonial usages. They smoked it in pipes, rolled it up into tubes made of corn husks, chewed it with lime, made it into cigars, and used it as snuff. Use was controlled and sparing; there was no chain smoking. That being the case, it seems unfair to blame addiction on the plant's solanaceous family tree.[45] Tobacco is a victim of the company it keeps, not the family it comes from.

In 1604, King James I, in his *Counterblaste to Tobacco*, declared smoking "loathsome to the eye, hateful to the Nose, harmful to the brain, and dangerous to the Lungs."[46] What a direct, no-nonsense statement! It puts the bland warnings on a cigarette pack to shame. But it was no more effective. In seventeenth-century England, it was the pipe that dominated the smoking market; later, snuff was all the rage, notably at the French court. In 1890, chewing was still the most popular way to use tobacco in the United States; the country boasted thousands of local brands and, in polite social circles, spittoons. Then came the cigar and improved table manners, followed by the mass-market cigarette. In its classy, well-advertised, cigarette guise, tobacco charmed a lost generation of flappers and bootleggers.[47]

Tobacco is not the only narcotic in the solanaceous family, much to the wholesome potato's embarrassment. Mandrake, henbane, and deadly nightshade (belladonna) are hallucinogenic, even poisonous.[48] In most countries, they are not deemed legal company. Given such relatives, is it so surprising the potato itself is sometimes considered dangerous? In fact, the potato's ill repute is not simply a matter of bad press. Below the skin's surface and around the eyes, a tuber contains small amounts of the glycoalkaloid solanine. Toxic levels build up when potatoes turn green. This happens if a tuber grows close to the soil's surface or is overexposed to the sun at harvest time. Peeling removes most of the solanine, and cooking breaks it down. So unless one eats green potatoes raw, the danger is minimal. By comparison, a potato plant's flowers, fruits, sprouts, and leaves have twenty times the glycoalkaloid level of its tubers. Consequently, the foliage makes a poor feed; even the woodchucks keep their distance.[49]

Mesoamerican farmers domesticated the solanaceous tomato. Andean farmers, by contrast, did not bother with the crop. They had a good substitute in the tree tomato. Common in Ecuador, the tree tomato is grown by small farmers at the same altitude as its relative, the potato. A tall, majestic plant with heart-shaped, purplish-green leaves, it bears but slight resemblance to the sprawling tomato. The taste of its reddish-orange oval fruits does call the tomato to mind, especially when squeezed into fresh juice, but hold the sugar. After the potato, Ecuador's "little orange," or naranjilla, is my favorite solanaceous plant. It grows on the eastern, misty slopes of the Andes, where rainfall is abundant. Naranjilla produces a tomato-sized, orange-skinned fruit with a deep-green, acidic pulp. When squeezed, it yields a tangy juice served fresh at breakfast. The orange-lime taste is unforgettable.[50]

A solanaceous plant's hallmark is the shape of its flowers. The leaves can be large or small, the plants tall or short, but the flowers are always star-shaped, symmetrical corollas of five petals. The stamens typically protrude from the corolla with the petals pushed back. Once examined mindfully, the family is easily identified.[51] Look at the flowers on a pepper plant, on tomatoes, or on eggplants; it is hard to miss the family resemblance. So too with common nightshade, riffraff like horse nettle, and the viney bittersweet, all solanaceous weeds of thickets and fields, common in the United States.[52] I confess special affection for the bittersweet, which grows wild, intertwined with the shrubbery in my backyard.

Given their close family ties, potatoes, peppers, tomatoes, and eggplants should not go into the same plot twice in succession. They draw the same nutrients from the soil, attract the same insects, and get many of the same diseases. As with marriage between first cousins, get a blood test first.

Regardless of how nutritious or notorious other family members might be, in its botanical diversity the potato takes a back seat to none of them. The potato clan itself, the genus *Solanum*, includes some 235 separate species, most of them wild. Their geographical range branches from the mountains of the central Andes south into Chile and north across Central America into Mexico and the southwest of the United States. Along the way, potatoes adapted themselves to the frigid Andean highlands, to subtropical rain forests, to semiarid valleys, and to temperate woodlands and pine forests.[53] From this rich diversity, the ancient farmers of the New World domesticated eight species. The Lake Titicaca basin is considered the crop's zone of origin, the place where domestication took place. Some 120 different wild potato species are found there, more than anywhere else in the Americas. The region's farmers still cultivate all eight species, including three famous for their frost resistance.[54]

The rice world is split between the short, sticky Japanese rices and the elongated, fluffy Indian types. Yet both are from the same rice species, *Oryza sativa*.[55] A comparable division exists in the potato world. The common spud, *Solanum tuberosum*, is divided into two subspecies: *Solanum tuberosum andigena*, which originated in the Lake Titicaca basin, and *Solanum tuberosum tuberosum*, which was domesticated independently in Chile.[56] Andigena types can have deep eyes and a starch content so high they break apart when cut. By comparison, tuberosums are smooth skinned and watery. True andigenas are adapted to short days, true tuberosums to long days.[57] In temperate zones, tuberosum types predominate. The andigena potato's redoubt is its homeland in cool Andean climates near the equator. There are exceptions. The equator cuts across the highlands of Kenya and Uganda, creating an Andean-like climate. Consequently, farmers can grow andigena types. The same is true in India's Ganges valley, where potatoes are a winter crop; the short winter days favor andigenas or andigena-tuberosum hybrids.

When the potato began its life as a world traveler, it did not bother with a passport. The result is a befuddlement of nomenclature that not even Interpol has managed to unravel. Had the potato applied for a passport in 1540, and had it done so under its common Quechua name, it would have traveled as *papa*. Today, the word *papa* prevails throughout Spanish-speaking America. The sweet potato, *Ipomoea batatas*, is an entirely different crop, adapted to a warm lowland climate. It is too bad the sweet potato did not get a passport under *camote*, its Mexican, Nahuatl name. In Latin America, *camote* (sweet potato) and *papa* (potato) never get confused.[58] This is not the case abroad, where confusion reigns.

How *camote* picked up the potato alias is still a matter of debate. Perhaps it comes from *batata*, the Carib-Arawak name for sweet potato, which was a popular crop in the Caribbean Islands in 1492. Early English sources used *batata* interchangeably for both crops. It is not difficult to get *potato* from *batata*. However it happened, English-speakers are stuck with the unhappy result. To distinguish between the two crops, the English language calls one of them "sweet." Thus, we have potatoes and sweet potatoes, both from the same linguistic *batata* root. The same usage shows up in Portuguese-speaking Brazil, a country that had trading ties to England in colonial times. Brazilians too call both crops *batata*, adding the adjectives *Inglesa* (English) and *doce* (sweet) to tell them apart. And let us not forget the nicknames. In English, the potato is a "spud," called such because of the digging fork used in its cultivation; lazy folk are "couch potatoes"; the stubborn are "potato headed"; and the nasally challenged are "potato nosed." A generation ago, teenagers in the United States honored the spud in a ritualistic dance called the "mashed potato." Potato and sweet potato nomenclature did not get mixed up in continental Europe; there are other confusions instead. When the potato crossed borders, it often got identified with a crop it resembled. To some, it looked like the wrinkled, subterranean truffle, a kind of earth "testicle." So disguised, the potato entered Germany, Poland, and Russia as some variant of *kartoffel*. In France, it became *pomme de terre*. Its name translated literally, the potato went to Holland as *aardappel* and from there to Dutch colonies in the Far East. It is even an "earth apple" in Iran. From the Danube down into the Balkans, the potato masquerades as "earth pear." In India, the potato goes by the generic Hindi word for tuber, *alu;* in China, where farmers grew both yams and taro root, the potato was dubbed "foreign yam" and alternatively, "foreign taro."[59]

Since the potato left Peru, it has acquired new names and new uses. A resourceful fellow, it found a way to get along with established crops from northern Europe's rye to Asia's rice. The potato's story, like its travels, is international.

CHAPTER 5

The American Potato

To judge from the frozen-food section at the supermarket, the modern potato is as American as apple pie. In the past, however, the noble spud so consorted with foreigners it seemed almost un-American. Irish immigrants were the first to make potato consumption an embarrassment. Other potato eaters followed: from Germany, Scandinavia, and eastern Europe. The newcomers ate more potatoes than bread; in fact, their potato intake was six times that of American citizens. The demand for potatoes expanded along with immigration. Between 1870 and 1910, potato production in the United States almost tripled, from 2.9 to 9.3 million metric tons. Per-capita consumption rose from sixty-three kilos to over ninety kilos, or almost two hundred pounds.[1]

Even so, most Americans were still partial to wheat. As immigrants sought acceptance, dietary conformity set in; they cut back on the potatoes and ate more bread. After 1920, restrictions on immigration reinforced the downward trend; new recruits no longer bolstered the country's potato-eating ranks. Potato consumption fell from its 1910 peak to about forty-five kilos per person in 1950.[2] In the meantime, affluent urban America was diversifying its diet. Families abandoned the potato's apparent monotony; busy housewives disliked how long it took to peel, cook, and mash the spud. The potato, in short, suffered from image and lifestyle problems.

What rehabilitated the spud was a positive relationship with the hamburger. And for this we must thank McDonald's. As a french fry, the potato is fat enriched and accepts an enhancer like ketchup gracefully.[3] In frozen form, the french fry gave the potato a new lease on life, this time away from home. Then came the microwave oven, which consolidated the potato's place in the family.[4] With a little coaching, even Dad could microwave a potato.

The potato's shifting place in America's eating habits reflected larger trends afoot. In 1900, over 60 percent of the U.S. population still lived in rural areas. In 1950, only 40 percent did.[5] The kind of potato people wanted changed, as did the kind of farm that produced it and how the crop got to market. When the U.S. Department of Agriculture took a potato census in 1909, they counted several thousand varieties, which they grouped into eleven categories. Today, just one variety, Russet-Burbank, accounts for half the country's production. In 1900, the crop came from small farms, few of which grew potatoes exclusively. The crop was seasonal, carted into towns from the surrounding countryside. Then urbanization and the railroad combined to change the pattern. Big cities created a mass market; the railroad made regional specialization practical.[6]

For potatoes, the most advanced state was Maine; within Maine, the most advanced area was Aroostook County. In 1900, the state's potato crop weighed in at 272,000 metric tons; thirty years later, the state produced 1.4 million tons.[7] Maine's short, cool, rainy summers; its loose, silty soils; and its links to the big-city markets of the Northeast combined to make the state the country's top producer. In 1923, Maine had potato yields that averaged almost 18 metric tons per hectare, far ahead of the national average of just 7 metric tons.[8] Production gravitated to larger, specialized farms. In 1935, just 2.2 percent of America's potato farms accounted for over half the nation's output; in 1945, this figure was up to 75 percent.[9] Maine concentrated on just two varieties, Irish Cobbler and Green Mountain. Its potato industry enforced high grading standards and built up a reputation for quality with consumers.[10] Other states followed.

Idaho built its reputation on Russet-Burbank, marketed as the "Big Idaho Baker." It targeted affluent urbanites willing to pay a premium for quality, uniformity, and dependability.[11] Today, no state even comes close to Idaho's production. For 1997–99, Idaho's potato output came to 5.6 million metric tons a year, compared to 4.1 million metric tons for second-place Washington state. Idaho's receipts topped $635 million. All together, the state accounted for 26 percent of a U.S. harvest valued at $2.7 billion.[12] Not only does Russet-Burbank grow exceptionally well in Idaho, but it also has a secure market, since Russet-Burbank makes a perfect french fry, at least in the deep frier at McDonald's.[13]

The story of Russet-Burbank's fame in the potato world is one of rags-to-riches success. In the 1840s, American potatoes were andigena types; as in Ireland, they fell prey to late blight. In 1851, as part of an effort to "reinvigorate" American potato stocks, the Reverend Chauncey Goodrich of Utica, New York, brought back a potato collection from South America. The samples included a long-day tuberosum from Chile that Goodrich dubbed "Rough Purple Chili." It set true botanical seed in New York state. From this self-pollinated seed, Goodrich bred Garnet Chili, which in turn begot Early Rose. In the meantime, Rough Purple Chili became a favorite with amateur breeders, displacing the andigenas. Even today, descendants of Rough Purple Chili can be found in the pedigree of almost every North American cultivar.[14]

Russet-Burbank takes half its name from amateur botanist Luther Burbank. In 1872, he spotted a seed ball on an Early Rose plant, which he had in his garden. Since Early Rose did not typically set fruit, this caught his attention. When the seed ball ripened, Burbank collected it and got twenty-three seeds. He planted the seed the next season, and all of it germinated. Of these twenty-three plants, one had higher yields and larger tubers than its parent, Early Rose. Burbank kept this one to try again the following year. It was a long, smooth-skinned, white-fleshed variety with a white skin. Burbank eventually sold the rights to a seed company, but the variety kept his name. At the turn of the century, the Burbank potato was popular out west. In 1914, a Colorado farmer noticed some mutated Burbank tubers with russet skins. He kept the clones.[15]

Dubbed Russet-Burbank, the newcomer took a while to catch on. In 1930, it accounted for just 4 percent of the U.S. market. Irrigation led to its rapid rise in Idaho. Growers discovered that when watered on a schedule, the plant produced "a beautiful, long, russeted potato." This was the "Big Baker" described above that growers marketed so ably. Until the 1950s, Russet-Burbank's redoubt was the fresh market. Processing has since changed the variety's portfolio.[16]

Big growers in Idaho worked out the blanching and freezing technology for potatoes. The prototype was the french fry, and the model potato was Russet-Burbank. It was the right potato in the right place at the right time. Low sugar content makes Russet-Burbank easy to store; its high dry-matter content and elongated shape make it ideal for processing. McDonald's chose Russet-Burbank for its french fries; the U.S. frozen-potato industry favors it almost exclusively. As a result, Russet-Burbank sets the standard by which all french fries are judged, not to mention crinkles and tater tops.[17]

Russet-Burbank has a dual purpose: it is good for the fresh market and good for processing. After a long decline, the practice of eating potatoes fresh has stabilized at about 23 kilos per capita, thanks largely to the microwave oven.[18] In 1980, only 10 percent of American households owned a microwave; a decade later, they were as common as refrigerators. The main thing Americans used the microwave for was baking potatoes.[19] And as likely as not, what Americans baked was Russet-Burbank. The potato, however, chalked up its biggest gains in the fast-food category, not as a fresh item. (See table 4.) Between 1980–82 and 1996–98, per-capita consumption of processed spuds rose from 30.6 kilos a year to 43 kilos. The frozen potato led the way, its per-capita consumption going up by over 50 percent, from 17.5 to 27 kilos. Meanwhile, canned spuds stagnated at under 1 kilo, chips held their own in the 7-kilo range, and dehydrated flakes went up from less than 5 to 8 kilos per person. When potato eating in all its forms is considered, total per-capita consumption increased from 52 kilos in 1980–82 to over 65 kilos in 1996–98. That included 6.3 million tons of frozen french fries a year, plus 5.8 million tons of fresh spuds and 2.3 million tons each of chips and dried flakes.[20]

The production, processing, and marketing trends sketched above benefited Russet-Burbank's reputation the most. Russet-Burbank became the paradigm for the ideal french fry and the ideal baked potato. Currently, Russet-Burbank makes up 74 percent of all Idaho's output; for Washington state, the figure is 41 percent; for Oregon, 43 percent. Nonetheless, it can be a troublesome spud. What makes Russet-Burbank so good is not its genetic makeup but the special environment it has found to mask its deficiencies. Without precision irrigation and just the right temperatures, its tubers end up knobby and deformed. This narrow range is hard to duplicate outside of Idaho and the Pacific Northwest. Only 26 percent of the crop in a stellar potato state like Maine is Russet-Burbank; only 7 percent of Michigan's is; and in New York State, Russet-Burbank is not among the state's top four varieties. Russet-Burbank does poorly abroad, which helps explain why the United States exports almost half a million metric tons of frozen fries a year. The largest market is in Asia, and Japan is the single largest customer. Without Russet-Burbank, a McDonald's french fry might lose the uniformity that is its hallmark. Or so many in the fast-food industry think.[21]

Given its storage and processing assets, no one worries much about how Russet-Burbank tastes.[22] But without hefty additions of butter, cheddar cheese, and sour cream, a baked Russet-Burbank is a dry, starchy affair. It has little disease and pest resistance, so growers douse it constantly with pesticides. In the semiarid west, its water-guzzling habits put it on a collision course with urban and industrial development. Nonetheless, it will be difficult to dislodge. The frozen-potato industry and its fast-food customers are hooked on Russet-Burbank; their equipment, marketing, and packaging are all geared to its peculiar characteristics. An improved variety like Butte, superior to Russet-Burbank in yields, disease resistance, and grading traits, faces an uphill battle. To an industry set in its ways, changing over to unfamiliar varieties, no matter how good they might be, is a costly step fraught with uncertainties.[23]

Russet-Burbank's clout blocks the way to hardier, better-tasting spuds. The food industry pays a premium price to get Russet-Burbank, and growers can afford the chemicals needed to protect the variety. So far, superior pest resistance has failed to sell America's potato growers on new varieties.[24] In developing countries, by contrast, farmers prefer resistant cultivars. Part of the reason is a less demanding market. Consumers eat their potatoes fresh and care less about their shapes and sizes. Overall, then, it's easier for Third World farmers to switch varieties. When they do, their pesticide bill goes down and their profits go up.[25] If American con-

Table 4. Average Per-Capita Potato Consumption in the United States, 1980–82, 1988–90, and 1996–98, in Kilos

Year	Grand Total	Total Fresh	Total Processed	Frozen	Processed Chips	Flakes	Canned
1980–82	52.4	21.8	30.6	17.5	7.6	4.7	0.8
1988–90	57.1	22.0	35.1	21.3	7.9	5.1	0.8
1996–98	65.1	22.1	43.0	27.0	7.3	8.0	0.7

Source: Compiled from National Potato Council, *1999 Potato Statistics Yearbook,* p. 46.
Note: 1 kilogram = 2.2 pounds

sumers got fussy about pesticides and less fixated on uniformity, an environmentally friendly potato might have a chance.

Still, Russet-Burbank no longer reigns unchallenged.[26] Norchip and Kennebec have done well as chippers, in part because the potato-chip sector shifts to newcomers more quickly than the frozen-food sector does. Thin-skinned, early-maturing Atlantic needs no hardening prior to processing.[27] The Golden Flake Company of Nashville, Tennessee, makes its chips from Atlantic, brought in fresh from Florida farms.[28] The table-stock market also takes on new varieties. Red Pontiac was introduced in the 1950s. Yukon Gold, popular in supermarkets today, was not stocked a decade ago. Potatoes labeled "from Maine" are often more recent introductions, such as Shepody or Ontario; those from Minnesota might be Norland.[29]

And now for the sad truth. The best french fry in the world is not American, and it is not made from Russet-Burbank. The title, at least in my view, should go to the Dutch, who sell fries cut from fresh tubers on every street corner in Amsterdam. The fries come with big globs of ketchup, mayonnaise, and mustard on top. If you want them plain, say so in advance. The most popular french-fry variety in Holland is Bintje. The English french fry is likewise a fresh affair. I never got a frozen chip with my fish anywhere, except in London. And let us not forget the French, who in revolutionary fervor planted potatoes in the Tuileries gardens. In 1837, after the monarchy's restoration, the royal family's chef perfected the double-cooking formula that created the "souffle" potato.[30] Since then, the royal souffle has become the common french fry. In Paris today, Greeks, Turks, and Algerians, not gourmet French cooks, make the best fries; they do so the old-fashioned way. They start with freshly cut-up, unfrozen spuds, which they precook in the deep frier and heap near the window to draw in customers. Then comes the final rendezvous with the lard. When an order comes in, a big scoopful of potatoes is cooked again, this time at a very high heat.

"French fries will not tolerate mediocrity; they either go limp and collapse, or they are marvelous."[31] Amen to that! Conversation is lethal to the American fast-food version of the fry; if not eaten immediately, it turns to rubber. This is not so in loquacious Europe, where patrons can chat, get into an argument, make out, and still find an edible french fry waiting. They can even take fries home and eat them the next day, a claim no frozen fry can make.[32]

I must confess a weakness for the potato chip, and not just in its Frito-Lay guise. The chip is an American product. It started out in 1851 as an upscale item at a fashionable resort in Saratoga Springs, New York. Making chips was at first a hand-frying, family affair with a well-defined neighborhood market. Only in 1926 did chips show up in waxed-paper bags that were ironed together. Later came cellophane, and then the continuous frier.[33] Through it all, the chip has adapted to circumstances better than the conservative french fry, which finds change threatening. The potato chip revels in the diversity of its dips, from dill and cream cheese to guacamole. And it is less prone to the perils of a one-size-fits-all mentality. A potato chip will shape itself into ruffles or reconstitute itself into Pringles without complaint. It accepts flavors added right in the bag, from barbecue and onion to vinegar and sour cream. In India, red curry is mixed into many a popular brand. The Calbee company of Japan makes a seaweed-flavored chip. Crispy "Cascade Style" chips from Washington state are thicker than regular chips and come without salt. Such boldness opens the door to spuds with a range of cooking characteristics; they need not measure up to the Frito-Lay golden standard. It's the sugar, after all, that gives a chip or fry its color. A low sugar content, as with Atlantic, makes for a light, yellow-gold chip; a high sugar content means a darker color. When a chip's destiny is to be mixed in with the spices, there can be more leeway on the sugar content.

Who makes the best potato chips? Is it Frito-Lay, Golden Flake, Charles Chips, Wise, or Great Value chips from the discount store? What flavor should chips be? There are competing standards and growing specialty markets. Kettle offers a "natural, gourmet chip" flavored with salsa, mesquite, yogurt, or green onion. Michael's Chips are unsalted, with "40 percent less fat." Politically correct Little Bear gets its spuds from farmers "who rotate their crops, replenish soil nutrients, and avoid toxic chemicals." Terra brand of Brooklyn, New York, makes a Yukon Gold chip in garlic, barbecue, and *au naturel* flavors. And don't forget Terra's purple chips, made from a purple-fleshed cultivar. But Terra has not stopped with the potato. Mindful of minority rights, it also chips sweet potato, taro root, manioc, and parsnip.[34]

So what is the potato's destiny? Just as urban demand, the railroad, and mass marketing changed the potato in America, similar trends elsewhere are likely to have comparable results. Rather than the mixed farming of the past, there will be more specialization. Consumption will be concentrated in the cities, not in the countryside. City dwellers will want a good-looking, well-groomed product. To gain market share, new cultivars will need to process well. Within countries, regions with the best climates and highest yields will win out over less favored areas. Such shifts are evident from Ecuador and Bolivia to Egypt, Uganda, and India. They reflect the multifaceted ways in which urban life rewrites the script for living; the city is unstoppable.

Nevertheless, there is still room for the potato to maneuver; Russet-Burbank is not the end of the story. In East Africa, farmers select new cultivars that fit the environment; they cannot afford to change the environment to fit the cultivar. Farmers in Egypt have stopped using pesticides, even though they produce for persnickety urban consumers. Where the potato is judged according to its customers' tastes, as with India's curry-flavored chips, varietal diversity has a better chance. Even with a french fry, other standards and ways of cooking are compatible with a quality product. The key is to keep the dynamics of the equation in mind, from consumption patterns and processing to product development and marketing.

1. Maca roots. Courtesy CIP, from CIP, *Pocket Guide to Nine Exotic Andean Roots and Tubers*, p. 20.

2. Ulloco diversity. Courtesy CIP, from Michael Hermann, *Andean Roots and Tubers*, p. 4.

3. Oca diversity. Courtesy CIP, from Hermann, *Andean Roots and Tubers*, p. 12.

4. Achira rhizomes. Courtesy CIP, from Hermann, *Andean Roots and Tubers*, p. 26.

5. Potato diversity. Courtesy CIP, from CIP, *Potatoes for the Developing World*, front cover.

6. Flowering potato plant with seed balls. Courtesy CIP, *International Potato Center Annual Report 1994*, p. 23.

7. Rustic seed bed. Courtesy CIP and PROINPA, from *CIP Circular* (June, 1994): 15.

8. Sweet potato roots, flowers, and foliage. Courtesy CIP Archives, from CIP, *Exploration, Maintenance, and Utilization of Sweet Potato Genetic Resources*, front cover.

PART III

The Andean World

CHAPTER 6

Andean Agriculture

N o landscape on earth compares to the Andes. The Spanish soldiers who besieged the Inca Empire in 1532 had firsthand experience in Europe and North Africa, in the Caribbean, Mexico, and Panama. But even to such seasoned veterans, the Andes had no equivalent; "the mountains were higher, the nights colder, the days hotter, the valleys deeper, the deserts drier, the distances longer."[1]

The organization of the Inca Empire was intimately intertwined with this landscape. Its tiered agriculture, ceremonial cities, roads, and terraces all reflected the unique character of the Andean environment. The skills of its engineers and craftsmen rivaled those of their peers in Europe. After the conquest, it was Inca architects, stonemasons, and artists who built the great cathedrals, monasteries, and palaces of the Spanish colonial world. For almost two centuries, the core of Spain's empire in the Americas was the heartland of the old Indian civilizations. Spain's success in the New World rested on the accomplishments of the Aztecs and Incas.[2]

To Spanish eyes, what mattered most about the Incas was the gold that adorned their jewelry and ceremonial objects, the gold that emblazoned Cuzco's Temple of the Sun. Yet, compared to the Incas' achievements in cosmology and architecture, in road building, in weaving, and in agriculture, the gold is but dross.

Inca ritual synchronized the seasons and crop cycles with the yearly movements of the stars. Inca engineers designed Cuzco so that the Temple of the Sun, ablaze in gold, stood at the heart of the city. The temple held the sacred image of the sun, an immense golden disk of rays and flames. The moon was cloistered with the sun in silver chambers; next to the moon stood the halls of the Seven Sisters (Pleiades), the planets, and the stars. From this astronomically aligned central point, Inca priests watched the heavens and tracked the year's agricultural cycle.[3]

Machu Picchu, the "lost city of the Incas," is a breathtaking site of symmetry, simplicity, and haunting beauty. No place on earth is quite like it. The sheer size, workmanship, and engineering that mark Andean architecture are remarkable. Inca masons built palaces and temples with three types of stone, each selected for its distinctive color, hardness, and weathering characteristics. To build the great ramparts at Sacsahuaman fortress, craftsmen cut gigantic blocks into irregular shapes, then bevelled the edges into decorative joints, skillfully worked to fit together exactly. The wall's geometry is as impressive as its massive size.[4]

The quality of Andean textiles was superior even to that of the great tapestries of Flanders. In the lowlands, weavers used primarily cotton; in the highlands, they used the long, fine wool of the alpaca. The finest blankets and cloth had a velvet softness, with deep, bright colors and a smooth, continuous seam. Weavers used dyes of exquisite scarlets, indigos, yellows, and blacks; they tinted the fibers before spinning. Made from flowers, herbs, and bark, the dyes were rich, true, and colorfast, superior to the tints used in Europe. Despite the dense population, people had ample food, warm clothing, and security. Along the great roads, the Inca stocked thousands of stone warehouses with fine cloth, blankets, grain, freeze-dried tubers, and salted meat.[5]

Two great north-south roads, one along the coast and one through the highlands, ran the length of the Inca Empire; roads also crisscrossed the mountains from the Pacific coast to the Amazon savannas. Every significant town was linked to the main thoroughfares. In all, the Inca network comprised an estimated 40,000 kilometers (25,000 miles) of roads. Impressive technically, roads were paved with flat stones and protected by retaining walls and earthworks. Engineers zigzagged them over the steepest slopes, using steps as necessary; where they found marshes, they raised the roads up on causeways; they spanned dangerous rapids and deep riverbeds with suspension bridges. The Inca had no wheeled vehicles. They designed the roads for traffic on foot, including pack trains of llamas. Way stations were strategically placed along the roads every mile. Trained runners carried information encoded in knotted quipus to every corner of the empire. Couriers made the Quito to Cuzco run, a distance of 2,400 kilometers (1,500 miles), in five days.[6]

We still admire Inca roads, the ruins at Machu Picchu, and the great fortress at Sacsahuaman. Andean pottery, ancient textiles, and gold artifacts are displayed in Lima's museums. Inca beliefs, aspects of daily life, and the nature of Inca political power can be reconstructed from the testimony of contemporary witnesses, such as the Inca nobleman, Garcilaso de la Vega, the soldier-chronicler, Pedro de Cieza León, or the Jesuit priest, Bernabé Cobo.[7] They can also be pieced together from archaeological remains. The greatest accomplishment of the Andean world, however, is the balance, flexibility, and sophistication of its agriculture.

Geared to every nuance of altitude and rainfall, the complexity of Andean cropping systems had no precedent in Europe. Even so, the Spanish chroniclers paid these systems but scant attention; Pedro de Cieza León, for example, virtually ignores the potato. What impressed him about Inca agriculture was the engineering behind it, the terraces and the irrigation systems.[8] Terraces varied in length, in width, and in the height between "steps." A first step can be over a kilometer long and sixty meters wide, the last just big enough to fit in a few rows of corn. The overall surface thus gained is great. Archaeologists have found terraces throughout the central Andes. Peru alone had approximately one million hectares of terraced land, of which about 40 percent is still in use.[9] Most terraces had irrigation; the water came from mountain streams farther up the watershed. Vital to agriculture, water was a symbol of supernatural power and vitality; its sources, from springs to the confluence of rivers, became sites for temples and shrines. The Inca capital, Cuzco, had extensive irrigation works, water-control projects, and fountains. Running water was piped to the city's palaces and temples and to the Sacsahuaman fortress.[10] Garcilaso de la Vega considered the Inca water system far superior to anything he saw in Moorish Spain. The distance the irrigation canals covered was greater, the channels were wider, and they carried more water.[11]

Irrigation works in Inca times (1438–1532) were typically state-sponsored projects.[12] Terraces and water systems, however, were basic features of Andean agriculture long before the Inca.[13] They are found in every ecological zone, from the altiplano to the Amazon basin. Over three thousand years ago, farmers grew potatoes and quinoa near Lake Titicaca at an altitude of 3,800 meters (12,500 feet). They did so in trenched, raised beds surrounded by water.[14] At about the same time, the Chimú culture of coastal Peru constructed an extensive irrigation network between the Chicama and Moche Rivers. Some eighty-four kilometers in length, the main channel winds through an uneven, difficult terrain, an engineering feat that even today would require a sophisticated command of hydraulics.[15] All totaled, Peru's coastal valleys alone had some three hundred thousand hectares of terraced land. On the eastern, lowland slopes of the central Andes—in the subtropical yungas— ridged fields, causeways, and platform mounds are common. Even on the margins of the Amazon basin, a region archaeologists once thought unfit for agriculture, ancient farmers built up elaborate earthworks to elevate their fields above the flood level of the surrounding savannas and wetlands.[16]

In the Andes, to go up or down 1,000 meters (3,280 feet) is to experience a radical shift in climate and hence in the type of crops that can be produced. In Inca times, a village in the semitropical yungas was a couple of days away from a village in the altiplano. The temperate valleys stood adjacent to both. This vertical compression allowed villages to advantageously exploit ecological diversity. Each tier produced what suited it best yet benefited from production in zones one or two ecological steps away. Bolivia's yungas, for example, are just a couple of tiers below the altiplano, but in terms of climate, the yungas are as far from Lake Titicaca as Maine's potatoes are from Cuba's sugar.[17]

From its core around Cuzco, the Inca Empire expanded north as far as Pasto in Colombia and south as far as the Maule River in Chile, a distance of some 5,440 kilometers (3,400 miles). From Pasto to approximately Salta, in northern Argentina, the Andes are above the Tropic of Capricorn in the torrid, equatorial zone. At such latitudes, the key to temperature is altitude; combined with rainfall, it defines the main Andean climatic tiers: the lowland valleys, or yungas; the temperate, intermontane valleys; the high mountain slopes; and the altiplano, or puna. (See table 5.)

Rain comes to the southern Andes from across the Amazon basin to the east, not from the Pacific Ocean to the west. Situated on the eastern slopes at between 1,000 and 2,300 meters (3,300 to 7,500 feet), the yungas are humid, subtropical, and densely forested. In this zone of undulating mountains and rolling hills, rainfall is heavy, with some precipitation almost every month. Daytime temperatures reach thirty degrees centigrade (eighty-six degrees Fahrenheit), but the nights are cool and refreshing. There is never a frost in the yungas.[18] In these warm, low-lying valleys, Andean communities grew sweet cassava, the starchy achira rhizome, sweet potatoes, the leguminous ahipa root, and chilies. Cotton for textiles came from the yungas, as did timber and forest products for highland construction.[19] And do not forget the highly prized coca leaf, common in the Andes long before the Spanish Conquest. Today, as in Inca times, the harvested leaves are dried in the sun and packed into bundles. Lime is then mixed in with the leaves for chewing, which releases a small amount of cocaine. The stimulation from the swallowed juice is probably less than that of a modern cigarette.[20]

A step up from the yungas, at between 2,300 to 3,500 meters (7,500 to 11,500 feet), come the midaltitude, intermontane valleys. The Incas, and later the Spaniards, built most of their cities in this temperate, mesothermic zone: Popayán and Pasto in Colombia; Quito and Riobamba in Ecuador; Cuzco, Huancayo, and Arequipa in Peru; Cochabamba and Sucre in Bolivia; and Salta in Argentina. The intermontane valleys have agreeable daytime temperatures of between twenty-two and twenty-five degrees centigrade (seventy-one to seventy-seven degrees Fahrenheit). Nights are bracing, with lows of between seven and ten degrees centigrade (forty-four to fifty degrees Fahrenheit); above 3,000 meters (9,800 feet), however, winter frosts are possible. The Spaniards preferred this zone for their haciendas. The climate favors Old World crops like wheat as well as cattle ranching and dairy farming. The zone's agricultural year is divided between a summer rainy season (December to March) and a winter dry season (April to November).

To the Inca, corn was the privileged crop, and they reserved the best land in the temperate valleys for it, including irrigated fields. Every ceremony and religious

Table 5. Verticality in the Andes, by Ecological Zone

Meters	Zone	Production	Characteristics	Feet
4,800 to 4,000	altiplano or puna	herding of llamas and alpacas bitter potatoes for chuño kañiwa, maca	periodic summer frosts constant winter frosts bittercold nights	15,700 to 13,100
4,000 to 3,500	high or upper valleys	andigena potatoes ulluco, oca, mashua quinoa, tarwi	frosts common in winter sporadic frosts in summer cold nights terracing and irrigation common	13,100 to 11,500
3,500 to 2,300	midaltitude mesothermic, intermontane valleys	dairy farming wheat, corn arracacha yacón, mauka	location of most Spanish cities winter frosts possible above 3,000 meters temperate, cool nights	11,500 to 7,500
2,300 to 1,000	warm valleys or yungas	achira, chilies sweet potatoes ahipa, cotton forest products coca leaf	abundant rainfall never frosts mild nights	7,500 to 3,300
under 1,000	savannas and forests	medicinal barks and plants, forest fruits, fish	abundant rainfall hot and humid fluvial	under 3,300

Source: The scheme follows Pulgar Vidal, *Geografía del Perú*, pp. 51–110, 208–10. In practice, this is only a rough guide to Andean geography; more complex subdivisions are possible.

ritual called for corn in some form. Compared to the lowbrow potato, corn was a high-status ceremonial food and a symbol of hospitality. Corn was also fermented into chicha. Since chicha was the everyday Andean beverage, people consumed it in enormous quantities, especially at ceremonies and festivals.[21] The Jesuit chronicler, Bernabé Cobo, complained bitterly that people "spend their days and nights drinking chicha excessively and dancing to the tune of harsh-sounding drums and songs." Besides corn, the temperate valleys were ideal for root crops such as arracacha, which is used today in baby foods; the sweet, fruit-like yacón; and the virus-resistant, rooty mauka plant.[22]

The high mountain valleys are at altitudes of between 3,500 and 4,000 meters (11,500 and 13,100 feet); they are steeper, narrower, and colder than the valleys below. For productive farming on the slopes, terracing is imperative. Otherwise, heavy summer rains erode the soils. Most of the ancient terraces in Peru's highlands are in the high mountain valleys. With the exception of La Paz and Potosí in Bolivia, few cities of consequence are found here. Spaniards left these colder, less hospitable highlands to Indian communities. The maximum daily temperature, even in summer, does not get much above twenty degrees centigrade (sixty-eight degrees Fahrenheit). Frosts are common during the dry, clear winter months, from May though August; they occur sporadically during the summer. Communities in the high valleys produce the widest assortment of potatoes, with frost-resistant types kept for the upper terraces. Other tuber crops include ulluco, oca, and mashua. To maintain soil fertility, Andean communities combined a long fallow with rotations to high-altitude grains like quinoa and kiwicha or to the leguminous tarwi, a lupine as nutritious as the soybean. In Inca times, llama caravans brought guano fertilizer up from the coast.[23]

The last step up for agriculture is the altiplano, or puna, an austere, windswept zone at elevations of between 4,000 and 4,800 meters (13,100 to 15,700 feet). In summer, daytime temperatures reach fifteen degrees centigrade (fifty-nine degrees Fahrenheit). In winter, it freezes most nights; in the summer, frosts can be expected periodically. In the southern Andes, most of the rain falls in the lower valleys; the altiplano rarely gets more than one thousand millimeters (thirty-nine inches), most of it between December and March. The Lake Titicaca basin is in the puna zone, but it is sheltered by a rim of mountains; this makes it warmer and wetter than the altiplano farther south. The mainstay of altiplano communities is herding llamas and alpacas. Llamas produce a coarse, stiff wool, which is woven into a rough cloth for cloaks and blankets. In Inca times, llamas were the beasts of burden; they were also kept for their meat, which was dried into *charque,* or jerky. Alpacas, by contrast, were kept almost exclusively for their fine wool.[24]

The zone's subsistence crop is the frost-resistant rucki-type potato plant. In some parts of the altiplano, frosts are possible three hundred days a year. Andean farmers domesticated the redoubtable rucki types, which comprise two distinct potato species.[25] Although bitter, the potatoes can be processed into a delicious, dried *chuño* for storage. To make *chuño,* communities sorted potatoes by size and left them exposed to freeze by night and defrost by day. When thus ripened for several days, rucki types lose their bitterness. The next step is to squeeze out the liquid, flatten the potatoes, and dry them in the sun. When stored in the altiplano, dehydrated *chuño* resists humidity, weevils, and moths for many years. In Inca times, it was a staple in the Andean highlands and much prized by communities in the lowland valleys. Even today, it shows up in every type of soup or stew.[26] Besides potatoes, altiplano communities domesticated the cold-hardy grain kañiwa, viable in protected areas at altitudes of up to 4,400 meters (14,400 feet), and the radish-like root crop, maca.[27]

The tiers described above are but rough guides to reality. Precisely where in the Andes one goes up or down these hypothetical steps makes a big difference. The great divide is between the wet Andes north of the Cajamarca Valley in central Peru and the drier ranges to the south. In Ecuador, for example, it rains heavily on both the Amazon and Pacific sides of the cordillera. Coastal Ecuador, especially the province of Esmeraldas, is awash in rain. South of Cajamarca, however, the Andes spread out, and the altiplano is higher. In the central Andes, the intermontane valleys to the west are much drier than those to the east. Along the Pacific coast, it is drier still. From Chiclayo, in Peru, almost to Santiago, in Chile, the coast is mostly desert. At Nazca, in southern Peru, the famous lines ancient people traced across the desert sands were still visible two thousand years later. They survived because it rains there rarely; most precipitation falls farther east, on the windward slopes of the Andes.[28]

The Andes reach their widest point near the densely populated Lake Titicaca basin; here, the altiplano is almost three hundred kilometers long and two hundred kilometers wide. What the rain clouds that blow in from the Amazon do not drop in the yungas, or temperate valleys, falls on the altiplano; there is little left for the parched, desert coast. What water gets to the coast comes from a few mountain-fed rivers. Without irrigation, agriculture in the coastal valleys is virtually impossible. In terms of altitude and rainfall, the tiers described above—yungas, intermontane valleys, high valley slopes, and altiplano—apply most directly to central Peru and Bolivia. The correlation, however, is at best approximate. Even in the central Andes, rainfall totals vary by valley, and every valley floor is at a different altitude. So the principle of verticality holds throughout, with microclimates staggered along the great north-south spine of the Andes.

In Inca times, towns located where tiers converged were the main centers. Tuber fields and cornfields could be easily reached above and below such towns. To exploit the tropics, Incas established colonies down in the yungas. Colonists sent cotton, timber, and coca leaves to the highlands, getting back a secure supply of corn, *chuño,* and jerky.[29] Today, the old irrigation systems and many of the terraces are gone. The demographic collapse that followed the Spanish Conquest, the imposition of new gods, and a different, hacienda-based production system have all taken their toll. Still, the ancient crops have endured in the communal fields of Andean highland farmers. In the central Andes, ties between llama herding on the altiplano and tuber production in the highland valleys are still intact; as in Inca times, llama dung enriches the potato plots. For their part, the yungas still supply the highlands with tropical produce, and the zone is still a destination for migrant highlanders.[30]

After the last ice age, agriculture started out in the warmer lowlands and along the coast. Archaeologists have dated crop fragments for arrowroot, manioc, squashes, and corn at between ten thousand and seven thousand years ago, long before such crops are found in the uplands.[31] The diversity of the Andean crops subsequently domesticated reflects the way altitude, rainfall, and latitude intersected to create novel situations. The result was an "archipelago" of climates, crops, and cultures.[32]

By Old World standards, Andean farmers were not really farmers at all. They had no plows or draft animals. To break up the soil, they used a *taklya,* or foot plow. It had a sharp blade of wood, stone, or metal fixed to a long pole; a foot rest above the blade; and a handle near the top of the pole. Plowing with a *taklya* was considered men's work. Using both hands, a farmer jabbed the blade full force into the ground; then he stepped up onto the foot rest, using his weight to work the blade into the soil as far as possible. Finally, he pulled back on the handle, turning up a large square of earth. Lines of turf were turned over back to back in opposite directions, creating parallel horizontal furrows, a technique that reduces erosion. After a pass with the foot blade, the women took over, working the soil further with hoes and clod breakers.[33]

The pattern in Andean agriculture is cultivation in small plots. So Andean farmers were, in fact, more like gardeners. On each terrace step communities planted a colorful collage of tubers, roots, grains, and legumes. Such diversity reduced the risk of a total crop failure. If a frost hit or a drought set in, the impact was different on each plot. Moreover, plot combinations created microecologies that impeded pests and diseases yet encouraged beneficial insects.[34] With small plots, Andean farmers were likely to spot novel variants of a familiar crop. This helped them match plants to microclimates. Adjacent tiers and thousands of plots stimulated genetic diversity. In the potato's case, Andean farmers ended up with eight domesticated potato species and over three thousand different varieties. The greatest diversity is still found in the highlands around Lake Titicaca and in the adjoining valleys, where potato types can be classified by the altitudes at which they grow.[35]

Andean farmers still safeguard this heritage in their fields.[36] Understanding how Andean communities organize agriculture is less romantic than searching for lost cities, but as a lesson in how to live harmoniously with the environment, it may be far more significant. The lost cities will always be there, waiting for archaeologists to find them. Andean crops and the agricultural know-how that sustains them may not. Urban migration now draws its recruits from even the most remote valleys, putting local knowledge of ancient crops at risk. In the cities, newcomers acquire different dietary habits. The old, time-consuming ways of cooking give way to fast foods. As a result, the demand for and knowledge of ancient crops is eroding. What ancient farmers achieved in the unique environment of the Andes could be lost forever.

CHAPTER 7

Lost Crops

Peru's national collection of three Andean tuber crops, ulluco, oca, and mashua, was stored at the Ayacucho agricultural station. In 1982, Sendero Luminoso terrorists attacked the station and threatened to burn down the storage facility. "They had their torches lit," remembered Ayacucho agronomist Carlos Arbizu, "but a campesino pleaded with them: 'Compañeros, this belongs to us, the campesinos of Peru, please respect it.' Thank God they did." But Sendero attacked again. This time, the head of the station told Arbizu to get the collection out of the station, to take it anywhere it might be safe. That very day, Arbizu loaded up the collection in his truck. That night, Sendero completely destroyed the storage building.[1]

Arbizu took his cache of tubers up into the highland valleys, to villages like Chunyag, situated at 3,600 meters (11,800 feet). "Wherever I went," he said, "I asked the villagers to plant the tubers and take care of them. They did. Soon, all the villages had gardens full of Andean tubers, little germ plasm collections spread out across the countryside." After the threat from Sendero diminished, Arbizu restored the collection, but the villagers kept samples of what they liked the best. Peru's ulluco, oca, and mashua had been saved, once again, by Andean farmers.

Today, Carlos Arbizu works for the International Potato Center (CIP) in Lima, Peru. Established in 1971, CIP is supported by contributions from private foundations, the World Bank, the United Nations, and a core of donor countries, including the United States. CIP's mission is research that helps poor countries grow more food in a sustainable way; its target crops are potatoes and sweet potatoes. Since 1991, it has added endangered Andean crops to its research agenda. It also spearheads work on mountain agriculture in the Andes. CIP promotes sustainable agriculture, organizes research networks, and trains scientists from around the world.[2] A member of the Consultative Group on International Agricultural Research (CGIAR), CIP is one of sixteen research centers focused on global food production problems.[3]

Arbizu, who grew up in Peru's remote central Andes, started collecting Andean crops twenty years ago. He studied at England's University of Birmingham under J. G. Hawkes, renowned potato researcher and historian. Today, Arbizu is responsible for the Andean root and tuber collection assembled at CIP. But he has not forgotten the lessons he learned at Ayacucho. CIP replants its collection of Andean tubers every year: 446 ullucos, 482 ocas, and 92 mashuas. It does so with the help of villagers from La Libertad, a town above the Mantaro Valley at 3,800 meters (12,500 feet). When planted in the high valleys, the crops have few insect or disease problems, so farmers do not need pesticides. Down a tier, at 3,200 meters (10,500 feet), however, viruses can multiply rapidly. La Libertad had already been planting oca and ulluco, but it had only a couple cultivars. "By working with us," says Arbizu, "villagers see for themselves how much diversity there is in oca and ulluco. When they find an unfamiliar variety they like, they pocket a couple small tubers as a sample, which is a good way to get more cultivars into circulation." When planted at the right altitude, Andean tubers are reliable and easy to grow. After planting tubers, farmers rotate to barley or quinoa, followed by a fallow.[4]

Today, CIP heads up an international effort to save endangered Andean crops, the so-called lost crops of the Incas.[5] CIP's list includes the above-mentioned oca and ulluco, plus mashua. All of these are tubers, and they are planted at approximately the same altitude as the potato; indeed, they can be found intercropped in the same fields. CIP's Andean agenda also includes maca, a starchy, high-altitude root crop, plus a set of roots from the warmer valleys below, namely arracacha, mauka, ahipa, yacón, and achira.[6] Research on the sweet potato, a root crop domesticated in the Amazon basin, is also part of CIP's research mandate, but not under the lost-crop rubric.

Most Andean tubers and root crops are unknown outside South America. Only the potato is widely dispersed today. Popular New World crops, such as tomatoes, peppers, corn, squashes, and most of the beans, reached Europe not from the Andes but from the Spanish Main or from lowland South America.[7] This is not because the Andean versions were genetically deficient. The explanation has as much to do with geography and the organization of trade as it does with biology.

The heartland of Andean crop diversity is in the highlands of central Peru and Bolivia. In colonial times, the central Andes were much more isolated from Europe than either the Spanish Main or Portuguese Brazil was. The Spanish fleets sailed directly from Spain to the port of Veracruz in Mexico or to Cartagena in Colombia. Later, they rendezvoused in Havana, Cuba, for the return voyage. From the beginning, exotic Caribbean and Mesoamerican plants entered transatlantic trade. Peru, by contrast, was off the beaten track, on the Pacific side of South America. The only direct sea passage was through the Strait of Magellan at the tip of the continent, a dangerous detour that added many thousands of extra sea miles to the trip. Because so many Andean crops reproduce from perishable, vegetative material, distance and isolation worked against them. The closest the Spanish fleets could get to Peru was Cartagena, on Colombia's Caribbean coast, or Portobelo, on the Isthmus of Panama. To get goods to these ports from upper Peru was a daunting logistical and administrative task.

Consider silver from Bolivia's mines. First, it had to be hauled down the mountains to the Pacific port of Arica, from whence it was shipped to the viceregal capital, in Lima. After cargo was cleared at customs, coastal shipping took it up to the Pacific side of the isthmus, from which it was taken overland by mule train to Portobelo, on the Caribbean coast. From there, it had to be shipped again, this time to Cartagena. To get anything from upper Peru to Spain took at least two years.[8] That being the case, it is a miracle the potato ever made it out. In fact, the first potatoes probably came from the Colombian highlands and not from Peru. Even today, the Colombian departments of Boyacá and Norte de Santander are big potato producers.[9] From either place, an enterprising potato could have made it down to Cartagena in a couple of months.[10]

Spain's commercial prominence declined after 1700. Fleets sailed irregularly, especially those with cargoes destined for Cartagena and ultimately Peru.[11] Not until the end of the eighteenth century did Spain permit free trade between all its ports and those of its colonies. Even then, the closest Atlantic port for exports from upper Peru was Buenos Aires, which is many arduous months of overland travel away.[12] Only when the Panama Canal opened in 1914 did Peru have direct sea links to Europe. Distance and isolation, however, are not the whole story. The conquest of Peru began in 1532, a generation after Columbus. There followed a decade of Inca resistance and civil war. By the time the Spanish crown consolidated its authority in Peru, the novelty of exotic crops and the "New World" had worn thin. Moreover, for purposes of genetic exchange, the viceregal capital was in the worst possible place. Pizarro built Lima down on the coast, far removed from the Inca heartland in southern Peru and Bolivia. Cortés, by contrast, built Mexico City over the ruins of the old Aztec capital, Tenochtitlán. The new city stood at the center of Aztec cultural dominance and demographic density. As a result, the exchange between cultures was more thorough in Mexico than in Peru, where Lima's isolation from Cuzco reinforced the cultural distance between Indian and Spaniard. Today, Mexico considers itself a mestizo nation of mixed cultures and identities. Mexican cuisine reflects this heritage, skewed as it is toward tortillas, chili peppers, and *mole poblano* sauce.[13] Mexican food is essentially Indian food and has been so since long before the onset of urbanization and the fast-food chain. The situation in the Andes is much more complex. Andean crops like floury corn, lima beans, potatoes, and squash crossed over into Spanish colonial cooking culture. But many crops did not, identified as they were with the highlands and a Quechua-speaking, alien world.

In contemporary Peru, Indian and European cultures "are still separate and unequal, and so too are their food crops."[14] But even among Andean people, the knowledge of ancient crops has eroded. A generation ago, Peru, Ecuador, and Bolivia were still predominantly rural societies. Today, only a third of Peru's labor force resides in the countryside; the figures for Ecuador and Bolivia are 29 percent and 44 percent respectively.[15] A modernized, market-oriented economy is not a hospitable place for the old crops. Many supermarkets do not stock them, and commercial food processors often ignore them. Nevertheless, Andean roots and tubers are a hardy, nutritious lot. They are discussed below, grouped by tier. (See table 6.)

The remarkable, radish-like maca root thrives from the altiplano to just below the snow line. It resists the zone's intense sunlight, fierce winds, and frigid nights. Maca survives at altitudes of 4,500 meters (14,750 feet), a habitat too bleak for even bitter, frost-resistant potatoes. Propagated from true seed, altiplano farmers surround the small maca plants with stone fences to keep out sheep and llamas. In size and shape, the maca plant resembles a large radish. Its colorful yellow roots are banded in varying combinations of purple and black; the flesh inside is a pearly white. Maca can be baked fresh or sun dried. It has a tangy butterscotch taste and aroma; dried, the roots become "brown, soft, and sweet, with a musky flavor."[16] The protein content in dried maca is 13 to 16 percent, superior to rice and wheat, and maca is rich in amino acids.[17] Fresh maca roots have lots of iron, calcium, and iodine. Maca advocates consider the plant medicinal. Described as "Peruvian ginseng," it allegedly improves mental and physical capacities, stimulates sexual drive, promotes fertility, and boosts the immune system. Pharmaceutical companies in Lima, such as Quimica Suiza, have invested heavily in maca research. In 1998, the company exported over $80,000 worth of gelatinized maca capsules to Japan, Europe, and the United States.[18] Maca's reputation on the health circuit has come just in time. In all of Peru, not much more than fifty hectares of maca is produced annually, and most of that is in just two departments, Junín and Cerro de Pasco.[19]

Ulluco, oca, and mashua tubers do best in the high mountain valleys. Farmers propagate them from tubers stored over the winter. Oca does not set true seed, not even under Andean conditions. Like potatoes, all these crops store well. Ulluco, oca, and mashua are thin skinned; along with potatoes, they show up in soups and stews. Mashua is rich in protein, which constitutes about 15 percent of its dry matter; for ulluco the figure is roughly 10 percent, and for oca 9 percent. All are good sources of carbohydrates. Ulluco, in addition, has a great deal of vitamin C, and oca is rich in calcium and iron. Ulluco, oca, and mashua are all frost tolerant and

Table 6. Andean Crops, by Selected Characteristics

Crop	Zone	Type	Preparation or Use	Propagation	Distinctive Color or Shape	Special Characteristics	Number of Andean Users	Time to Maturity (in months)	Accessions Held by Andean Germplasm Banks[a]
maca	altiplano	root	baked fresh	seeds	radish-shaped roots; banded in purple and black	13–16% dry-matter protein content; stimulates sexual drive; promotes fertility; boosts immune system	<1 million	7–9	33
ulluco	high valleys	tuber	boiled; used as a puree in soups and stews	tubers	stubby to round splotched tubers; glossy, heart-shaped leaves resembling nontuberous begonias	smooth, thin skin with watery flesh rich in vitamin C	30 million	6–8	2,476
oca	high valleys	tuber	boiled, baked, fried, or freeze-dried	tubers	colorful, cylindrically shaped, indented tubers; shamrock-shaped leaves that resemble columbine	firm flesh; bitter types can be freeze-dried into chuño; rich in calcium and iron	15 million	6–9	3,899

Table 6. *Continued*

Crop	Zone	Type	Preparation or Use	Propagation	Distinctive Color or Shape	Special Characteristics	Number of Andean Users	Time to Maturity (in months)	Accessions Held by Andean Germplasm Banks[a]
mashua	high valleys	tuber	boiled, baked, or fried	tubers	sprawling nasturtium; oval, indented colorful tubers	12–15% dry-matter protein content; calming effect; reduces sex drive; diuretic; high yielding	<10 million	6–8	1,032
arracacha	midaltitude valleys down to the yungas	root	boiled; added to soups and stews	stalks	large, conical storage roots	used in baby foods; an ingredient in soups and stews; orange-fleshed types rich in vitamin A	30 million	10–14[b]	577
mauka	midaltitude	root	boiled or fried	cuttings from roots or stems	large cylindrical yellow-skinned roots	pest and virus free; rich in calcium, potassium, and phosphorus	<100,000	12–18	119

ahipa	yungas	legum-inous	eaten raw as a root; fresh vegetable used in salads and stir fry	seed	large, top-shaped root; white, crispy flesh; mildly sweet taste	8–18% dry-matter protein content; fixes nitrogen; matures rapidly	<100,000	6–9	69
yacón	midaltitude valleys down to the yungas	root	fresh fruit	crown offsets	crunchy, sweet, watery roots	unmetabolized sugar for diabetics	<1	6–7	433
achira	yungas	rhisomic	boiled or baked; root processed into starch	rhizomes	large, edible canna rhizomes	thrives in warm, rainy valleys with saturated soils; large starch granules	<1	8–12[c]	362

Sources: Compiled from crop descriptions and tables in Vietmeyer, *Lost Crops of the Incas*; Herman and Heller, eds., *Andean Roots and Tubers*, pp. 6, 8; and Arbizu, Huamán, and Golmirzaie, "Other Andean Roots and Tubers," pp. 40, 42.

[a]Many of these may be duplicates.

[b]Immature roots can be dug in 4 to 8 months.

[c]Rhizomes can be left in the ground almost indefinitely.

relatively pest free. Like potatoes, however, they are notorious for virus buildup, which reduces yields. All have adapted to the short day length of tropical latitudes, much like the original potato types that reached Europe centuries ago.[20]

I first spotted ulluco in an outdoor market in Cochabamba, Bolivia. Well, not exactly. I first thought the apricot-sized yellow tubers with reddish-purple splotches were an unfamiliar but colorful potato variety. The women who ran the ulluco concession soon set me straight. Compared to potatoes, ulluco has smoother, thinner skin; the tubers are smaller—some more round than oval, others elongated. If not more varied in color than potatoes, they are much brighter, with strong shades of rose, orange, red, and magenta. Inside, the flesh is yellow to white. Boiled in their skins, they make a smooth puree that can be added as a thickener to soups and stews. Ulluco's glossy, heart-shaped leaves look like those of nontuberous begonias. The plant spreads out close to the ground.[21]

Cylindrical in shape with deep eyes, oca tubers remind me of small pine cones. Three or four oca tubers fit easily into the palm of one's hand. Attractive and colorful, oca is sold in mixtures of creamy white, greenish yellow, pale pink, and deep scarlet red. Compared to ulluco, which tends to be watery, oca has a firmer flesh; in addition to being boiled, it can be baked and fried. The shamrock-shaped leaves of the bushy oca plant resemble columbine. There are both sweet and bitter types of oca. Left in the sun, the glucose in sweet varieties doubles. Bitter oca types can be processed, like potato *chuño*. The oca is soaked in water and left overnight to freeze; the water is stamped out prior to sun drying. Ulluco can be similarly processed.[22]

Mashua is the highest yielding of all the tubers. With minimal care, harvests from this sprawling nasturtium reach twenty to thirty metric tons per hectare. Since a single plant can produce four kilos of tubers, a few rows suffice for home consumption. Like oca, mashua tubers are oval and indented, but they are smaller, with one end typically stubbier than the other. (See fig. 1.) The colorful white to yellow tubers are streaked and spotted in orange and reddish violet. Mashua can be boiled, baked, or fried. A standard in folk medicine, it has a calming effect, reputedly reducing the sex drive. Mashua is also used as a diuretic for cleansing the kidneys. When intercropped with other tubers, it seems to deter insects and reduce bacterial diseases.[23]

Farmers like to plant small plots of ulluco and oca near their potatoes. Finding mashua is much harder. A market study from Ecuador showed that even in Quito only 20 percent of the consumers interviewed knew what mashua was, much less ever purchased it.[24] In rural areas, people were embarrassed to say they ate it, since city folk identify its consumption with Indians and backwardness. To a lesser extent, ulluco and oca also suffer from a negative image. As Andean farmers become market oriented, they cut back on traditional, less popular, crops.[25]

"Putting samples in a germ plasm bank is but one aspect of protecting a crop's

Fig. 1. Pedro Cruz with mashua plants. Courtesy CIP, from Hermann, *Andean Roots and Tubers*, p. 18.

diversity," Patricio Espinosa told me. "We also need to preserve the knowledge Andean people have. Each variety has its own taste and cooking characteristics. To reverse the downward trend, we have to match the appropriate varieties with specific uses."[26] Espinosa and his Ecuador-based research team want to bolster urban demand in the country's main cities. With backing from CIP and a five-year grant from the Swiss Development Agency, they sampled consumers by age and social class, asking them which Andean crops they knew, how often they bought them, and how they prepared them. The research team followed the crops back to the countryside, looking at how farmers in selected villages produced, stored, and marketed them. The result was a series of basic studies, a cookbook, and a television series entitled "Cook with Class," which showed how to prepare Andean specialties.[27]

The dispersed parcels characteristic of Andean farming make crop estimates notoriously unreliable. To get a more accurate assessment, the team worked back from consumer samples, calculating how much ulluco and oca had to be produced to meet urban demand in Quito and Cuenca, both highland cities, and coastal Guayaquil, the country's largest city. The team concluded that production in Ecuador greatly exceeds official figures, especially for ulluco. Annual per-capita consumption in the three cities averaged over eleven kilos, implying an annual production of at least 32,000 metric tons. As for oca, only consumers in Quito and Cuenca purchased significant amounts of the crop; overall production was estimated at about 5,000 metric tons. Mashua consumption is marginal in the cities, but it still persists in the countryside, especially in the Quechua-speaking central sierra. In this zone, people claimed to consume it regularly, usually in soups and stews.[28]

In Ecuador's sierra, rural communities distinguish between ulluco, oca, and mashua varieties by the size, color, and hardness of the tubers produced. They also know which varieties grow the fastest and the best way to use each one. Quito's urban consumers, by contrast, are much less sophisticated. They complained that ulluco had too much gelatinous mucilage, as if this were an inherent trait of the crop. Rural people, by contrast, can gauge a tuber's mucilage content by its color and shape.[29]

"Andean crops will flourish," said Espinosa, "when urban consumers start to value them." The project's interviews showed that housewives simply did not know which varieties to select and how to cook them. After "Cook with Class" aired on television, demand for oca and ulluco increased in the supermarkets. Andean tubers, the program noted, were not only nutritious but healthy, as farmers produced them without using pesticides. The project started passing out recipes at the supermarkets and using taste tests to educate consumers about the varying mucilage content in ulluco varieties. The challenge now is to work out a dependable marketing scheme, so that communities can specialize and consumers can get a quality product.[30]

Compared to the tubers, Andean root crops are produced at lower altitudes, in the milder, generally wetter, mesothermic valleys. This temperate group includes arracacha, mauka, ahipa, yacón, and achira. Of these, arracacha is the best known. In the survey discussed above, 97 percent of consumers in Quito reported household arracacha use; for Guayaquil, the figure was 91 percent, and for Cuenca 68 percent. Annual per-capita consumption in the three cities was put at 8 kilos, 8.9 kilos, and 2.7 kilos respectively.[31] Similarly, in Venezuela, Colombia, and southern Brazil, arracacha is a popular item available in the produce section of most supermarkets. A member of the carrot family, the arracacha plant has large, conical storage roots resembling parsnips. (See fig. 2.) Roots tend to be close to the surface, like those of irises. Aboveground, arracacha puts out large, fleshy stems that look like celery stalks, except that they are round and less watery. The crop is propagated vegetatively. Farmers detach the fleshy stalks from the rootstock, cutting them back and trimming the leaves. The honed stalks are dried a few days before replanting. Roots take many months to bulk; harvests start a year after planting. The crop can be left in the ground for up to sixteen months, which is fortunate as it

Fig. 2. Harvested arracacha roots. Courtesy CIP, from Hermann, *Andean Roots and Tubers,* p. 22.

stores poorly and has a short shelf life. Arracacha is usually boiled fresh; it has a smooth texture and a nutty taste, not unlike that of butternut squash. It is also added to soups and stews. Andean consumers prefer white-fleshed varieties; Brazilians favor orange-fleshed types high in carotene. Since arracacha is noted for its digestibility, Nestlé-Brazil uses it in baby foods and instant soups. A small amount of fresh arracacha meets the daily requirements for ascorbic acid, vitamin A, and calcium; however, it has virtually no protein.[32]

Mauka is disappearing. Ask a hundred Andean farmers what it is, and ninety-nine will not know. The crop, in fact, was not even noticed by ethnobiologists until the 1960s. Mauka is an ancient crop, and only a few groups of Andean farmers in isolated departments of Ecuador, Peru, and Bolivia still grow it. Mauka prefers a somewhat cooler climate and higher valleys than arracacha does. Its large, cylindrical, yellow-skinned roots protrude aboveground, as big as zucchini squashes. Propagated with cuttings from both stems and roots, mauka can be harvested in a year, but it does much better if left in the ground to mature for two. The white-fleshed roots are typically boiled or fried. Mauka is rich in calcium, phosphorus, and potassium; its 7-percent protein content is high for a root crop. Mauka is pest free and does not carry viruses.[33] Apparently, it has a protein that impedes virus replication.[34] So far, CIP's germ plasm collection contains just five mauka accessions.

Fig. 3. Ahipa plant with roots. Courtesy CIP, from CIP, *Pocket Guide to Nine Exotic Andean Roots and Tubers,* p. 19.

Ancient farmers of the Americas domesticated the only leguminous root crop, the yam bean. Ahipa is the Andean species; jícama, available at many U.S. supermarkets, is a Mexican species. Although cultivated extensively in pre-Inca times, ahipa's redoubt today is Bolivia, where production is concentrated in the rainy, temperate valleys of the eastern Andes. Brown-skinned, with a white, crispy flesh, each ahipa plant yields a single root that weighs half a kilo and is shaped like a top. (See fig. 3.) The nitrogen-fixing ahipa is upright and compact, rather than sprawling. Compared to the late-bulking arracacha and mauka, rapidly maturing ahipa yields a marketable crop in just four months. Farmers replant each year using true seed. Ahipa is rich in carbohydrates, high in protein, and low in calories. Some 11 to 14 percent of the roots's dry matter is composed of protein, a percentage which is comparable to cereal crops like wheat and corn and much better than the potato. Ahipa roots are eaten raw or sliced up for salads. Ahipa's mildly sweet taste and appearance are reminiscent of the water chestnut. Ahipa can also be boiled or stir fried. Although botanically a vegetable, in Bolivian markets it is sold by the fruit vendors. An adaptable plant, ahipa flowers, produces roots, and sets seed irrespective of day length.[35]

Like ahipa, yacón doubles as a fruit and prefers the moist Andean valleys to the east; it is also unrestricted by day length and low in calories. But it grows much taller, usually over two meters, reaching maturity in six to seven months. Storage roots as big as sweet potatoes grow out from yacón's thick, rhizomic crown. (See fig. 4.) To replant the crop, farmers use offsets, which they break from the crown. Yacón is a popular treat, its crunchy, sweet fruit as refreshing as watermelon. The plant's range is more extensive than ahipa's. Andean farmers from Ecuador into Bolivia plant it for household consumption. Yacón roots are dried in the sun for a few days to increase their sweetness. They keep well and are peeled before eating. Fresh tubers are mostly water and fructose sugar, stored as insulin, which the body does not metabolize. Yacón is ideal as a fresh fruit for diabetics and weight watchers; in Japan yacón is used to produce a natural sugar substitute.[36]

A species of tall, colorful canna, achira looks almost exactly like its ornamental garden relative. In achira's case, however, the huge, starchy, underground rhizomes, or corms, are edible. From the West Indies to Chile, achira shows up sporadically in local markets. Only in Ecuador and Peru, however, is it a substantial crop. A robust plant of warmer, rainy valleys, achira thrives in saturated-soil conditions that would rot most root crops. Down in Ecuador's yungas, at between 2,200 and 2,500 meters (7,200 and 8,200 feet), farmers can plant the rapidly growing achira all year. It can be harvested within six months or left in the ground almost indefinitely. Farmers propagate the plant from rhizome tips that grow from the corm. Achira tolerates heavy soils and has few disease and pest problems. Corms can be cooked fresh, much like potatoes or manioc. They make for a nutritious animal

Fig. 4. Yacón roots. Courtesy CIP and Gigi Chang, from CIP, *International Potato Center Annual Report 1994,* inside front cover.

feed, especially when the plant's stems and leaves are mixed in. Achira's most important, large-scale use, however, is in starch production. Corms are 22 to 25 percent dry matter; of this, 12 to 16 percent is starch.

In Ecuador's Patete district, in the yungas of Tungurahua Province, making achira starch is a local industry. The washed corms are shredded in mechanical graters, forming a sticky mass that is pressed though a sieve. The waste that cannot be strained is recycled into the fields, bolstering the soil's organic matter; the paste that goes through is removed and mixed with water in large basins or cement tanks. The lighter, fibrous pulp rises to the surface, and the heavier starch falls to the bottom. Achira has the largest starch granules ever measured. When cooked, the glossy transparent starch is easily digestible. In Ecuador, achira starch ends up in a wide assortment of local products from bread and pastries to soups, sauces, fillings, prepared meat products, candies, and beverages; it adds to texture, helps with gelling, increases adhesion, and is a natural preservative.[37]

To what extent are the worthy crops above accurately described as "lost"? Except under poetic license, they are certainly not lost crops "of the Incas." Domestication preceded the Incas by many thousands of years. Nor were these crops lost by the many generations of Andean farmers who nurtured, selected, and replicated them down to modern times. Some so-called lost crops enjoy a large following, even in the cities. Andean consumers of ulluco and arracacha, for example, are put at thirty million, and consumers of oca at fifteen million. Arracacha has a large following outside the Andes, notably in Brazil. Nonetheless, large-scale use and genetic diversity are not the same thing. Despite Brazil's widespread arracacha consumption, production is based on just one yellow-rooted clone. Restricted to just a couple varieties, genetic diversity can decline as farmers phase out older but less popular types. So when a crop's future is being assessed, sheer numbers are not always the best guide.

Mashua use is put at just under ten million households. That seems reassuring on the surface but is less so when closely scrutinized. Most consumption is by rural households, with the notable exception of the department of Boyacá in Colombia, where mashua is a popular item of mass consumption. As the rural population migrates, demand for mashua will most likely drop. Worse still, yacón, maca, and achira are eaten by less than a million people, almost all of them from rural areas. So the future does not bode well for these crops. Achira, however, may hold on as a source of starch for food processing, both within and outside the Andes. In Vietnam, farmers grow more achira than any place in the world. The crop is made into starch for transparent noodles. In this form, it substitutes for a more expensive noodle made from mung-bean starch. Finally, ahipa and mauka are consumed regularly by less than one hundred thousand people, mostly in rural households. According to sources at CIP, only a few hundred farmers still cultivate them.[38]

How lost an Andean crop is, then, depends on many factors. If farmers stop growing a crop before a good sample of its genetic diversity is collected, it may be lost as a food crop permanently, reduced to a botanical curiosity. According to Carlos Arbizu, collections of maca, mauka, and ahipa are rudimentary; no one knows for sure how much diversity is already lost. Andean crops are at a great disadvantage in scientific terms too. Interest in their agronomic potential is recent, dating back to the 1960s and 1970s. Serious collection began in the 1980s and early 1990s. Research on how the crops can best be produced, harvested, and utilized has just started. In most cases, nutritional information is rudimentary. Little is known about how adaptable these crops are to other microclimates and latitudes. Will mauka still be virus resistant outside its customary niche? Can Andean crops be improved? For global food crops like rice and wheat, agronomists have a knowledge base that makes improvement possible. For Andean crops, the key questions are often unasked and even more often unanswered.

CHAPTER 8

Wild Potatoes

From the yungas down into the floodplains of the Amazon basin, native peoples grew manioc, yams, sweet potatoes, and yautia. All told, South America is home to over twenty-five domesticated root and tuber species. With respect to both climatic range and the botanical families represented, this is the greatest diversity found anywhere in the world.[1] The roots and tubers are just a start. Andean farmers domesticated the robust, protein-rich grains quinoa and kañiwa, both typically planted at high altitudes in rotation with potatoes. A step down comes the amaranth grain, kiwicha. The lost-crop list likewise includes eleven lesser-known fruits, such as the giant Andean blackberry and the capuli cherry, plus several solanaceous fruits in the potato family: the orange-like naranjilla plant; the sweet-tasting, cucumber-like pear melon; and the large, reddish-purple tree tomato. The Ecuadorian coconut palm grows at elevations as high as 2,800 meters (11,000 feet). There are two "lost" pepper species, two squashes, and three types of beans, including tarwi. The seeds of the high-altitude tarwi are 40 percent protein and 20 percent oil, as nutritious as soybeans.[2]

Also on the lost list is the potato. Well, not the popular, common spud, whether in its long-day tuberosum form or in its short-day andigena guise. In 1997, only 144 of the 3,527 accessions in CIP's collection of cultivated potatoes were classified as *Solanum tuberosum tuberosum.* The largest holdings, some 75 percent, were varieties classified as *Solanum tuberosum andigena,* the potatoes common to the Andean highlands. Now, no one disputes the popularity of the common potato (*Solanum tuberosum*), whether in its tuberosum or andigena form. But Andean farmers domesticated seven additional species of potato, which differ from tuberosum-andigena subspecies in fundamental respects.[3] (See table 7.) Some of these novel fellows occupy habitats too cold for *S. tuberosum;* others overlap with it in its range. Their starch, protein, and vitamin-C content differ from the common spud, as do how they taste and how they can be processed. The exotic species have bright, diverse colors; they are often deeply indented and come in more varied shapes, such as cones and crescents. The multicolored potato clusters CIP features on its publications draw on this eye-catching impact. By comparison, *S. tuberosum* makes a less striking display.

Ajanhuiri and the rucki potato types can withstand the altiplano's frosts. Most varieties produce bitter clones, which get processed into *chuño*, but blue ajanhuiris can be sweet and floury. Knobby pitiquiña, with its long, cylindrical tubers, has a nutty taste and is popular with Bolivian farmers. Limeña has a deep-yellow flesh and is ideal for soups. Traditionally, Andean farmers intermixed pitiquiña and limeña in their fields, along with common potatoes. Small and irregular in shape, phureja types are adapted to the warmer, moist slopes of the Andes. Lacking dormancy, they start to sprout as soon as they are harvested. Phureja potatoes are noted for their heat tolerance and late-blight resistance. All told, these robust but neglected species have tremendous potential as potatoes in their own right. In addition, CIP scientists draw on their genetic traits to improve the quality and hardiness of the common spud.[4]

Table 7. Number of Accessions Maintained in the Gene Bank at CIP, by Cultivated Species

Cultivated Species	Accessions	Ploidy Level
Ajanhuiri *(S. ajanhuiri)*	10	Diploid (2n=2x=24)
Limeña *(S. goniocalyx)*	48	Diploid
Phureja *(S. phureja)*	170	Diploid
Pitiquiña *(S. stenotomum)*	268	Diploid
2x Hybrids	56	Diploid
Chaucha *(S. chaucha)*	97	Triploid (2n=3x=36)
Rucki *(S. juzepczukii)*	31	Triploid
Rucki *(S. curtilobum)*	11	Hexaploid (2n=5x=60)
Andigena *(S. tuberosum)*	2,644	Tetraploid (2n=4x=48)
Tuberosum *(S. tuberosum)*	144	Tetraploid
4x Hybrids	48	Tetraploid
Total	3,527	

Source: Adapted from Huamán, Golmirzaie, and Amoros, "The Potato," p. 23.

CIP's collection of cultivated Andean species initially had more than 15,000 varietal accessions. With the potato, however, appearances are deceptive; the plant is notorious for its "environmental plasticity."[5] Potato plants selected in different places may look like unique varieties, but if they are grown together under identical conditions they often turn out to be duplicates. Using both field trials and molecular analysis, CIP gradually pared the collection down to about 3,500 accessions.[6] Even so, the work required to maintain it is daunting. With rice, seeds can be put into cold storage for decades. With the potato, tubers only last a year; then another cycle starts. Each year the collection is replanted in the highlands, then harvested, sorted, and stored. Duplicates are kept on hand in Lima, Peru. For added security, the accessions are also held as tiny plantlets in test tubes. (See fig. 5.) Under controlled light and temperature, these in-vitro sets can be stored for two years. The test-tube collection is itself duplicated and stored in Quito, Ecuador. As a last resort, CIP keeps true seed from nonsterile species under deep freeze. So far, however, the best way to maintain the collection's trueness to type is to plant it.[7]

Impressive as it is, the diversity of the domesticated spud is just a start. Backing up the eight cultivated species is a vast gene pool that comprises 235 wild potato species. No domesticated food crop has so large a clan of extant wild ancestors.[8] And no one has done more to locate these intrepid progenitors than Peruvian breeder and taxonomist, Carlos Ochoa.

The first assignment Ochoa had, which was back in the 1940s, was in the Mantaro Valley in central Peru. After a severe frost destroyed the valley's potato crop, Ochoa wanted farmers to restock with more frost-resistant types, but no one knew where to get them. In those days, Peru's breeders worked on improving commercial crops such as wheat; no one paid much attention to a lowly subsistence crop like potatoes. So Ochoa began experimenting. He started from scratch, educating himself about the habitats, distribution, and characteristics of the parent potatoes he crossed, both cultivated and wild. Eventually, he developed six new potato varieties, some of which are still popular in Peru's highlands. The work made a lasting impression on the young agronomist. To shore up the potato's eroding genetic base, Ochoa set out to find, preserve, and describe as many wild species and rare cultivated potatoes as he could.[9]

Over the next forty years, Ochoa identified over 80 new wild potato species, or a third of the total known to date. Three are named in his honor. And he has saved untold "lost" varieties from extinction. In the beginning, he often worked alone and at his own expense. Later, he had backing from CIP. He has stalked the high valleys of the Andes from Colombia to northern Argentina for the telltale sign of a wild potato in bloom. He has braved the blustery spring weather of the Andes, not to mention bandits, earthquakes, and erupting volcanos. Thanks to Ochoa's efforts, CIP now has over 1,500 accessions from 112 different tuber-bearing, wild potato

Fig. 5. CIP's potato collection: tubers, in-vitro tissue culture, and true potato seed. Courtesy CIP Archives, from *CIP Circular* (June, 1986): 6.

species. For his "pioneering work in potatoes," Ochoa has received awards from Brown University, the Organization of American States, and the Inter-American Institute for Cooperation in Agriculture.[10]

A wild potato is a survivor that has kept its heritage alive without domestication. It has withstood a thousand droughts and El Niño rains; has reproduced despite the weevil and the tuber moth; has lived with viruses, laughed at late blight, and outdistanced bacterial wilt. Wild potatoes are thus a treasure trove of robust genetic traits. The only problem is to figure out which traits go with which wild species. That is what makes Ochoa's work so important. Finding a wild potato is just a start. The real achievement is telling its unique story: where it comes from; how much rain, frost, and drought it is used to; which pests and diseases live in its neighborhood; what its stems, leaves, flowers, and seed balls look like. And remember, with a potato plant, what is underground is a crucial factor. Are the tubers round, long, flat, or irregular? What color is the flesh and skin? Are the eyes deep or shallow? In his latest book, Ochoa describes Peru's wild potato species. He includes hundreds of his own photographs and drawings, plus tables and color illustrations. Taken together, his tomes on South America's *Solanum* genus are a potato breeder's bible.[11]

To create spuds less addicted to modern farming's chemicals, tapping the wild potato is good place to start. And do not forget the unexploited diversity of South America's eight cultivated species, from the bitter, high-altitude ruckis to the humidity-loving phurejas, not to mention the multifaceted andigenas. For saving wild potatoes, we are in debt to Carlos Ochoa. For the diversity of cultivated spuds, we are in debt to Andean farmers, past and present.

In 1520, Andean farmers cultivated seventy different plant species, which is comparable to the combined totals for Asia and Europe.[12] Andean farmers still treasure this inheritance. I saw this for myself when I visited Bolivia's potato-research station at Toralapa.

The station is situated in a narrow valley at 3,400 meters (11,150 feet) in Cochabamba Province. There I learned about the work underway: projects to reclaim old varieties, seed production in protected beds, and on-farm trials for frost resistance. In the afternoon, we went to visit farmers and potato plots. Don Carlos Huascar has a small farm off a dirt road higher up the mountain slopes, at 3,800 meters (15,000 feet).[13] Short, thin, and wiry, he had on rubber boots, tattered jeans, and a heavy gray sweater. Adjoining his adobe cottage, colorful dahlias and daisies were massed. Don Huascar has separate plots for raspberries, artichokes, and lima beans, plus several dispersed potato fields. He said he plants "fifty different kinds of potatoes, of all shapes, sizes, and colors."[14] Interspersed among the potatoes, he has twelve different types of oca, plus ulluco and mashua, the latter inherited from his father. With the November rains, the first tubers to go in are the ocas. He plants all his tubers in mounded rows. Don Huascar said he keeps the soil fertile by adding sheep manure and leftover garden refuse to it. He uses no fertilizer, no insecticides, no fungicides.

Don Huascar was building a little terrace in front of his house with stones taken from the field. "I like the view," he said. I stopped taking notes and turned around to look down the valley: rocky crags, eucalyptus groves, tuber fields in shades of green, the barley golden brown. This mountain world, I thought, was so silent and sad, so heartbreaking and serene. Against such a landscape, I felt the truth: that we are made for the world, that the world is not made for us. In the high Andes, rural families face an impoverishment measured by high infant mortality and semiliteracy. But doesn't the modern world live in an ignorance more profound, in a solitude deeper still? The city's pace blots out everything. There are no other voices. We are so lost. But in the Andes, his true home, Don Huascar is never lost.

PART IV
Potato Projects

CHAPTER 9

No Easy Answers

I t took me a decade to get to Bolivia. The first try was in 1980 by boat from Puno, on the Peruvian side of Lake Titicaca. Before I bought the ticket, right-wing paramilitary squads unleashed a campaign of terror in the country's capital, La Paz. I stayed in Peru and went to Cuzco instead. I tried again in 1983, this time from Asunción, in Paraguay. I was writing about community health projects and was headed for the medical school at the University of San Simón, in Cochabamba. I had a ticket on Eastern Airlines. The flight was cancelled. A strike by Bolivia's oil workers made refueling impossible; the airports were closed.

Bolivia is the only country in South America that managed a successful nationalist revolution during the Cold War. In 1950, in prerevolutionary Bolivia, some 70 percent of the population was engaged in agriculture. Most of the rural workforce consisted of Aymara- or Quechua-speaking campesinos. Large haciendas with one thousand hectares or more constituted just 6 percent of all farms but owned 92 percent of Bolivia's cultivated land. By comparison, small holdings of less than five hectares accounted for 60 percent of all farms but held a mere .02 percent of the land.[1] Landless campesinos with labor obligations to the haciendas made up most of the rural population. In return for their work, they could tend small plots on their own.

After agriculture came mining, which earned half the country's foreign exchange. Opposition to the mining and hacienda-owning oligarchy came from the Mine Workers Federation, led by Juan Lechín. It formed the core of the Nationalist Revolutionary Movement (MRN), which enjoyed widespread middle-class support. The party's candidate, Victor Paz Estenssoro, won the 1951 presidential elections, but the military seized power to forestall an MRN government. Insurrection followed. In La Paz, miners and their urban supporters defeated the army after three days of heavy fighting. From the cities, the rebellion spread to the countryside. Armed campesinos broke into the haciendas, burned work records, and killed or terrorized the owners. Within a year, the old hacienda system lay in ruins. Dispossessed landlords never received compensation. An agrarian-reform law was decreed officially in 1953, but land distribution had already occurred directly through seizure. Overall, about half the country's landless campesinos ended up as small farmers. Bolivia's land redistribution was the most extensive and enduring carried out anywhere in South America.[2]

Paz Estenssoro's left-wing government nationalized the mining sector, increased social services, and raised wages. The government printed the money to cover the costs. The rapid inflation that followed wiped out the savings of the middle class and estranged them politically from the MRN. Paz Estenssoro was reelected in 1960, but a military coup toppled him in 1964. There followed a long sequence of military regimes opposed tooth and nail by the labor unions. In the prolonged standoff that followed, social services, mining, and agricultural production deteriorated. The country's debt increased, and so did the drug trade.[3] When Paz Estenssoro, at age seventy-seven, showed up again as president in 1985, the country was near collapse. Its $3.4 billion foreign debt almost equalled its gross domestic product (GDP), the mining sector was near collapse, and the annual inflation rate was heading for a disastrous 24,000 percent. The result was a breathtaking change of course. The Paz Estenssoro government ruthlessly slashed public spending, opened up the economy to foreign investment, cut tariffs, fired two-thirds of the country's thirty-three thousand miners, and then sold off the mines. The official unemployment rate hit 25 percent. In desperation, impoverished miners headed for the tropics to recoup their fortunes as coca farmers. But for the moment, the budget was balanced and inflation under control.[4] When I finally made it to Bolivia in 1990, things had settled down. I flew in from Chile, where I had spent a week in Ñuble Province visiting rice farms.[5] In Bolivia, CIP had joined a national effort to shore up the country's potato production. Called the Potato Research Project (PROINPA), this national effort had long-term financial support from the Swiss Agency for Technical Cooperation (COTESU); most of the agronomists and research scientists involved came from Bolivia's Agricultural Research Institute (IBTA).[6]

The project's goal is to improve Bolivia's potato-production technology. In 1990, it was still in its fact-gathering phase, sponsoring workshops in potato-growing areas. Given the merciless budget cutting of the 1980s, agricultural research was at a standstill. Facilities had deteriorated; scientists had to abandon projects, thereby losing years of work. Consequently, PROINPA had to start almost from scratch. That being so, the project was determined to diagnose the situation accurately so that its subsequent work would pay off. I headed for Sucre, capital of Chuquisaca Province, for a PROINPA-sponsored three-day workshop on seed production.

"Seed does not lie," said workshop participant Julio Laredo. "It can only give back the quality it has." Laredo, a stocky, blunt man in his early forties, was in charge of seed certification for Chuquisaca Province.[7] The potato is Bolivia's most important food crop. Annual per-capita consumption nationally exceeds 100 kilos per person; for the country's potato farmers, consumption is closer to 150 kilos, which is as high as potato consumption gets anywhere in the world.[8] During the 1980s, Bolivia's potato made a poor showing. For the years 1985 to 1989, production averaged 710,000 metric tons, a drop from the 755,000 tons registered a decade before, in 1975–79. In the meantime, the country's population had increased from 4.3 million in 1970 to 6.4 million in 1990. The drop in the potato harvest occurred even though farmers devoted roughly the same acreage to the crop. Average yields fell from 5.5 metric tons per hectare in 1975–79 to 5 tons for 1985–89. These yields were far below those in both Ecuador and Peru, which averaged 8.7 and 8.6 metric tons respectively.[9]

Ancient Andean farmers had domesticated potatoes in the Lake Titicaca basin. How could it be that in its zone of origin the potato's yields could be so low? That was the question the workshop was expected to answer. To most of the experts there, the culprit was the poor quality of the seed stocks farmers used. At harvest time, farmers set aside tubers for the next crop, at least 2 metric tons per hectare planted.[10] Constantly replanting the same recycled seed stock can lead to a buildup of virus diseases. To keep seed healthy, it has to be selected and stored carefully. When sanitary precautions are not taken, yields spiral downwards. The solution is to get disease-free, high-quality seed to farmers. With good seed to start with, yields go up dramatically. Almost any CIP analysis of Third World potato production states the clean-seed premise.[11] Applying this premise, however, is not as simple as it looks.

Farmers in the United States and Canada started to use certified seed in 1915. To qualify, the seed had to be as free as possible from seed-borne diseases, including viruses. This required independent inspection by an unbiased third party, such as the local extension agent. Reports from inspectors certified that potato seed "had been selected, grown, and rogued for diseases under supervision."[12] There followed a series of studies by the U.S. Department of Agriculture to determine the merits of seed certification. In 1924, 11,627 reports filed from twenty-seven states and eight Canadian provinces showed an average yield increase of 3.1 metric tons per hectare.[13] Since potato yields then averaged 7.8 metric tons, certified seed boosted production by almost 40 percent.[14] Between 1925 and 1935, production of certified seed in the United States and Canada almost trebled, rising from 155,000 metric tons to 449,000 tons.[15] Today, commercial growers buy fresh seed stocks regularly. Even the home gardener can get certified seed. Behind the seed stand decades of research at land-grant universities. Today, growers can select from a pool of improved clones with better stress tolerance, disease resistance, and cooking characteristics.[16] That being the case, why not transfer American-style seed programs and their superior varieties to Bolivia?

A research method is one thing, its products quite another. In Bolivia's highlands, farmers produce several species of potato, but the watery, American-style tuberosum is not one of them. Not only is the climate too cold for tuberosum, but disease and pest problems, seasonal day length, and rainfall patterns all differ greatly from those in Idaho or Maine. Tuberosum does have a niche in Bolivia, but it is down in the warmer valleys and lowlands. Cultural practices likewise show sharp contrasts. In Bolivia, growing potatoes is a labor-intensive, community-centered effort. In Cochabamba Department, a typical farmer has about three hectares, only one of which ends up in potatoes. On this plot of land, farmers plant spuds with different tastes, maturities, and uses, including the rucki types for *chuño*.[17] A potato farm in the United States, by contrast, is an agribusiness venture that can cover hundreds of hectares. Farmers specialize in a single variety, such as Russet-Burbank. In short, a potato-research agenda that fits the United States, or for that matter, Tunisia or Uganda, is unlikely to suit Bolivia. Like it or not, for Bolivia to improve its potato prospects, it has to work out its own priorities. It cannot borrow them from somewhere else.

Workshop participants agreed that Bolivian farmers needed better seed and a more systematic approach to production. André Devaux, however, was concerned about the magic-bullet mentality. "We have to be careful not to build objectives from the top down," he said. "It seems simple enough. Produce a lot of seed, get it to farmers, and the job is done." Devaux was PROINPA's codirector. A CIP agronomist from Belgium, he had worked for several years in Rwanda before his assignment to Bolivia. "The problem is," he said, "poor farmers will never spend money on expensive, certified seed. What we need are affordable, practical technologies that promote seed quality."[18] In Devaux's view, improving seed selection and storage was a good place to start. In the meantime, before tackling a seed program, PROINPA had lots of fact-finding ahead. It needed to know which potato types predominated at each tier, from the altiplano down into the temperate valleys and yungas. Its priorities had to reflect the climate, pests, and disease problems of each zone. In the past, farmers in the highest valleys specialized in seed, as the cool climate reduced disease incidence and pest pressure; in turn, they supplied farmers at lower altitudes. Was the old seed system still in intact?

To answer these questions, PROINPA sent out interdisciplinary teams to selected potato-growing districts. The teams visited potato plots, divided them roughly into five parts, and sampled the plants in each. They labeled samples by district, altitude, and community and sent them back to project laboratories for disease and pest analysis. The teams asked farmers about the potato-production cycle, about which varieties they produced, and about which problems they considered the most serious. They collected data on rotations and potato-production practices; they got estimates from farmers on yields, costs, profits, and marketing. Using this information, they ranked project priorities. For each plot, they drew up a plan of action that specified the results expected.[19] These plans now guide the project, from its breeding and pest-control objectives to crop production and farmer-focused seed multiplication. At every step, farmers are involved in testing new technologies.[20]

To bolster potato production, PROINPA's core strategy had to include seed production. That was already clear in 1990. After the Sucre workshop, I spent a week at PROINPA headquarters in the city of Cochabamba, which is Cochabamba Department's capital. Prior to the economic and political troubles of the 1980s, Cochabamba had been Bolivia's top potato-growing department.[21] PROINPA inherited and restored the research station at Toralapa. It was situated on the edge of the altiplano at 3,450 meters (11,300 feet), a three-hour drive from the city of Cochabamba. I went there to meet Nelson Estrada.

"CIP's research has focused on tuberosum potatoes," said Estrada.[22] "To make an impact here, better yields from andigena potatoes is the place to start." Born in Colombia, CIP breeder Estrada had helped transfer andigena traits such as late-blight tolerance and nematode resistance to the common, tuberosum potato.[23] Snowy-haired and wrapped in a poncho, he was at home in the highlands. Estrada had an infectious enthusiasm and a devotion to Andean potatoes. Small farmers in the Andes prefer andigena types, he explained. "They have more dry matter, more protein, and a higher starch content." For a sticky puree, starchy andigenas cannot be beat. Andigena potatoes are less prone to spoil when left out in the sun, and seed quality holds up longer. By comparison, tuberosum potatoes are thin-skinned, store poorly, and spoil easily. Andigena plants are usually taller than tuberosums, with strong, thick stems. Many andigenas have deep eyes, which discourages peeling them raw; for a spud boiled in its jacket or baked, the trait is not a drawback.

In Bolivia's highlands, andigena potatoes predominate; they are more diverse genetically than the tuberosums. CIP's germ plasm collection, for example, has 2,644 andigenas compared to just 144 tuberosums.[24] (See table 7.) Some potato experts argue that the tuberosums do not even merit a separate classification. Under mass selection, long-day potato plants (tuberosum types) can be derived from short-day plants (andigena types), hence the contention that farmer selection in Europe led to a long-day andigena-type potato. Until late blight devastated Europe's potato crop in the 1840s, andigenas held sway.[25]

PROINPA's on-farm research showed that frost and dry spells do the most damage in the highlands. At 3,000 meters and above, Bolivia has more potato acreage (some 100,000 hectares) and more small potato farmers (some 150,000) than in any other zone.[26] Across this area, the chance that frost will damage the crop in any given year exceeds 30 percent. Part of the problem is exposure. Andigena potatoes take at least 150 days to mature. By comparison, tuberosum types in the lowlands can be harvested in just 90 days. Moreover, the timing and severity of frost damage is unpredictable. Even though spending money on fertilizer or better seed can mean higher yields, farmers know that a heavy frost can wipe out the gains. Of course, they have time-tested ways to reduce the risks. Plots located on the slopes, for example, are less prone to frost damage than those on flat ground. In areas notorious for frost, Bolivian farmers grow the frost-tolerant rucki types rather than andigenas. Rucki potatoes, however, must be freeze-dried, as they are too bitter to be eaten fresh. Hence, for a commercial crop, as opposed to a crop for household consumption, the highland farmer's best option is andigena. To shore up production and profits, farmers need reliable frost tolerance in andigena varieties. The problem is relevant far beyond Bolivia's borders. Of the 400,000 hectares of potatoes cultivated in the Andean highlands, 70 percent are threatened by frost.[27]

In Bolivia, the mercury never drops below freezing during the day. Frosts come on clear, cold nights after a sunny day, usually between 4:00 A.M. and 6:30 A.M. According to Estrada, what the project needs is a nonbitter potato plant that withstands temperatures as low as -3 to -4 degrees centigrade (25 to 27 degrees Fahrenheit). This alone can increase yields by 20 to 30 percent. Since frost damage is worse during dry spells, the project has to combine both frost and drought tolerance. Estrada explained that Bolivia has a collection of some six hundred indigenous andigena cultivars. Screening this material is a top priority. In fact, some andigena cultivars can already take a light frost of -1 degree centigrade (30 degrees Fahrenheit). Another approach is to cross andigenas with frost-tolerant rucki types without transferring rucki's high glycoalkaloid content. Frost tolerance can also come from wild potatoes. CIP's wild collection contains fifteen hundred accessions from one hundred and twelve different wild potato species. So far, CIP has identified ten wild species with high frost tolerance; *Solanum acaule,* for example, resists temperatures as low as -7 degrees centigrade (19 degrees Fahrenheit).[28]

Making a direct cross between cultivated potato species is bedeviled by ploidy level. The common potato, whether andigena or tuberosum, is a tetraploid, with forty-eight chromosomes. Unfortunately, most of the other domesticates are diploids, with twenty-four chromosomes. Even for simple inherited traits, patient backcrossing is the rule. More challenging still is crossing the common potato with its wild diploid relatives, plants that have valuable, robust traits. To do this, breeders can generate haploid plants, which have half the number of chromosomes of the domesticated parent plant. Once this is done, they can cross a common potato with a diploid species, including wild relatives. Another alternative is to use 2n pollen grains from a wild, diploid plant. Such pollen has twice the number of chromosomes, which permits a cross with a tetraploid plant. When these strategies fail, breeders try protoplast fusion and embryo rescue.[29] Estrada had preached the frost-tolerant merits of wild species for years. His method was to first cross a diploid wild potato with a diploid domesticated one, such as *Solanum phureja*. Then he backcrossed to andigena types.[30] Estrada practiced what he preached. In its quest for a potato that could stand up to Andean frosts, Estrada's team crossed native Bolivian cultivars and wild species.[31]

Despite its obvious importance, frost tolerance is but one of many factors that impede Bolivia's potato production. PROINPA's diagnosis showed that whether it was breeding objectives, pest problems, or rotation systems, the project's priorities had to be based on the climatic zone in question. (See table 8.) For the highlands, frost and drought tolerance are high on the list.[32] For the temperate valleys, frost drops out but drought tolerance remains. In the rainy yungas, however, droughts are rare. As for disease, late blight cuts across zones; the highlands escape bacterial wilt, but the valleys below do not. Nematodes are troublesome in cooler zones above the yungas. All zones must grapple with the formidable potato weevil. The highlands are spared the tuber moth, which likes hot weather, but the yungas are not. To lead the attack against such foes, the project has a set of complementary, farmer-friendly strategies.

On the front lines against frost, drought, and bacterial wilt are improved varieties. As for pests, crop management is the key. Whether it is battling weevils, the

Table 8. Production Difficulties, by Tier

Meters	Zone	Insects	Diseases	Climatic Stresses	Feet
4,000 to 3,000	altiplano and high valleys	weevils nematodes	late blight	frost drought	13,100 to 9,843
3,000 to 1,500	temperate valleys	weevils nematodes tuber moth	late blight bacterial wilt	drought	9,843 to 4,921
1,500 to 900	yungas	weevils tuber moth	late blight bacterial wilt		4,921 to 2,952

Source: Adapted from IBTA-PROINPA, *Informe Anual Compendio 1993–1994*, p. 60.

tuber moth, or nematodes, the project has simple, inexpensive strategies that can greatly reduce crop damage. Across all zones, PROINPA emphasizes proper seed storage and selection. To protect the soil's fertility and build up its organic content, the project promotes fallows, the incorporation of plant residues, and rotations to leguminous "green manure" crops. What results, however, is by no means a uniform plan applied universally across districts. The agronomists assigned to each place must sort out the priorities. Even if the items listed for districts in Cochabamba, Chuquisaca, and Tarija Departments are the same, the rank order can be different. In the highest zones, for example, the cool climate slows down late blight, so its control is not paramount.

The scheme is intricate, but so is the reality that the project must confront. PROINPA built up its objectives systematically from direct field research. Obviously, it presumed a factor like frost tolerance was a key to success in the highlands, but it did not necessarily know if farmers would concur. There are few mistakes more fundamental in agriculture than fixing what farmers do not think is broken. From the start, in its diagnosis of problems and in its plan of action, the project listened to farmers; where relevant, it sought their direct participation.[33] As self-evident as this strategy looks on paper, in practice it is a lot of extra work. Whether the problem is seed production, breeding for frost tolerance, or farmer participation, there is no easy answer. The secret lies in combining the ingredients.

CHAPTER 10

Rustic Beds

L a Paz snuggles into a steep mountain valley. The airport is above the city on the altiplano, elevation 4,200 meters (13,800 feet). It was 6:00 A.M. An orange-red sun tinted the snow-capped mountains; below were Lake Titicaca, fields, and llama herds. It left me breathless. Fortunately, the passenger next to me had *sorojchi* pills for altitude sickness. I took one, got off the plane, and waited anxiously for my luggage; I had only half an hour to catch the Cochabamba flight. Time and distance are not what divide us now.

I had not visited the PROINPA potato project for six years. The first time, I was interested in planning; this time, I wanted to see the results. I had lunch with André Devaux, the projects codirector. A new government, he told me, was busy decentralizing state agencies. Gonzalo Sánchez de Lozada had won the 1993 presidential election with 35 percent of the vote. In Bolivia, when no candidate gets a clear majority, Congress brokers the election. Since each region has its favorite son, getting a majority is virtually impossible; consequently, wheeling and dealing with Congress is the only way to end up president.[1] Bolivia is really several countries patched together. There is La Paz Department, with the Lake Titicaca basin and the capital city; Cochabamba Department, whose humid valleys straddle the yungas; Chuquisaca Department to the south, with its picturesque colonial capital, Sucre; and farther south still, Tarija Department, which borders Argentina. Across the mountains, in the humid, tropical lowlands, lies the country's largest department, Santa Cruz. Add to its climate the business mindset of the city of Santa Cruz, and the result seems like another country. Together, these five departments—there are a total of nine—have roughly 75 percent of the country's population and most of its big cities.[2] Regional loyalties are strong, and all-weather roads few. Under the best of circumstances, getting from one department to the next is a challenge; during the rainy season, it is often impossible.

Sánchez de Lozada's decentralization scheme proved cumbersome in practice. PROINPA's approach to the potato was already decentralized by climatic zone and by department. The impending reorganization created uncertainty and confusion; no one knew how agricultural research could be more decentralized than it was already. The Bolivian Institute for Agricultural Technology (IBTA), responsible for a chunk of PROINPA's budget, had not received funds for months; salaries were in arrears. To make matters worse, the leadership of IBTA, in La Paz, changed with disturbing regularity. Since the PROINPA project started in 1989, IBTA had gone through six directors. André Devaux was understandably preoccupied: "What the development textbooks skip is how to handle highly politicized public agencies."[3] Amen to that.

The next morning, I left for PROINPA's research station at Toralapa. Back in 1990, getting a frost-tolerant potato to farmers was the cornerstone of the project's breeding strategy. I wondered what had happened. Nelson Estrada, CIP's frost-tolerance expert, had retired just a few months before. Willman García, a young Bolivian who had worked closely with Estrada, brought me up to date.[4] He sported a woolen jacket and a colorful scarf; the air was clear, cold, and bracing. We headed for the frost-simulation unit.

For the altiplano and high valleys, Bolivian farmers need potatoes that combine frost and drought tolerance with resistance to late blight and nematodes. Such a variety could double production. The challenge is to get the traits into high-yielding cultivars that taste good enough to satisfy both Andean farmers and urban consumers. That is a tall order. As García explained, the first step is to combine frost and drought tolerance; additional traits can be selected for subsequently. Regardless of which variety gets selected, however, it has to pass muster across the board. "What good is a frost-tolerant potato prone to late blight?" García asked. Not much, of course.

Selection for frost tolerance starts by crossing two promising parental plants. Each cross generates a hybrid family. The progeny are twice screened for frost hardiness in the simulation unit: when seedlings are about six weeks old and then again at twelve weeks. This permits an initial selection. In 1994–95, PROINPA screened 20,000 hybrid plants from 138 families. The two frost tests netted just 761 promising plants, or 3 percent of the total. These plants in turn are checked for drought tolerance and late-blight resistance.[5] García estimated that between 1.0 and 1.5 percent of the hybrid plants so generated go to field tests. Farmers try out the PROINPA selections in their own plots and in their own way; at harvest time, they evaluate the results. The final test is how the tubers cook and taste.

We got in the jeep and went to Koari District, not far from the station. Farmers were evaluating some fifty Toralapa varieties, primarily for frost tolerance. A few nights before, the mercury had plunged down to -2 degrees centigrade (28 degrees Fahrenheit). Some potato plants had collapsed, their leaves wilted. Others stood tall, impervious to the frost. Willman warned me not to judge everything by externals. An attractive plant can put too much energy into its appearance and not enough into tuber production. Some damaged varieties, by contrast, resprout and recuperate rapidly. "They may look terrible after a frost," he said, "but might still produce high yields."

Frost, I recalled, once hit my potato patch back home. I simply cut off the wilted stems and let them recuperate. Of course, that was the only frost of a long growing season. In Koari, by contrast, the potato runs a gauntlet of six to eight frosts a year. Consequently, frost tolerance is of obvious importance to farmers. Interspersed among the varieties being tested were Waych'a, a commercial favorite sensitive to frost, and a bitter, frost-tolerant rucki, used for *chuño*. When campesinos walk by the test site, they can readily compare the newcomers with the familiar Waych'a and rucki types. Thus, the field trials double as demonstration plots. At harvest time, Koari farmers collectively evaluate the results and make selections. Such participation paves the way for new varieties, since farmers already know something about them.

It can take ten to fifteen years to breed a new variety, from the first cross to its official release and multiplication.[6] Just selecting good parents and generating families can take two years. PROINPA already had frost-resistant varieties released for on-farm field trials. I asked García how they managed this so quickly. "Of course, we worked hard," he explained, "but most of all, it reflects Nelson Estrada's experience. He had such a good eye. He knew instinctively which parents went together, and he chose judiciously from their progeny. Because of him, we avoided many mistakes."

By the 1998 season, PROINPA had four varieties on line that had fared well in five districts of Cochabamba and La Paz Departments. On-farm trials showed that they withstood frosts down to -3 degrees centigrade (26 degrees Fahrenheit) for two hours. The favorite was Illimani, which farmers preferred even to the popular Waych'a. They liked it for its deep eyes, its uniformity, and its starchiness. Yields were between 35 and 39 metric tons per hectare. This is far better than the frost-tolerant bitter ruckis; even under the best conditions, rucki yields hover between 2.5 and 5 metric tons.[7]

Frosts threaten potatoes throughout the highlands, not just in Bolivia. Does that mean that every Andean country has to breed its own frost-tolerant cultivars? To prevent such duplication of efforts, CIP encourages national programs to coordinate their research agendas. The Andean Cooperative Potato Research Program (PRACIPA) is an example.[8] PROINPA takes the lead on frost tolerance; member countries share in the results. For its part, PROINPA benefits from the research its network partners undertake. This is invaluable, as PROINPA cannot afford a special project on every potato disease and pest problem in Bolivia. In addition to having access to the research of its Andean partners, PROINPA can draw on the work of larger, regional networks coordinated by CIP.

Consider late blight, the scourge of Bolivia's temperate valleys and yungas. In 1996, the project released six varieties with late-blight resistance. Most of them were tuberosum-andigena hybrids with an international pedigree; they came from research programs in Mexico, Colombia, and India.[9] The same kind of exchange applies to nematode resistance or to bacterial wilt. PROINPA does not have to start from scratch. It can obtain promising cultivars from CIP or from other national programs. The material selected comes as a virus-free plantlet in a test tube, sent to PROINPA according to quarantine protocol. From the plantlet, PROINPA takes stem cuttings, which are used in multiplication. Such introductions are first screened on PROINPA's experimental plots and later by farmers. Under no circumstances does PROINPA release a variety directly, regardless of its reputation elsewhere. A cultivar that resists late blight in India may succumb in Bolivia's environment.

By now it was lunch time, so Willman García and I joined the rest of the staff in the station's spartan cafeteria. It had stone walls, folding wooden chairs, and metal tables. The fare was simple but suited to the mountain climate: chicken soup and noodles; a plate of rice heaped with carrots; lima beans; and potatoes boiled with a beef bone for added flavor. Finally came lime-flavored gelatin for dessert and pots of steaming coca-leaf tea, a popular and legal highland specialty.

The diplomatic crisis at IBTA did not set back the project's timetable on frost tolerance. Shoring up potato production, however, is a battle on many fronts. At the Sucre workshop in 1990, participants blamed Bolivia's low yields on pest-ridden, diseased seed stocks. When it started, PROINPA did not have clean seed ready for farmers. So it looked for interim strategies to improve seed quality. Managing the tuber moth is an example of this.

A serious pest in the temperate valleys, the tuber moth gets into the seed stocks farmers store for the next crop. Better management is the key to controlling the insect. Tuber moths can easily get into harvested potatoes, especially when they are left in the fields overnight. The pest lays its eggs in the eyes of the tuber, where they are nearly impossible to spot. The next morning, farmers end up storing infested tubers. Later, when the larvae emerge and start to eat the seed stocks, farmers douse them with toxic insecticides. The project, however, had cheaper, more effective alternatives. Tubers can be kept in dense piles between layers of crushed eucalyptus leaves, a natural deterrent. For extra protection, farmers can shake up the tubers in a bag containing a biological powder. The bioinsecticide is made from virus-infected tuber-moth larvae; one hundred grams is enough to cover twenty-five kilos of potatoes. The project's Toralapa facility can produce six metric tons of the powder a year.[10]

The upper valleys may escape the tuber moth, but not virus diseases carried inside a tuber's cells. A virus-infected potato generates virus-infected progeny; yields fall. "Imilla Negra used to be a popular variety in the altiplano," Gino Aguirre told me, "but virus infections got so bad farmers almost abandoned it."[11] Aguirre had on rubber boots, dungarees, and a bright woolen sweater. He worked on seed multiplication and kept track of the station's germ plasm collection. This included Andean tubers: 480 ocas, 130 ullucos, 55 mashuas, and potatoes—1290 accessions representing seven of eight potato species. Most of the potatoes were andigena types, but the collection also had a large subgroup of ruckis.[12] The accessions are kept in cold storage or in vitro; they are also planted in the station's fields. We headed for the lab, where Aguirre explained how PROINPA generates virus-free seed tubers.

The project starts with a microscopic, virus-free meristem cutting in a test tube. As this tiny shoot grows, additional cuttings are taken, rooted in a medium, and then transferred to protected seedbeds. We examined some of the beds. Aguirre pointed out the short-statured ruckis; they had tiny violet-sized flowers and small, erect leaves that help protect them from hail. The taller andigenas had thick stems with larger, more colorful flowers and broader leaves. The densely spaced potato beds produce small, virus-free seedling tubers, which PROINPA returns to the villages that supplied the originals. The communities involved then multiply the cleaned-up seed stock and distribute it. As of 1998, the project had returned twenty-four native cultivars, virus-free, to Andean communities, five of them the bitter rucki types.[13]

I asked Aguirre how the project decided which varieties to clean up. "We consider a cultivar's social and economic role in a community," he said. "Rucki types occupy a special niche in the altiplano's production system, so including them is critical." Even when cleaned up, however, rucki types still have low yields. Consequently, the project keeps an eye open for promising andigenas. There is no sense introducing new varieties when local ones will do. A notable example is Gendarme, or "Chejchi," an andigena collected in La Paz Department. The skin of this semierect, nematode-resistant variety is splotched with purple, red, and yellow; it has a cream-colored flesh and tolerates frost down to -3 degrees centigrade (26 degrees Fahrenheit) for three hours. The main drawback of this variety, true of many andigenas, is late maturity; for Gendarme, the time between planting and harvesting, depending on altitude, is from 150 to 180 days. Nonetheless, virus-free Gendarme seedlings have taken hold in the highlands.[14]

Willman García took me back to Cochabamba that night; I had to catch an early flight to Tarija the next morning. Despite all of the hard work, PROINPA's stock of new varieties is just a start. The challenge is to get them into circulation. That means generating seed stocks in quantity, a real challenge in a country as climatically diverse as Bolivia. To make sure I understood this, Devaux had arranged a visit to Tarija, a department with potatoes grown from the altiplano down into the yungas.

"How many varieties PROINPA says it has is irrelevant," said Jaime Herbas in his calm, matter-of-fact way.[15] True enough. In the United States, research programs churn out fancy new cultivars all the time; few of them end up with a market share of any significance.[16] Herbas headed up the PROINPA team stationed in the city of Tarija, the departmental capital. He was giving me a thorough orientation for a trip the next day to the altiplano and the Iscayachi research station. "Who needs seed, when, and which varieties? No single answer applies here," said Herbas. In Tarija's yungas, as in the San Andrés district, common tuberosum types predominate. Situated at 1,950 meters (6,400 feet), this region has a subtropical climate that permits two potato crops a year. Farmers are commercially minded, producing for urban markets. The main problem is getting enough good seed. Tuberosums store poorly in the humidity. To make matters worse, most of the district's soils are infected with bacterial wilt. When clean seed is planted in such soils, yields hold up; when farmers start off with infected seed, however, yields drop sharply. Consequently, the district's farmers need clean seed regularly, seed that has not been exposed to bacterial wilt. For the San Andrés district, the crux of the problem is the seed supply; there is no lack of demand.

When PROINPA did its on-farm diagnostic, it realized farmers in San Andrés already had a variety they liked, the tuberosum Americana. Introduced in the 1950s, Americana matures in just ninety days. Its culinary qualities make it a favorite with housewives in the city of Tarija. Eventually, however, Americana seed stocks had so degenerated that low yields forced farmers to give it up. After interagency meetings in 1991, PROINPA took on the task of cleaning up Americana. Three years later, virus-free seed was reaching farmers.[17]

For the potato farmers of San Andrés, the solution is certified seed production by a national seed program.[18] Demand is high, as farmers renew seed stocks regularly, and it is variety specific. A formal seed program needs volume and rapid turnover. To reduce production costs, certified seed is multiplied several times prior to its release; farmers use the seed for a season and then restock. For farmers in the highlands, in contrast, a formal seed program will not do. In the upper valleys, farmers like to recycle their seed stocks for many seasons. The thick-skinned, starchy andigenas keep better than tuberosum types, especially given the cool, dry winter conditions. Moreover, highland farmers plant multiple varieties, not just a couple. When they do buy seed, it is usually in small quantities. All these factors make a formal seed program in the highlands a losing battle.[19]

"Seed production is viable," Herbas explained, "if small farmers do it themselves." In fact, Andean farmers have been swapping seed at markets for centuries. The traditional system has its own division of labor. Communities at lower elevations prefer seed from the altiplano, which is less prone to disease and pest problems. The strategy at PROINPA is to build on what farmers already do, helping them do it better. Cleaning up seed provides the prototype. When the project returns an old favorite like Imilla Negra to a community, the virus-free tuberlets must first be multiplied for distribution. That involves farmers in seed production, albeit on a small scale. The challenge for PROINPA is to expand and enhance this work.[20] My orientation complete, we agreed to leave early the next morning for the Iscayachi station.

Situated in a warm, pleasant valley, Tarija is an unhurried town with a well-preserved, attractive square. While waiting for Herbas to come by with the jeep, I walked to the market on the edge of town. Dilapidated pickup trucks lined the road. It was dawn, and truckers were already unloading. The perishable tomatoes, grapes, and peaches came packed in crates. Tougher items, like the potatoes, carrots, and onions, were heaped in the street. Everything had to be sorted, bagged, and reloaded into pushcarts. The market's potato section had piles of white-skinned tuberosums and purple, oblong-shaped andigenas. I found some speckled ulluco and creamy-yellow oca. I stopped at a kiosk for some *café con leche*. The proprietor gave me a cup of boiling milk to which he added a couple drops of a thick, coffee syrup. The drops hardly changed the milk's color, but they gave it a rich coffee taste.

The Iscayachi station is situated at 3,350 meters (11,000 feet). To travel in Bolivia is to zigzag up, down, around, and then up again. The jeep climbed over the main pass above the city. In Tarija's high valleys, cattle and sheep grazed on the tough native grasses. In the distance, herders had piled up rocks into circular corrals. The criollo breed favored here is kept mostly for its meat. The sheep herds are in decline, due to competition from synthetic fibers. We started down. Reddish-orange flowers dappled the roadsides; across the valley lay green, terraced fields. Campesinos wore ponchos dyed in deep shades of blue, red, and purple, bright colors that ward off the gloom of a rainy day. The narrow, muddy road to Iscayachi continues on to the Argentine frontier. In Bolivia, as in most of South America, the borders get the worst roads. Moving people, merchandise, or produce across them had never been a priority. Along the Uruguay River, which separates Brazil from Argentina for almost 700 kilometers (420 miles), there is but one bridge.

The Iscayachi station works on potatoes, lima beans, and barley. Hernán Cardozo, a Bolivian, was in charge of the potato program.[21] At the time of my visit, he had just completed a seed multiplication course at CIP headquarters, in Lima. The weather was overcast and blustery. Cardozo was anxious to get started, as heavy rains make the dirt roads muddy and treacherous.

The station multiplies cuttings that it gets from PROINPA. Once rooted, the cuttings are transferred from the greenhouse to protected beds. The densely planted beds yield small tubers. These tubers go to communities for multiplication in rustic seedbeds. In this fashion, farmers can produce their own quality seed. The project had beds in five communities last year and in sixty this year. "We could have had a hundred," Cardozo said, "but that was too many to supervise." Every week, technicians go to each community and check out the seedbeds. "Close supervision is required at the start," Cardozo said. "The first year a farmer has a bed, no matter what, we must not let it fail."

Made of local material, usually stone or adobe, a typical rectangular bed is about 2 meters wide and 10 meters long. Situated so that it faces north into the summer sun, the bed has a back wall that is almost twice as high (.85 meters) as its front wall (.45 meters). To make sure the seed produced is sanitary, the old soil is removed to a depth of twenty to thirty centimeters and replaced with a substrate of manure, sand, and clean soil. The bottom of the bed is usually below ground level. At night, when there is danger of frost, or during a hailstorm, the bed is covered by mats of straw, old fertilizer bags, or eucalyptus branches wired together. Wooden frames covered with heavy plastic can be used but are twice as expensive. The beds are small enough to be watered manually as needed. The walls protect tender plantlets from high winds and temperature extremes. A bed's estimated cost, including labor, is fifty-five dollars. Considering that a bed lasts many years and a family does most of the work itself, this cost is minimal. Given the high density in the beds, farmers must use fertilizer, but it pays off. Two to three kilos of seed planted in a bed yields around fifty kilos, a multiplication rate in excess of 1 to 15. If multiplied out again in open fields the next season, farmers net enough to plant a hectare.[22]

The farmers of Iscayachi District are seed producers. They plant in December for an April harvest. The seed is stored pending the spring crop at lower altitudes. Seed planted in protected beds costs farmers' customers about 15 percent more than seed produced the old way. High-quality seed from the beds, however, adds at least two metric tons per hectare to overall yields. That more than compensates for the added cost.[23]

Cardozo and I broke for lunch at the station: soup with rice, beans, and potatoes. I sat with Javier Franco, who had accompanied me from PROINPA headquarters in Cochabamba. A Peruvian and *eminence gris* of CIP's nematode research, he had a decade of experience in Bolivia. Franco was easygoing when it came to the potato but critical when it came to development schemes. He explained that under Bolivia's decentralization plan, no one knew which agency was supposed to write the checks. Project agronomists had gone unpaid from October to January; the money just sat in La Paz. Fortunately, business can be done on trust in Tarija;

otherwise, the salary crunch would have crippled work all along the line. The discussion veered from the muddy roads to politics. In this case, discretion was the better part of valor, so I listened: "the roads curve not just because of the terrain, but so they can pass by the right properties"; "one minister starts paving a road and the next one stops the project; the equipment just sits by the side of the road." I had heard it all before, just with different metaphors. In the United States, the roads are great, but when it comes to educating a child or providing health insurance, we are all curves and unpaved roads.

Lunch over, I went with Hernán Cardozo and Javier Franco to the village of San Lorencito.[24] Since it was off the beaten track, we took the four-wheel-drive truck. Cardozo never passed anyone without offering them a lift, which was usually to the next settlement.

In San Lorencito, houses and seedbeds are made from river rocks and fieldstones, the district's ubiquitous building materials. With an artistic eye, builders had set the stones in intricate, balanced designs. Most houses faced each other across courtyards shielded from the wind and cold. Irrigation canals brought potable water up alongside each house, from whence it was channeled to outdoor sinks and privies. To reduce the threat of fire, cooking was relegated to a separate building.

Don Alberto showed us his seedbed. It was close by so his family could water the plants and cover the bed at night as needed. From outside, the bed seemed too shallow, but when I looked in from the top, I realized that the bed's floor was at least half a meter below ground level. Digging out the soil here and replacing it was a good use of space, as the ground is otherwise too rocky for gardens. Beds were surrounded by a second, higher wall to keep out the pigs, sheep, and cattle. Even the colorful dahlia gardens had stone walls around them.

San Lorencito's farmers were multiplying three PROINPA varieties, Gendarme, Revolución, and Desirée. Compared to frost-damaged plants I had seen in nearby fields, these verdant, healthy-looking plants pulsated with energy. To cover the bed, Don Alberto uses transparent plastic sheets stapled on wooden frames. They can be attached to latches set into the stone, secure even in gale-force winds. Don Alberto had improved on PROINPA's technology. He spaced his cuttings at twenty-four per square meter, twice the recommended density. He showed us the results: great piles of smooth, pink-skinned Desirée, separated by size. Most were large enough to satisfy commercial farmers; the smallest ones could be replanted in the seedbeds for another multiplication. "The farmers here are very happy," Cardozo said as we left. "They are inventing things; they are proud." After the potatoes, families plant cabbage, lettuce, and onions in the beds. It was mid-March, already too late for such crops in open fields but not too late in beds protected from the frost. Most of the production goes to household consumption.

I asked Cardozo how often farmers had to change the soil in their beds. He said that the station changed the soil every year, regardless. Campesinos, however, might get by for several years; it all depended on nematodes.

According to Javier Franco, the most serious nematode pest in the highlands is *Nacobbus aberrans,* false root knot nematode. It infects the potato plant's root system, forming little knot-like nodes the size of rosary beads. The embra is within the node and lays its eggs there. When infestation is high, the potato's root system is less dense, stems are weak, the plant turns yellow, and yields fall. Although nacobbus does not directly damage tubers, it gets into the skin. When tubers from an infected area are harvested and then planted somewhere else, the pest spreads. Hence, to produce sanitary seed, the soil must be nematode free.

In the past, altiplano farmers planted potatoes in the same spot only once in seven years. Today, fallows are much shorter. The result is mounting nematode problems. In the highlands, the potato doubles as both a subsistence crop and a cash crop; it is irreplaceable. PROINPA's approach is to reduce, not eliminate, the nacobbus population. "What we want is balanced management in the field," Franco explained. "Killing off a specific nematode type is unnecessary, probably impossible, and certainly undesirable. The likely consequence is the upsurge of a new strain even more difficult to combat."

To develop user-friendly technology, Franco and his PROINPA team work closely with altiplano communities. After the potato harvest, farmers need to collect and burn leftover stems and damaged tubers, as nematodes hide out in the debris. Leguminous crops like lima beans and tarwi, both adapted to cold climates, reduce nacobbus build up. That is also the case with barley, a popular altiplano cereal crop. It acts as a trap: the nematode can enter the plant's roots, but once there, it cannot complete its life cycle. Sensitive to soil temperatures and humidity, nacobbus populations have peak seasons. By changing planting dates by a couple of weeks, project agronomists hope that farmers can avoid the worst periods.[25]

Franco and I went to inspect fields where nematode-control projects were underway. Franco pointed out plots that had potatoes last season but had been rotated to lima beans this season. When planted densely, this leguminous, nematode-unfriendly plant raises soil temperatures, reducing the pest population. The Iscayachi station had also introduced rapidly maturing lima-bean varieties. Farmers get high prices in Tarija for an early harvest of lima beans, as the crop is still out of season. After the harvest, the nitrogen-rich lima-bean foliage is incorporated into the soil as a green manure. For comparison, farmers leave selected plots fallow. When they return to potatoes, they can see how much more productive the nitrogen-enriched plots are. Since lima beans are a staple in the altiplano, the rotation already fits into the established crop cycle.

So too does a rotation to quinoa, a grain native to the altiplano. Franco and his team were calculating how effectively quinoa reduced nematodes after potatoes. I went for a look. Quinoa has a tall stalk with small, round grains covering the large tassel at its top; Spanish chroniclers called quinoa the rice of the Incas. Production here was for domestic consumption only. Also recruited to the project's antinematode campaign is the leguminous Andean lupine, tarwi. Its showy blue flowers cheer up even the most dreary landscape. Exceeding a meter in height, the plant has pods that are all attached to a single stalk.[26]

Some potato varieties can resist the nacobbus nematode. An example is the multicolored andigena, Gendarme, which is also frost tolerant. Nonetheless, Franco cautioned against pushing new varieties on reluctant farmers, who are fussy about shape, color, or taste. A thunderstorm brewed; we could see it heading our way. Even so, I could not resist inspecting the potato fields. Timed to altiplano day length, andigena varieties flower and seed prolifically. I had seen potatoes flower back home, but nothing like this. Here, the potato displayed its secret life as an ornamental unabashed. To judge from the seed balls, which covered more mature plants from head to toe, potatoes in the Andes enjoy an untrammeled sex life.

Even though it was early March and still summer, it got cold fast, with winds and hail sweeping across the valley. I scurried back to the truck for my overcoat. Too late. Cold and wet, we stopped at the first cantina. I sat next to Franco. I asked him if he thought the enthusiasm over the seedbeds was justified. "What makes the rustic beds so important," he emphasized, "is their flexibility. In the high Andes, formal seed production is not the solution. Communities want to multiply their favorite varieties, not just those PROINPA recommends. If they can start with clean seed, plant it at the right altitude where nematodes are not so bad, seed quality will hold up for a very long time. Farmers can see with their own eyes that the beds work. If they can collect some rainwater, or divert water from the canals by their homes, they can follow potatoes with other garden crops. The beds open up a lot of options. Farmers never worked with cuttings or seedling tubers before, now they do."

"What does development mean?" I asked Franco rhetorically.

"Let campesinos run their own lives and communities," he said. "We can offer new technology, but they must decide how to use it. NGOs think they have the answer, CIP thinks it has the answer, the World Bank thinks it has the answer, but only the people affected have an answer. Doing things the way other people want them done is not an answer, and it is not a good definition of development."

After the 1997–98 season, PROINPA took a tally. Some 1,236 families in 129 altiplano communities had a total of 1,376 seed beds. They produced over forty-three metric tons of high-quality seed.[27] The men build the beds, but the women often end up managing them. That's because so many men migrate to the cities seasonally for work. In 1997, of the 790 participants in PROINPA's seedbed training programs, 252 were women.[28]

CHAPTER 11

Potato Weevils

E cuador's geography divides the country into three zones: the Pacific coast, the sierra, and the eastern jungle.[1] The tropical coast includes the country's largest city, Guayaquil. The breathtaking ascent from the coast to the sierra goes from sea level to rugged passes in excess of 4,500 meters (14,800 feet). Then comes the drop down into the central valleys. Quito, Ecuador's capital, is situated in the sierra at an altitude of 2,800 meters (9,000 feet). Across the cordillera to the east begins a descent into the jungle, or *oriente,* where most of Ecuador's oil is found. Compared to its immediate Andean neighbors, Ecuador has much less territory in the Amazon basin. By the terms of the 1942 Rio Protocol, Ecuador lost its territorial claims in the Amazon basin to Peru, an agreement it subsequently refused to accept. Maps made in Ecuador still include the lost territory.

The Andes narrow as they go north into Ecuador; the central valleys are lined by volcanoes. Within a corridor that is just 240 kilometers (144 miles) from south to north, the snow-covered volcanic peaks of Sangay, Tungurahua, Cotopaxi, Antisana, and Cayambe loom over the sierra's towns, glowing in the moonlight. At 5,895 meters (19,340 feet), Cotopaxi is the highest, its perfectly shaped cone visible from Quito's suburbs. This array of volcanoes is unequaled anywhere in the Americas, except perhaps for the Pacific Northwest's "ring of fire," formed by glacier-covered Mount Baker, Mount Rainier, Mount Hood, and Mount Saint Helens. The highest of these peaks is Mount Rainier, which at 4,392 meters (14,410 feet) towers over Seattle. In Ecuador, that is not even high enough for snow.

The Andes come closest to the Pacific Ocean at Tulcán, near Ecuador's border with Colombia; here, the distance between the eastern and western cordilleras is less than fifty kilometers. Beyond the Pacific slopes are the rain forests of Esmeraldas Province. In Peru, the coast is mostly desert. The misty mountains of Ecuador are part of what geographers call the "páramo Andes," in contrast to the "puna Andes" farther south.[2] Ecuador's lush sierra includes Carchi Province, which gets so much rain farmers can plant potatoes all year round. Higher up, the foggy páramos are marshy, natural water collectors, a far cry from Lake Titicaca's thirsty altiplano.

The equator passes through Ecuador's cordilleras, creating what is called an "intertropical convergence zone." Turbulent weather systems and fronts are rare; instead, climate results from localized weather systems.[3] In the central valleys, temperatures change from sunrise to sunset but not seasonally; during the day heat builds up and then rises at night. Thus, the higher ends of the valleys tend to be cooler and get the most rainfall; the lower ends are hotter and drier, a few almost like deserts.[4]

Compared to the climate of Bolivia's highland valleys, the climate of Ecuador's sierra is milder; the frosts are fewer; and those that occur are less severe.[5] In Ecuador, farmers rarely plant crops above 3,500 meters (11,500 feet). Basins and valley bottoms get occasional frosts, but inversion—the tendency of cold air to fall and warm air to rise—protects the slopes. Above 4,000 meters (13,100 feet), Ecuador has no densely populated villages, fields of bitter potatoes, or llama herds. *Chuño,* the freeze-dried potato specialty of the southern Andes, is unknown in Ecuador.[6] The Inca introduced alpaca and llamas, but they never flourished on Ecuador's páramos. As for highland agriculture, it never relied so heavily on irrigation and terracing. Most pre-Hispanic irrigation was confined to the drier valley floors, mainly for corn; as for terracing, there is little evidence of stone-walled embankments on mountain slopes.[7] Agriculture in Ecuador never had the tight organization characteristic of Inca imperial rule in Peru. Apparently, the rainier climate, abundant game, and low demographic density made elaborate hydraulic schemes unnecessary. Compared to the stingy altiplano, Ecuador is a land of plenty.[8]

The Inca conquest of Carchi Province was beaten back by the fierce resistance of a federation of Cara. In the end, the Inca ruler Huayna Capac (1493–1525) accomplished through treachery what force of arms could not—or so legend says. He lured Carchi's unarmed and unsuspecting warriors to peace talks. Fully armed Inca troops, hidden from view, waylaid them, hacked them to pieces by the thousands, and threw them into the lake. To this day, the place is called Yaguar Cocha— the Lake of Blood. To destroy their territorial base, the Inca resettled the Caras elsewhere, sending in loyal subjects as replacements.[9] Not long after Huayna Capac's death, the conquistadors toppled the Inca, and Ecuador passed to the Spanish Empire. Carchi's indigenous population never recovered.

In 1954, land ownership in Ecuador reflected the country's colonial past. Less than fourteen hundred haciendas held over 2.7 million hectares, or 45 percent of the country's acreage. Three decades later, land-reform projects had changed this picture considerably. In the sierra, over 1 million hectares had been distributed to one hundred thousand farm families and communities. Ecuador still had haciendas, but their owners now possessed hundreds of hectares, not thousands.[10] This is not revolutionary by Bolivian standards, but it is by Brazilian standards. In Brazil, a military regime (1964–85) subsidized big estates and export agriculture. By the time civilians took over again in 1985, land reform was out of fashion.[11] Ecuador's approach was cautious enough to prevent a coup on the right, as happened in Brazil, and radical enough to forestall guerilla warfare on the left, as happened in Colombia.

Oil helped ease Ecuador's transformation into an urban society. In 1970, 55 percent of its six million people still lived and worked in the countryside. A decade later, the balance had shifted decisively to the cities.[12] In the meantime, Ecuador cashed in on the world's volatile oil markets. Production increased from a total of 1.4 million barrels of crude oil a year in 1970 to 80 million in 1979, of which it exported 46 million barrels.[13] This is not much by international standards, but it is sufficient to make a big difference in a small country. Oil prices cooperated, increasing from $2.40 a barrel in 1972 to $35 in 1980; the country's oil revenues followed suit, rising from $323 million to over $2 billion. Between 1972 and 1980, Ecuador's economy grew at an average rate of 9 percent a year. In 1980, Ecuador had over $800 million in monetary reserves.[14]

The oil boom helped modernize the country's infrastructure, from paved roads to new government buildings and rural electrification. Jobs in construction, textile factories, urban services, and the oil sector pulled labor from the countryside. Today, only 29 percent of Ecuador's 12.6 million people are classified as rural.[15] Even so, the agricultural sector is still the country's single largest employer. The rural labor force is disproportionately concentrated in the highlands, where land reform had the greatest impact.[16] In Carchi Province, which had few Indian communities left, mestizo farmers benefitted the most.

"No province in Ecuador sells more potatoes than Carchi's commercially minded farmers," explained Wilson Vásquez. "Average yields are thirteen metric tons per hectare, the highest in the country." A thin, serious man in his forties, Vásquez had spread out a topographic map of Ecuador on the table. We met at the Ministry of Agriculture in Quito; Vásquez worked with the potato program at the National Agrarian Research Institute, or INIAP.[17] Most of Carchi's potato farmers plant at least ten hectares of the crop annually, which is a lot by Ecuadorian standards. Carchi's cool climate and dependable rainfall are great for potatoes. In 1995, Carchi's farmers grew 40 percent of all the country's spuds, but on just 25 percent of the acreage devoted to the crop.[18] They can plant potatoes almost anytime; most of them are improved, long-duration andigenas that take five to six months to mature. Chola, a red-skinned variety with a creamy-yellow flesh, is the most popular.

I asked Vásquez why it takes andigenas so long to grow. In India's Ganges valley, the winter potato crop is ready in seventy-five days. Vásquez explained that the higher the altitude the longer it takes a crop to mature. In Ecuador, the combination of high mountain valleys, cool temperatures, and short days slows down plant growth. At 3,000 meters (9,840 feet), for example, maize needs eleven months. With potatoes, traditional andigena landraces need seven to eight months. Improved andigena hybrids, which are crossed with rapidly maturing tuberosums, mature two months earlier. Potatoes are produced exclusively in the sierra; the country's wet, marshy coast is for rice and bananas. "Ecuador's potato zones are distinct sociologically," Vásquez noted, returning to the map. In the central sierra, Chimborazo Province produces 20 to 25 percent of the potato crop, but farms are smaller and yields less.[19] Chimborazo's campesinos belong to traditional, Quechua-speaking communities. Farmers rarely plant more than five hectares of potatoes; yields average about seven metric tons. Farmers in the central sierra prefer a white-skinned potato with cream-colored flesh. In the south, in Cañar and Azuay Provinces, small plots predominate. Campesinos rarely have more than a couple hectares. In this zone, purple-skinned Bolona, with its reddish-tinted flesh, is preferred. Compared to farmers in the north, who sell their crop to truckers, farmers in the south produce for domestic consumption. They have little or no capital to improve production.

My orientation complete, I headed for the potato research station at Santa Catalina, outside Quito. The station is the headquarters for FORTIPAPA, a potato project set up in 1991 by INIAP, the Swiss Development Agency, and the International Potato Center.[20] Given its focus on generating farmer-friendly technology, FORTIPAPA has much in common with PROINPA, its sister project in Bolivia. Like PROINPA, it works on pest control, does participatory research with farmers, and produces seed in protected beds. In 1998, for example, farmers from eight communities in five provinces planted and evaluated over forty new varieties. In Cañar Province, the project helped select and clean up the popular cultivar Bolona.[21] FORTIPAPA's breeding program drew on Ecuador's collection of old potato landraces. For example, it has tested 131 entries of the domesticated Andean species *Solanum phureja* for late-blight resistance. Phureja is adapted to warmer valleys and is renowned for its disease resistance.[22]

The key to success, according to INIAP agronomist Héctor Andrade, is "systematic testing and focused technology transfer." Director of INIAP's potato research, Andrade brought a fresh approach to old problems. Although INIAP had released many improved varieties, its selections had little impact. "The breeding scheme we previously used in Ecuador," Andrade said, "took too long and cost too much. The traits selected for reflected what researchers wanted more than what farmers needed or what consumers preferred. Consequently, with support from FORTIPAPA, we changed strategies." To improve selection, INIAP involved farmers and consumers early on. The variety Fripapa is the result.[23]

Ecuador had a growing market for precut french fries. The challenge was to find a variety of high processing quality. To accomplish this, INIAP did three years of participatory research. It worked with Ecudal, a food-processing company. "Ecudal told us the kind of potato they wanted," Andrade said. "We took varieties that met their standards to farmers, who tried them out and told us which ones they found acceptable and why. Then we went to consumers for taste tests." INIAP ended up selecting Fripapa, an oblong, rose-colored, yellow-fleshed variety. An andigena-tuberosum hybrid, it matures in just five months and resists late blight. By the time Fripapa was selected, Ecudal had processed it, farmers had grown it successfully, and consumers had already said they liked it. INIAP has now distributed virus-free seed tubers to farmers for local multiplication. Ecudal provides technical assistance, quality control, and most important of all, a secure market. Frito-Lay has likewise teamed up with INIAP to test and select local cultivars for processing.[24]

"To make sure new varieties get accepted, we work hard to understand consumer demand," Andrade emphasized. In Quito, per-capita potato consumption is 120 kilos a year.[25] Consumer studies by FORTIPAPA show that the most popular way to prepare the spud is to cook it in soups; next comes the home fry, and then the mashed potato.[26] Taste tests, including that for Fripapa, reflect the utilization hierarchy. Consumers considered how well potatoes cooked, plus their color, consistency, and flavor.[27] FORTIPAPA also checked out how much consumers recognized differences among varieties. At least half of those interviewed in Guayaquil, Quito, and Cuenca, the country's three largest cities, correctly distinguished between the old favorite, Chola, and the improved newcomer, Gabriela.[28] Knowing how households cook potatoes, plus which varieties they prefer and why, is a crucial piece of marketing information. It helps the project decide which varieties should be cleaned up for seed multiplication. So far, improved INIAP varieties account for about 40 percent of the urban market.[29]

According to Charles Crissman, Carchi's profit-minded farmers respond quickly to market changes. He says that in the past decade, Carchi had switched varieties almost completely. An agricultural economist from CIP and former Peace Corp volunteer, Crissman had spent the last three years tracking pesticide use in Carchi.[30] He and I were headed for Tulcán, the provincial capital, near the Colombian border. During my ten days in Ecuador, I was visiting all of FORTIPAPA's regional offices. On my first visit to Ecuador, over thirty years before, the 250-kilometer trip to Tulcán from Quito had taken all day; this time, it took just three hours. Better roads now allow Carchi's farmers to specialize in potatoes for the Quito market. They produce a dependable, high-quality spud, good enough for both household consumption and for processing. Carchi farmers have the know-how, they have the capital, and they have the pesticides.

Modern agriculture and pesticides are twinborn. ("Pesticide" is a generic term that includes both insecticides and fungicides.) To control insects, weeds, and diseases, farmers worldwide use 2.5 million tons of chemical pesticides at a cost of between $20 and $25 billion a year. Japan, Europe, and the United States rank as the biggest users, applying 10.8 kilos (23.8 pounds), 1.9 kilos (4.2 pounds), and 1.5 kilos (3.3 pounds) of active ingredients per hectare of farmland, respectively. Averaging 220 grams (7.7 ounces) of pesticides per hectare, developing countries are still far behind. With over half the world's agricultural output, such countries account for roughly 25 percent of pesticide consumption. In Ecuador, half of the pesticides applied go to the country's primary export, bananas. In the sierra, particularly in Carchi, the potato crop gets a heavy dose.[31]

To gauge pesticide use, Crissman headed up a FORTIPAPA research team that conducted a detailed survey in Montufar, a canton in the heart of Carchi's potato-producing region. Over a two-year period, the team worked intensively with forty farmers, who planted a total of 320 parcels of potatoes. They tracked the date a pesticide was used, parcel by parcel, and whether farmers applied just an insecticide, just a fungicide, or a mixture of both.[32] Given the time, cost, and labor that spraying crops entails, farmers typically combine them into powerful "cocktails." They apply the mixtures all at once in a single pass though their plots. From a farmer's point of view, the habit has much to recommend it. Fields are rarely contiguous; instead, Carchi farmers have parcels under cultivation at different locations up and down the steep mountain slopes. Lugging equipment from one field to the next is demanding enough, not to mention making two passes when one will do.

As for insecticides, the results showed that farmers used chemicals with a total of 19 different active ingredients; of these, 2 products, carbofuran and methamidophos, made up 47 percent and 43 percent of the total, respectively. As for fungicides, the products used had 24 active ingredients, of which mancozeb accounted for 81 percent of the total.[33] On average, farmers made seven pesticide applications, with mixtures of 2.46 insecticides or fungicides in each. This added up to 1.5 kilos of active pesticide ingredients per hectare.[34] The canton's rate of pesticide poisoning, 171 per 100,000, is one of the highest ever recorded. Tests of spray operators showed a "consistent pattern of central nervous system damage, due to pesticide residues absorbed through the skin."[35] A subsample of 174 farmer applicators revealed that almost 25 percent had experienced an acute pesticide-poisoning episode in their lives. Chronic dermatitis was twice as high and skin lesions more common than in a control group. Particularly alarming was the reliance on the highly toxic carbofuran, which is strictly controlled in the United States, and organophosphates such as methamidophos, which destroys an insect's nervous system.[36] The insecticide works the same way on humans. Compared to the incidence found in town families from nearby San Gabriel, the incidence of "delayed reaction time" in the Montufar sample was a standard deviation above the "normal" town population; the incidence of severe "nervous system disorder" was three standard deviations above the normal population.[37]

To protect farm families, the first line of defense is prevention. An attractive brochure on "Rules of Pesticide Use" is published by USAID and CARE-Peru. Most pesticide absorption is though exposed skin, especially on the hands, legs, and back. The rules say to use protective clothing such as rubber gloves, boots, and jackets; to wear a face mask; not to apply pesticides against the wind; not to eat or drink until the work is completed and one has changed clothes; to be sure to wash up well with soap and water; not to keep pesticides in the house; and to be sure to dispose of old containers safely. As for the form and quantity applied, the brochure says to read the instructions on the label and heed their warnings. Farmers are directed to be careful not to mix pesticides and fungicides together.[38]

Except for wearing rubber boots, not many Carchi farmers follow the rules. As for myself, I must admit to violating most of these rules in my garden at one time or another—not on the potatoes, which I grow pesticide free, but on the roses and fruit trees. I wonder how many liters of Round Up, bottles of Fungisex, or cans of Diazanon spray Americans keep in the basement, out in the garage, or even under the kitchen sink. To get Carchi's farmers to apply pesticides safely, warnings on labels and free brochures are not enough. Whenever possible, FORTIPAPA personnel make the pitch for safety. But are farm families—not to mention day laborers even farther down the totem pole—likely to invest in the recommended safety equipment? Without extension and community-focused education, it is unlikely that caution will trickle down to the rank and file.

The best way to protect farmers from pesticide risks is to reduce reliance on the potato's chemical arsenal. In the sierra, farmers spend 3.5 times more on chemical concoctions to protect potatoes than they do to safeguard wheat and barley and 6.7 times more than they spend on corn or beans.[39] So cutting back on the pesticides farmers dump on the potato is a good place to start. In Carchi, the main target of insecticides is the voracious Andean potato weevil. From Colombia to Bolivia, the weevil is the worst high-altitude pest, wreaking havoc in potato fields. When it comes to rotations or varietal selection, researchers need to incorporate the knowledge and experience farmers have. But when it comes to insect cycles, farmer knowledge often breaks down. The Andean potato weevil is a case in point. Only the damage caused by late blight rivals the damage this pest inflicts on potatoes in the Andes.

The adult weevil hides during the day. It comes out at night to eat the potato plant's foliage and to mate. The female lays its tiny eggs on field debris at the base of the plant. Over a female's four months as an adult, she lays over six hundred eggs, laying the largest quantity from December to March, which is the height of the potato season in the Andes. After a month, the larvae hatch and tunnel into the soil. The crescent-shaped white larvae give the pest its local name, *guzano blanco,* or "white worm." The worms bore into the tubers, leaving behind a telltale prick on the skin's surface that is difficult to detect. They live inside the tubers for six weeks, eating away, after which they bore their way out and down into the ground to a depth of between twenty and thirty centimeters. Thus protected in the ground, the larvae build their delicate cocoons, in which they pupate. The newly formed adults hibernate in the soil until the rains wake them up. Then they break out of their cocoons and head for the surface, looking for innocent potato plants to infest.[40]

They must find potatoes. The weevil is a fussy, monophagous pest; it will not eat anything else. If farmers rotate crops, the adult pest walks to adjoining fields. Hence, starting with clean, weevil-free seed is not enough. Farmers have long recognized the white-worm, larval-stage insect as a potato menace. These worms are easy to see in Carchi's black, volcanic soils, and farmers have long saturated the ground with chemicals, trying to kill off as many as possible. What farmers did not know is that the adult weevils and the white worms are, in fact, the same pest; they are just at different stages of development.[41] Armed with this vital entomological detail, farmers began to substitute knowledge-intensive strategies for pesticides. "The key to controlling the weevil," explained Fausto López, "is to reduce the adult population before it reproduces." López headed up FORTIPAPA's substation at San Gabriel, the seat of the canton of Montufar. An agronomist, he received additional training in community organization and on-farm research at FORTIPAPA's office in Chimborazo Province.[42] He took us to see how farmers use the new pest-control technology. Carchi farmers are the best educated in the country; many have completed high school. To judge from the pick-up trucks plying the muddy mountain roads, the potato is a profitable venture. We stopped at a farm up the valley at about 3,000 meters (9,800 feet). We were supposed to inspect varietal trials, but what caught my attention were the tree tomatoes. Over two meters tall, the plants have orange-red fruits the size of large, oval-shaped plums. I ran to catch up. Farmers came over from their fields to join us. López did a little evangelizing, reminding them that "reducing pesticide use is everyone's job" and that "we all must work together, not each lost in his own world."

To reduce weevil incidence, farmers use homemade traps, which they put out in the fields a month before planting time. The bait consists of foliage from volunteer potato plants that spring up in old fields. The farmers dip the leaves in a solution containing two grams of acefato per liter of water, which is a minimal dosage of an insecticide much less toxic than carbofuran. They lay out these traps every ten meters, cover them with a square of old cardboard, and put a rock on top of each so that it does not blow away. Adult weevils are attracted to the bait, feed on the leaves, and die before they can lay their eggs. When the toxicity wears off, which occurs after eight to ten days, the bait is changed. The traps are replenished regularly until the seed tubers start sprouting. Thereafter, farmers spray only if they find enough adult weevils around the base of the stalks to justify it. According to Fausto, "This is an enormous change; many farmers no longer spray at all." Studies show that the traps reduce insecticide use by 70 percent. Foliar insecticides, by contrast, cost at least forty-five dollars per hectare and are frequently applied to little effect.[43]

The new technology is not just for mestizo farmers; in Ecuador's central Andes, this new technology is standard in twenty indigenous communities in four cantons of Chimborazo Province.[44] And it is not just for Ecuador. Many of the practical aspects of weevil control were first worked out elsewhere. Some components require no pesticides at all. In Peru's highlands, for example, kids hold contests to see which team can catch the most weevils. They go to the fields after dark with flashlights and lanterns, picking weevils off the ground and putting them in plastic sacks. The weevil's most vulnerable point, however, is during its eight-week pupal stage. The slightest injury to the cocoon kills the pupating insect. By working the soil early, farmers bring the cocoons to the surface, where the sun and frost kill them. And if the cocoons escape the elements, they are unlikely to escape the chickens. Peru's weevil-control program—backed by CIP, funded by USAID, and administered by CARE-Peru—targets 3,500 peasant families in 117 Andean communities in four departments. So far, the project has trained seventy-six pest-control extension workers. Some 70 percent of the campesinos in project districts can describe the weevil's life cycle, can recognize pesticide risks, and know how to select less toxic chemicals. Based on community participation and feedback, the project designed its own training materials. Its bulletins, brochures, slides, and videos now circulate throughout Peru's highlands. The impact that weevil control has is cumulative. After four years, without using any insecticides at all, pilot-project farmers in Peru's Huatata District reduced the percentage of weevil-damaged tubers from 44 to just 8 percent; in three years, Chilinpampa's farmers reduced it from 60 to 14 percent; and in two years, Aymara District reduced the percentage from 38 to 25 percent.[45] CIP follow-up studies calculated the net gain to farmers at $241 per hectare. Yields went up, and with weevil damage much reduced, what farmers sold fetched a higher price.[46]

Carchi's commercially minded farmers cannot be bothered with catching weevils at night. Adding up production costs such as seed (10 percent), fertilizer (20 percent), machinery (12 percent), pesticides (12 percent), and labor (29 percent), a typical Carchi potato farmer already invests between $2,000 and $2,500 per hectare of potatoes produced. The net profit per hectare on the 320 plots that Crissman studied came to about $525.[47] Despite labor costs, it is much cheaper to put out traps with an insecticide-treated bait than to spray potato plots repeatedly.

Insecticides are more destructive ecologically than fungicides. Nevertheless, what fungicides lack in firepower they make up for in the sheer quantity applied. In Carchi, farmers spray against late blight, the preeminent potato fungus, six to eight times a season.[48] With weevils, farmers can count how many critters they have per square meter and calculate whether an additional spray is worth the cost. With late blight, on the other hand, there is less margin for error. A little bit of late blight can quickly become too much. By the time the symptoms show up, it is often too late. Consequently, farmers spray defensively and regularly. They use protectants that keep the fungus from growing or systemic fungicides that kill the spores. Trying to get farmers off the fungicide treadmill is not easy. The safest path for FORTIPAPA is to get varieties with greater late-blight resistance to farmers. This lets them control the fungus with fewer sprays. Fripapa, the variety INIAP selected for processing, has durable late-blight resistance. For the popular but susceptible varieties, FORTIPAPA calculates the timing and minimal number of sprays required to keep the disease at bay. Excessive spraying, especially when farmers mix in systemic fungicides, can induce the fungus to mutate, which can result in new, fungicide-resistant strains. Fewer, more effectively timed applications both save farmers money and reduce the rate at which the fungus mutates.[49]

By midmorning, a drizzly rain made the muddy roads treacherous. Farther up the hillside, work teams were harvesting a two-hectare parcel of potatoes. We decided to stop. The slope was too steep for a tractor, much less a truck. Walking up the horizontal rows was like going up a gigantic staircase, each step a meter above the next. We made our way to the top and went over to talk with the owner.

A solid, daunting woman in her fifties, Doña Marta had wrapped herself in ponchos and sweaters; she wore the traditional black-felt bowler hat and long, laced, leather boots.[50] On this misty, cold morning, she had been there since 4 A.M. Doña Marta looked over her potato plots like a general surveying a battlefield. I asked about her workforce.

"Workers get morning coffee, a lunch, plus 2,000 sucres [3,860 sucres = one U.S. dollar] for every quintal [one hundred pounds, or about forty-five kilos] harvested," she answered. Doña Marta estimated that a good worker averages twelve bags a day. For potato sacks, she uses recycled, heavy-plastic fertilizer bags that cost her 500 sucres apiece. When a bag is filled, it is weighed and sewed closed. Doña Marta put the total cost of harvesting a quintal at 3,200 sucres. This includes paying the organizer of the work teams. When a trucker shows up, she bargains over the selling price per sack. Sometimes, she pays for loading the truck, sometimes not. The trucker pays her in cash. I had no doubt that Doña Marta drove a hard bargain.

In Carchi, the buoyant demand for labor has helped small farmers supplement their income. Since potatoes are a labor-intensive crop, it takes the equivalent of 157 days of labor to plant, weed, protect, and harvest a hectare of them.[51] Everybody knows which corners to go to in San Gabriel to hire workers. The various work teams have organizers who bargain over the cost of each job. Apparently, the whole family sometimes pitches in. The team that Doña Marta contracted consisted of five families that worked together—men, women, and children. They kept up a steady but unhurried pace. The youngest kids chased each other up and down the terraces.

"I have planted potatoes since I was eighteen, since my father died," Doña Marta said. I asked her what she did to get a good crop. She answered, "Start with good seed, use fertilizer, protect the crop." Doña Marta calculated that this particular field used 40 quintals of seed. "When the seed was new," she said, "the yield was 800 quintals; now it is down to 450, so it is time for me to renew the seed." She plants potatoes in horizontal rows; as the plants grow, soil is hilled up around them, which helps keep insects and the late-blight fungus from getting at the tubers. As for plant protection, she believes in using strong chemicals that do the job. She thinks today's pesticides are not as strong as they once were.

Her field was so overrun by weeds, it did not look like a potato patch at all. Once the plants set their tubers, however, weeding is beside the point. Farmers can leave the crop in the ground for several weeks, letting the foliage die off almost completely. Doña Marta had also stopped spraying against late blight. The misty rains do not leach the fungus down to the tubers. Nonetheless, during the growing season, she had sprayed the field a dozen times. This was the third time she had planted potatoes here in succession; it was time to rest the land. With the profits from her potato crop, Doña Marta planned to buy a couple of cows and convert the field into a high-quality pasture. For Carchi farmers, buying cows is like putting money in the bank. They get their money back with interest. Selling the milk provides a small but steady income. If the demand for meat goes up, farmers can sell their cows at a profit. Such at least is the classic economic explanation.[52] Doña Marta's explanation was less complicated. "I just like having cows," she said. "It makes me feel like a real farmer."

In another plot, Doña Marta said, she had ulluco, oca, and lima beans. "When I was a girl, we grew some mashua, but not any more. My mother used to cook it, hot and spicy in stews. Or she added sugar, and we ate it like a sweet. Mother said mashua was good for my health, but I no longer remember why," she said, with a sad smile.

The sun broke through the clouds. In Carchi, there are no seasons. Harvested plots stand next to parcels freshly plowed. Barley fields ripen alongside decayed corn stalks ready for burial. Cows graze with their calves. Beyond the hedgerows and eucalyptus are the potato fields, blanketed in a dozen shades of green. Here in Carchi's valleys, Mother Nature does not keep secrets: childhood, impetuous youth, maturity, and old age—each has its place, and each is but a reflection of the other.

CHAPTER 12

Watersheds

"The problem with the cows is that they are in the wrong place," said Oswaldo Paladines. "They should be on the hillsides, not down in the valleys." Tall, hair thinning, in his fifties, Paladines headed up Ecuador's Fund for Agricultural Development (FUNDAGRO), an organization that works on rural training and extension.[1] He had three decades of experience in agricultural work, including a stint at the International Center for Tropical Agriculture (CIAT) in Cali, Colombia.[2] We had met over coffee in El Angel.

The old hacienda, Paladines explained, typically had acreage on the valley floor, where it was warmer, plus land at various altitudes up the mountainsides to the páramo.[3] The hacienda owners planted vegetables in the warmer bottomlands, then tubers like potatoes in cooler zones higher up. Finally, where slopes were the steepest, they had their pastures. Ecuador's land-reform laws gave hacienda owners first choice of which lands they would retain and which they would divide up. Naturally, they chose the land most easily exploited—namely the valley floors and the lower surrounding fields. The campesinos got the hillsides. The perverse result is the opposite of the old hacienda system. Now, ranching is done on the valley floors, and crop production is on the steep hillsides, precisely where it should not be. Where rainfall is heavy and the soils are poor, there is terrible erosion.[4]

The harm done by eroded soils, depleted forests, and contaminated water cannot be undone by potato research, no matter how good it is. The CIP-backed Consortium for the Sustainable Development of the Andean Ecoregion—a group know by its Spanish acronym, CONDESAN—recognizes this obvious truth.[5] Created in 1994, the Consortium pulls together expertise on conservation, agriculture, and community action. Members include community groups, nongovernmental organizations (NGOs), farmer associations, universities, government agencies like

FUNDAGRO, and research centers like CIP.[6] Protecting the biodiversity of Andean crops is one of the Consortium's objectives. It also works on erosion control, the use of water resources, and ways to make agricultural production sustainable. Consortium projects include alpaca grazing in the altiplano, the construction of rustic beds for household vegetable crops, and watershed conservation.[7] National research programs in Colombia, Ecuador, Peru, and Bolivia are all Consortium members. In Ecuador, the Angel River valley in Carchi Province is a Consortium research site.

I asked Oswaldo Paladines why there was a site at the Angel River. "Serious erosion has yet to set in," he answered, "so we still have a chance to preserve the watershed. It makes no sense to select a site where anything we do is too little too late." Carchi rarely has destructive downpours; a wet drizzle is the rule. Consequently, erosion is less advanced there than it is in other provinces. FUNDAGRO also had prior experience in the zone, mostly working with farmers on small-scale milk production. For all these reasons, the Angel River valley was a good choice.

The Angel River is the life of the watershed. It rises in the páramo and provides irrigation throughout the valley, even to communities beyond the watershed. From a technical, water-use standpoint, the system is outmoded. Almost 60 percent of the water carried is lost before it reaches a farmer's field. The system's deficiencies, however, are as much social and structural as they are technical. The valley has some sixteen semi-independent irrigation systems, all of which branch off from the main channel, which is some forty kilometers long. There is conflict over water rights all along the way. To upgrade and maintain the system, all the communities involved will have to cooperate. Unfortunately, farmers have invaded the upper reaches of the valleys. Land that once formed a natural buffer between the cropland below and the páramo above is being worked, causing erosion that silts up the irrigation canals. As topsoil and pesticides are washed into local streams, the quality of the water is deteriorating.

Most of Ecuador's irrigation is for fields on the upper valley slopes, those closest to a water source.[8] The Angel channel is unusual, as it goes from the páramo all the way through the high valleys and down to the flatlands by the Mira River, at 1,500 meters (4,920 feet). The Mira River valley is warm enough for orchards of tropical fruit trees like papaya and avocado. Farther up the Angel valley, as it gets cooler, farmers plant beans and peas. Nearing the town of El Angel, at 3,400 meters (10,480 feet), the main crop is potatoes, and higher up still come the lima beans, quinoa, and pastures. And last, at 4,000 meters (13,120 feet), is the páramo. I had come to El Angel with CIP economist and FORTIPAPA director, Charles Crissman. The Consortium has offices in El Angel—large, cold rooms with cement floors and no heat. The rooms double as meeting halls; folding chairs are stacked up in the corners. Currently, the project is tabulating the region's population density and migration patterns. On the wall is a map of the watershed and its irrigation system. The páramo is legally protected, which slows down but does not halt agriculture's intrusion into the buffer zone. Since the páramo is the irrigation system's source, I wanted to see it. We took the project's jeep.

It was a pleasant, sunny day in El Angel. Up on the páramo, however, it was windy, cold, and dank. At night, temperatures plummet down to -5 degrees centigrade (23 degrees Fahrenheit). A clear day with a powerful sun can send the temperature soaring to 25 degrees centigrade (almost 80 degrees Fahrenheit)—but not today. I put on my gloves and scarf and pulled down my ski cap. A blustery gray sky, thickets of scruffy bushes, and clumps of tough, greenish-brown grass give the páramo its forbidding and austere beauty. In the distance, through the fog and mist, we could make out the marshy lake that is the source of the valley's water. The Consortium, Crissman told me, is documenting just how much water collects on the páramo. Using a fog trap, they have recorded accumulations as high as 30 millimeters (1.2 inches) in twenty-four hours. Over the course of a month, they measured precipitation on a total of twenty-two days.

We headed back down. A muddy road is no obstacle to crossing the flat páramos. But going down, mud makes the mountain roads dangerous. We were lucky that the drizzle had not settled into the valleys. It was a Monday, market day in El Angel. Even though El Angel is a small place, its weekly street market dwarfs the produce sections of the largest supermarkets. Everything is fresh, in season somewhere, either in the central valleys or down on the coast. I made my way slowly through the narrow streets set aside for market day, notebook in hand. I spotted some white and yellow oca tubers, shaped like stubby pine cones, and two types of ullucos: round ones with smooth red skins and elongated ones, some of them pink skinned, others creamy white. The spice stalls had anchiote, a red dye, plus anise, cinnamon, and aromatic medicinal teas made from dried leaves and bark. From the temperate valleys of Riobamba came grapes, apples, pears, and raspberries. From the cool central valleys came lettuce, cabbage, turnips, and cauliflower, plus carrots, beets, and radishes, not to mention peas, lima beans, onions, and garlic. Vendors heaped up the peppers in vast piles sorted by size, color, and potency, including in their piles the tiny red ají, which is ground up for use in hot, pungent sauces. There were piles of large-kerneled floury maize, dark-red tomatoes, and bright yellow squashes.[9] From the warm valleys came sweet potatoes, starchy manioc roots, and the carrot-like arracacha. Oranges and apples, peaches and pears, plus watermelons, papayas, and avocados were from the temperate tiers. From the tropical coast came the papayas, bananas, pineapples, and passion fruit, plus the round, greenish-yellow guava fruit, the cannonball-sized zapote, taxo for juice, and the spiny green guanabana.[10] The only crop I missed that day was the potato, which has its own market day each Sunday.

It was noon; we headed for El Angel's only restaurant, which is wedged into one room of an old colonial-era mansion. The walls had once been a deep, cobalt blue; roughly hewed, squared-off beams braced the ceiling. The paint was chipping, the tables were wobbly, and the outhouse had a squat toilet. But the avocado soup, the lima beans, and the rice with tomatoes, peppers, and fried eggs made it all worthwhile. Given such abundance, it is hard to believe that the twenty-first century will usher in scarcity and want. Compared to Asia, South America has ample natural resources and is not densely populated. Even so, the future of Andean agriculture is uncertain; farming the slopes can do irreparable damage. I asked Charles Crissman if mountain agriculture was viable. "Only with terracing," he said, "and only with crops matched to the soils and terrain. Otherwise, erosion will destroy the land's fertility, forcing families off the land."

A collective discussion followed. Ways to gear terracing to the needs of small farms had already been worked out. Farmers can plant tough barrier grasses between plots to break the rainfall; they can shore up hillsides with fruit trees or with rapidly growing timber harvested piecemeal for firewood. The overall social benefits make terracing worthwhile. Unfortunately, on any particular farm, the costs often exceed the short-term benefits; land conservation is the kind of investment that requires a long-term mentality. Ecuador could back hedgerows, contour plowing, and reforestation projects—if it valued them. The government subsidizes highways, cooking gas, and rural electrification. So why not erosion control? "Defending a country's natural resources," Crissman argued, "ought to rank as a public good equal to a highway."

After lunch, Crissman and I left for Tulcán, but not before I set up a meeting with Oswaldo Paladines for the following week in Quito. This time, he and I met at FUNDAGRO headquarters, in a spacious, carpeted office. I was critical of how much countries spend on roads, a service for the rich. "It is easy to criticize infrastructure projects and subsidies," Paladines said. "We need roads to link up markets, and the subsidies on cooking gas have public benefits." The subsidies made 15-kilo canisters of gas so cheap that even rural families switched from wood to gas, which has helped save what is left of the country's forests and wood lots. "Rural development," Paladines said, "is as much a matter of social structure and education as it is technology." To reinforce his point, he went over FUNDAGRO's analysis of the Angel watershed and its farmers.

How the economy works, he said, reinforces the urban-rural divide. All the big companies and all the capital are in Quito. The money earned on Carchi's haciendas is not reinvested there; it is drained away and used elsewhere. The old hacienda class, of some thirty to fifty families, has left for Quito. Anyone with an idea or money to invest goes to the big cities. The result is tremendous inertia and apathy in the towns. In El Angel, there is not a single new house or bank.

In the countryside, farmers work their land as individuals, not as members of communities. This reflects Carchi's mestizo roots, as opposed to the mindset of Ecuador's indigenous culture. Some 78 percent of the farmers in the Angel watershed are small landowners with an average of 3.5 hectares of land. The annual income from such a farm is $1,100, to which should be added the $800 a year earned by working as a day laborer, for a total of $1,900. A middle-sized farm averages 7 hectares, with an annual income of $3,500, plus $400 earned at manual labor, for a total of $3,900. Large farms average 13 hectares with average incomes of $6,600; owners of such farms do not work for anyone but themselves. For Carchi overall, the years of school its children complete is above the national average. Children of small farmers typically finish grade school; for the children of middle-sized farmers, high school is the norm. In the case of large farms, some children even end up with university degrees. Although farmers in the project's work zone differ by income, they all live in run-down houses. That is because they put education ahead of home improvements. They realize their children will have to earn a living in the city. They want them to leave; they know dividing the land up further is not viable.

A family's capacity to generate income is directly related to how much land it has; the greater the investment it can make in production, the greater the access to technology, credit, and technical assistance. FUNDAGRO estimates that better production technology can increase small-farm incomes by 40 percent. That seems like a lot, but it only comes to an additional $440, which is not much in absolute terms. On the other hand, with good capital and good management, middle-sized and large farms can easily double their income. But such farms constitute only 22 percent of the total. Small farms make up the majority. The small farmer eats well, but additional subdivision is out of the question; it means complete impoverishment. For small farmers, a new potato variety is not the answer. They have few resources and no market power. Attempts to form cooperatives have failed. Nonetheless, the small holder prevails everywhere in the country. Eventually, consolidation of small farms is inevitable. For now, farmers must use what they have to the best of their ability, and that means education and technical assistance. It does not matter whether public or private agencies provide extension, as long as the work is done conscientiously and efficiently.

"We need to add aspects to rural development that are not based so exclusively on farming," said Paladines. For the Angel watershed, ecotourism might provide an extra boost. Tourists visit Tulcán and go back to Quito the same day. A more imaginative approach would promote the region's picturesque towns or hiking in the countryside. "How do we visualize the future of hillside reserves and protected environments?" Paladines asked. "The páramo above El Angel is part of a special reserve, but who will enforce the regulations, or work on maintenance? Is this not the kind of job rural associations should be paid to do? Together, new initiatives are possible, and not just in agriculture."

Carchi is an exceptional province. In most of Ecuador's sierra, traditional communities are still strong. Land reform gave them a new lease on life. In Paladines's view, however, Ecuador has not come to terms with its indigenous communities. "Should they be self-governing enclaves or should they build nice houses with bathrooms and forget the past?" he asked.

Good questions, but who has the answer?

CHAPTER 13

Multiplication

Agronomist Fausto Merino had an answer: applied research and on-farm training that would help communities farm better. Merino headed up the agricultural station in Riobamba, capital of Chimborazo Province. He had over twenty years of experience with extension, research, and farmers. With respect to the station's potato program, Merino worked on community-based seed multiplication.

Ridges cut across Ecuador's cordilleras. Riobamba is less than two hundred kilometers from Quito on a paved, well-maintained highway. Nonetheless, the trip takes almost four hours, as the mountain valleys in between the two cities separate them like steps on a ladder. To the south of Quito, Los Chillos valley is green and wet. But past the páramo and Cotopaxi volcano, the valleys are sunny and dry. From Salcedo to Ambato, apple, pear, peach, and cherry orchards line the road. Once across the ridge that protects Ambato to the south, the road narrows and winds down into lush, green countryside. If snow-capped Chimborazo is visible, the vista across the valley is unforgettable. Soaring 6,310 meters (20,700 feet) into the sky, Chimborazo is higher than any peak in North America, including Alaska's Mount McKinley.

I had heard of Riobamba's FORTIPAPA director before. I had met his students everywhere. They told me that Fausto Merino took inexperienced recruits like themselves and shaped them into knowledgeable researchers, trainers, and extension specialists. By the time I got to Riobamba, I was expecting a drill sergeant of sorts, not the quiet, unassuming man I met. He had an unhurried way of being in a hurry.

Merino had a slide projector set up in the training room that adjoined his small office.[1] During 1996, the station had organized some thirty-one short courses, training sessions, and practical demonstrations both in Riobamba and surrounding towns. Over eight hundred people participated: campesinos, community leaders, seed producers, women's groups, technicians, and students. Sessions lasted a day and focused on a specific activity such as seed multiplication, weevil control, soil conservation, or on-farm trials.[2]

According to Merino, accurate information is the cornerstone of successful technology transfer. How can a community-based project make an impact without knowing its customers? Priorities must reflect the reality of the farmers and communities a project serves. For small-scale agriculture, the relevant facts concern farm size, credit access, cropping patterns, and production technology. No two places are exactly alike. In agriculture, the details often matter the most. What follows is the project's diagnosis of local potato problems, which Merino illustrated with charts and slides.

Chimborazo Province has approximately 365,000 inhabitants; 65 percent live in rural areas. Of the rural population, 45 percent is a Quechua-speaking Indian population organized into legally recognized communities of seventy to one hundred families. The region also has cooperatives and farmer associations. In the 1970s, hacienda owners and campesinos battled over land reform. Recipients got land, most of it on the hillsides, but little by way of credit or technical assistance. Extension hardly existed; improved technology did not get applied to small farms. In Chimborazo Province today, a small farm is defined as one that has between half a hectare and 2.9 hectares; such units constitute about 60 percent of all farms. Middle-sized farms, which account for about 20 percent of the total, have between 3 and 10 hectares. Large farms have more than 10 hectares and constitute about 20 percent of the province's holdings. The fragmentation of holdings is a serious obstacle to improving the region's agriculture.

Climates in Chimborazo range from subtropical, at 1,200 meters (4,700 feet), to cool and temperate, at 3,600 meters (11,800 feet). For potatoes, sandy, black volcanic soils are the best. The tubers have less disease and pest problems at elevations above 3,000 meters (9,840 feet). The potato is the staple and chief commercial crop for both small farmers and indigenous communities. Per-capita consumption in Chimborazo's rural communities is about 250 kilos a year.[3] Farmers plant the main potato crop from October through December. On the leeward, western slopes—the Pacific side of the cordillera—rainfall is concentrated seasonally; the winter months, July and August, are cool and wet, and only one potato crop a year is customary. On the windward, eastern slopes—the Amazon side of the cordillera—rainfall is better distributed; farmers often sow a second potato crop in May and June.

How long it takes a potato crop to mature depends on the variety planted, on altitude, and on the distance from the equator, which effects day length. A traditional variety like the purple-eyed Uvilla needs nine months. At altitudes between 2,300 and 3,200 meters (7,550 to 10,500 feet), an improved andigena variety like Gabriela takes six months to mature; early-maturing Esperanza takes five months. Go up another 500 meters, and each will need another six weeks.

Of a farmer's total production, Merino noted, moving on to another slide— this time a pie chart—17 percent is held over for seed, 26 percent goes to domestic consumption, and 57 percent is sold. Unfortunately, farmers are caught in a vicious cycle of low-quality seed and poor harvests. When farmers keep their own seed, no matter how bad it is, at least it is free. Farmers will only restock as a last resort; only about 3 percent do so each year. At community meetings, training sessions, and farmer field days, project agronomists explain how important it is to start with fresh, clean seed that has been multiplied only once or twice. "Of course, farmers already know this," Merino said. "The sticking point is where to get clean seed at a reasonable price." When farmers finally discard their old seed, they restock at local markets with tubers actually intended for consumption. The problem is, vendors often mix varieties of the same color together. Potato stocks sold in the markets are often infected with viruses. This is all right for cooking but not for use as seed. Sometimes the potatoes have weevils. In short, the quality and longevity of the tubers farmers get are minimal. To get high-quality seed, they would have to go to Quito or to Carchi Province, a trip of several hundred kilometers.[4]

Once the project has diagnosed a district's potato situation, it works out a strategy to improve it. The work is dynamic and interactive, involving research, training, extension, and crop production. The project does not impose its technology; it responds to demands from communities and farmers, usually as a result of demonstration projects. Working with farmers, the project explains new production technology, not just for seed multiplication but for a cluster of potato problems that effect seed quality.

Merino explained that until the Riobamba station got backing from FORTIPAPA and CIP, it did not have the budget and staff needed to work systematically on seed production. Now it does, and farmer training is a key component of its work. The project chooses representative districts and targets specific communities within each. The first step is to decide which varieties to multiply. For this, the project sets up evaluation plots in selected communities. "How farmers assess a variety carries more weight than what a researcher thinks," Merino said. "After all, it is the farmer who takes the risks and does the work."

During on-farm testing, farmers conduct three field evaluations: one a couple weeks after planting, one at flowering time, and one at harvest time. What are the criteria farmers use? After planting, they consider how vigorously and rapidly the seed tubers sprout. At flowering, farmers compare the number of stems, how quickly the plant has matured, and its disease and pest resistance. A plant's size and overall shape are likewise significant, as such factors can impede or facilitate weeding and spraying. At harvest time, farmers compare the size and shape of the tubers, the color of the skin and flesh, how deep the eyes are, and the overall yield. When the final selection is made, market acceptance and taste are key considerations. After three planting cycles, on-farm trials narrowed down the contenders from thirteen varieties to four: Fripapa, Margarita, Santa Isabel, and Rosita, all of which showed some late-blight resistance.

In the meantime, the project had not neglected related production problems. To reduce weevil damage, it introduced rustic traps that used potato leaves dipped in insecticide as bait. By using varieties more resistant to late blight, the project showed farmers that they could reduce fungicide applications from fourteen to two. The project promoted fertilizer use, especially of phosphates, and stressed soil management to counter erosion. Merino estimated the return to farmers on such investments at 300 percent.

A new variety means nothing unless farmers can get enough seed to plant it. "We wanted farmers to take over seed production," Merino said. "For this to happen, we had to show that so doing is worthwhile." The project is not trying to set up a formal "seed system." Its objective is to promote the use of basic seed by as many communities as possible. Not every community can take up seed production. A village has to have favorable climatic conditions and internal social cohesion to be successful. Choosing collaborating communities wisely, Merino explained, strengthens the demonstration effect. A series of bad choices can lead to costly failures that set a project back. In selecting its partners, the project chose communities at high elevations where potato production has fewer disease and pest problems, sites where drought is unlikely and frosts infrequent. To facilitate the project's work, a collaborating community had to have adequate road access. The project wanted populous communities with a lot of potato acreage but with low yields, places where clean seed could have an immediate impact. The project assessed how keenly a community wished to participate and how trustworthy it was; it wanted to work with united communities, not with those divided internally or at odds with their neighbors.

The seed-production project began in six strategically located villages. In each place, the project had to select and train one or more promoters (*promotores*), who become, in turn, the extension agents for the community, the links between the project and farmers. "A promoter should be experienced in potato production," said Myriam Trujillo, "with leadership skills recognized by the community; they must be open to change, have an entrepreneurial mentality, and good social skills; they must have the community's trust." A dark-haired young woman in her twenties, Trujillo started out as the project's official secretary but ended up as one of its best community organizers.[5] She helped set up community assemblies, group reunions, and workshops. A good seed-production program, she said, needs solid community organization and capable leaders. The keys to success are motivation, training, and participation.

The seed project started with a group of fourteen promoters. The group got extensive training on all aspects of seed production, from seedling health to protected beds and rapid multiplication. Trujillo helped communities set up a legal organization that could receive funds and monitor expenses. Each community-based organization, in turn, has groups that work on storage, local seed distribution, and marketing. Everyone in the organization has some responsibility. Producing seed is more demanding than producing a regular potato crop, Trujillo explained. The community starts with disease-free seed tubers that it gets from the Santa Catalina research station outside Quito. The seed costs 600 sucres a kilo (3,860 sucres = one U.S. dollar). "That is to make sure the community respects it," Trujillo said, "to make sure it treats a seed tuber differently than it does a regular field tuber." Promoters keep track of each aspect of seed production, including pesticide expenditures. They do a lot of work and cannot be expected to work for free; usually, the community compensates them with a portion of the seed harvest.

My orientation complete, we broke for lunch—well, dinner. In Ecuador, as in most of South America, the main meal comes at noon, not at night. Meals are not gulped down mindlessly, and the food is not all heaped up on one plate. Eating a hearty soup, carrots mixed with lima beans, roast beef, mashed potatoes, freshly baked bread, goat cheese, papaya with limes, and coffee—it all takes time. When Merino and I finally headed out in the jeep to visit project sites, it was almost 3 P.M. We went through Cajabamba and then started up the gravel road to Guacoma. We did not get far. The gas pump clogged, and the jeep stalled. We pushed and pried, but to no avail. So we hitched a ride to Cajabamba, where Merino got a taxi back to Riobamba. He promised to return with the master mechanic, *el maestro.* I stayed behind, keeping vigil.

Cajabamba is a busy town. The sierra's central highway and the coastal road to Guayaquil intersect here in a tumult of exhaust-spewing buses, overburdened trucks, and crowded taxis. The buses are the most interesting. The long-distance ones that plied the Quito to Guayaquil run were the newest; some even had air-conditioning, which is a real plus on the coast. For the Cuenca route south, the roads are second class and so too are the buses. Hand-painted and named after saints or the Virgin Mary, they listed from side to side, in need of all the divine protection they could muster. Even so, they were a cut above the aging, ragtag minivans that worked the gravel pathways to a hundred destinations scattered across the upper valleys. Whenever a vehicle pulled in, a hawker jumped out, yelling destinations and looking for passengers. I kept my ears pealed for "Riobamba," just in case. After Merino left, I realized I had neither the office's address nor its phone number.

My jacket was still in the jeep. To keep warm, I stayed in the sun, changing look-out positions as the shade intruded. An hour passed. Bored of bus-system analysis, I took up watch at the cantina on the far corner. Finally, and to my relief, I spied Merino, accompanied by a driver and a mechanic. We went back to the jeep. The mechanic came armed with hoses and gadgets. We swapped vehicles. With dusk approaching, Merino wanted to get to at least one community.

Guacoma, a village of about thirty families, is situated at about 3,000 meters (9,840 feet). This community had formed the Bethlehem Seed Association. Its members had total of thirty-four hectares, which they farmed collectively.[6] The entire village was waiting to greet us. The Quechua women, their hair in traditional braids, had colorful woolen skirts, shawls, and black hats. The men, their hair closely cropped, had on heavy sweaters and baggy pants with the cuffs tucked inside their rubber boots. The teenagers went in for jackets and sneakers. The children, giggling shyly, hid behind the adults. We met the promoter, Sr. Pablito; he had dressed up for the occasion in a reddish-brown poncho with thin, blue stripes. Like most community leaders, he spoke both Spanish and Quechua. He looked about thirty years old and had completed two years of high school.

Sr. Pablito explained that the promoter's job is to monitor seed production. He keeps track of each variety the association plants, calculating how much the community spends on each step, from planting to harvesting. Both the project and the community contribute to seed production, and both share in the harvest. The project supplies the initial virus-free seed tubers, plus fertilizer, insecticide for the weevil traps, and fungicides for late-blight control. Community members provide all the labor. Sr. Pablito, prompted in Quechua by the group around us, named each task that the community did on its own: "It prepared the soil, sowed the crop, did the weeding, mounded up the soil around the plants, put out the weevil traps, sprayed against late blight, harvested the crop, and selected seed for storage." The distribution of the seed harvested reflects how much of the cost the community itself covers. The first year, the project-community split was sixty to forty; this year, all the community needed from the project was orientation, so it expected to keep all the seed it produced.

We went to look at some seed plots. The promoter tracked costs in a thick, hardbound, legal-sized pad, each sheet of paper ruled and divided into little squares. Everyone called it the *biblia*, or bible. It recorded how much the community had spent to rent a tractor, how much seed went into each plot, and the spacing between rows. Each plot was represented by a large square that took up almost an entire page. Within each square, Sr. Pablito wrote down in a clear, precise script the name of the variety planted, how old the seed was—that is, whether the seed was fresh from the Santa Catalina Station or had already been multiplied—the amount of fertilizer used, plus the date and number of pesticide applications. "Before, we did not pay enough attention to the seed we used," Sr. Pablito said, "but now, thanks to the project's training and advice, we have learned how to manage production, which helps the community." With Merino beside me, this was certainly the right thing to say.

Frosts can strike unexpectedly, so it is important to protect tender seedlings. For this, the Guacoma community had built a double-sized protected bed with walls made of mud mixed with straw. A central wall divided the bed into two equal rectangular parcels, each about 2.5 meters wide. Sheets of heavy plastic rolled up on poles made it easy to cover the beds quickly when frost threatened. In Bolivia, over two thousand kilometers to the south, protected beds face north into the summer sun. In Ecuador, which is on the equator, the sun is directly overhead. Consequently, the walls can be the same height on all sides, and the beds can face in any direction. So far, Chimborazo's highland communities have built over thirty such beds.

Guacoma's association started its seed production with virus-free tubers, most of which it planted densely in protected beds. None of this seed is sold for consumption; it is all held over for a second multiplication. For this, the tubers go directly into the community's main potato fields, which are higher up, at 3,400 meters (11,150 feet). Sr. Pablito suggested we go up for an inspection; half the community piled into the truck. To show us which varieties were in each seed plot and which controls had been applied, the promoter brought along the *biblia*.

Sr. Pablito noted the contour plowing and hilling, which minimized erosion. We got out of the truck and headed for the potato plots. The association had a hectare of red-skinned Gabriela and white-skinned, pink-eyed Esperanza. The remains of weevil traps could still be seen in the fields. Sr. Pablito estimated that they had put out about six hundred traps per hectare. He told me to select a couple of plants at random. The tall, erect plants were almost ready to be harvested. The Gabriela plant selected had a total of forty-eight tubers; I counted them myself. Some were small, but that is acceptable for seed production. The smallest could be recycled to a protected bed. The Esperanza plant yielded thirty medium- to large-sized tubers. The largest were too big for seed, so they would be sold commercially. I asked Sr. Pablito how this compared with the yields the community used to get. "We never had yields like this before," he said. "Maybe four or five usable tubers. And they were not so large. Many were diseased or infested with insects, so we threw them away." Back home in my own garden, a multiplication rate of one to twelve is the most I ever averaged. Most seed projects assume a multiplication rate of one to ten.

Twilight is short in the Andes. Once the sun drops behind the peaks, the bright, golden glow quickly dissipates into night without the formality of sunset. Suddenly it was dark and time to go. I could barely make out the quinoa fields and the barley plots farther up the hillsides. The lights of neighboring villages shimmered across the valley. Rural electrification is one of the few fringe benefits Ecuador's oil boom brought to the country's villages.

On the way back to Riobamba, I asked Merino if seed production paid for itself. He groped around in the glove compartment, pulled out a small notebook, and turned on the inside light. He explained that a multiplication rate of one to fifteen was more than sufficient. In Shobol, another project site, the community had completed a second multiplication. It planted 40 quintals (1 quintal = 100 pounds, or about 45 kilos) of Gabriela on 1.4 hectares and harvested 874 quintals. The multiplication rate was one to twenty-two. The split was roughly sixty to forty; 524 quintals went to the community, and 349 quintals went back to the project for distribution elsewhere. With Esperanza, the community sowed 20 quintals on seven-tenths of a hectare and harvested 326 quintals, for a multiplication rate of one to sixteen. The return on the investment over costs was 180 percent for Gabriela and 57 percent for Esperanza.

No doubt. But was this enough to explain the zealous embrace of seed production I had seen at Gaucoma? Seed is only one element in the potato-production equation. What was I missing? It took me awhile to fully appreciate the miracle inherent in potato arithmetic. At work was a multiplier effect on a vast scale, a return per potato invested that not even a hedgefund could match. The next day, I finally saw what was happening firsthand.

The Guabug Seed Association has fifty-three members and a total area of fifty-six hectares; it had worked with the FORTIPAPA project for three years.[7] Merino took me there early in the morning. We pulled inside the gates of what used to be an old hacienda. The run-down buildings now belong to the community, which uses them to store farm equipment. We met the promoter, Sr. Julio. A quiet man in jeans, a sweater, and sandals, he seemed especially proud of Guabug's rustic, diffused-light silo. (See fig. 6.) "We used to store potatoes in a dark musty room," he said. "The potatoes sprouted, but without light they ended up weak and stringy. When kept outdoors like this, sprouts are short, fat, and strong; they can be broken off and planted."

I finally got the picture. Using sprout cuttings, multiplication is rapid, and the rate is many times higher. I had seen this technology in CIP's tissue-culture laboratory a decade before; I had just never seen it in the field.[8] I presumed such techniques were for scientists, not farmers. Diffused-light storage silos had changed all that.[9] Guabug's silo had four levels, or "shelves," staggered half-a-meter apart from bottom to top. Its capacity was twenty-four hundred kilos. The silo was constructed from local materials: straw and bamboo for the roof, eucalyptus lumber for the frame, and woven straw mats for the shelves. Construction had begun the previous year; this was one of five diffused-light storage projects. The next year, com-

Fig. 6. Diffused-light storage silo, Huancayo, Peru. Courtesy CIP, from Booth, Shaw, and CIP, *Principles of Potato Storage,* p. 38.

munities built an additional twenty-one such silos. Seed is separated and stored by size: the largest tubers, of seventy to ninety grams, go on top; the smallest, of thirty to fifty grams, on the bottom; in between go middle-sized tubers, of forty to sixty grams. To separate potatoes by size, whether for seed or for sale, the association uses chain-linked grates with holes of different sizes. Villagers start with the largest grate; they heap up potatoes on top and sift. What does not fall through is set aside. Then they take the second grate, with middle-sized holes, and repeat the process. The remainder is small potatoes.

The popularity of diffused-light silos was a direct consequence of sprout technology. In Guabug, the association planted sprouts in protected beds, which greatly extended the useful life of a seed tuber. Tubers taken from storage have strong, thick sprouts. The sprouts are broken off from the "mother tuber" and planted in a small raised seedbed; the mother tuber goes back into storage for resprouting. (See fig. 7.) Women in the community have specialized in this step, which requires agility and patience. I was in luck, for a women's group was preparing a seedbed that morning. The raised bed of fine soil was about 1.5 meters long, half a meter wide, and twenty centimeters deep. Prior to planting, the women firm up the soil with a flat board. They had about two kilos of medium-sized seed potatoes in an

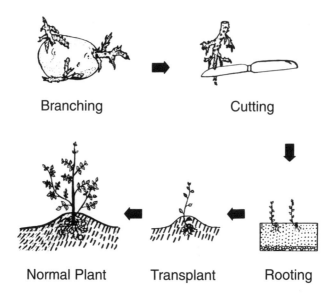

Branching   Cutting

Normal Plant   Transplant   Rooting

Fig. 7. Rapid multiplication from sprouts. Courtesy CIP, from CIP, "Sprout Cuttings," slide 26, p. 10.

orange plastic container. They were taking sprouts from Santa Isabel and Fripapa tubers, two of the project's new, late blight–resistant varieties. One of the women took a stick, making small holes in the soil about three centimeters apart. Then she carefully broke off a sprout from the tuber and pressed it into the ground until two-thirds of it was buried in the soil. She continued until she filled the row and then started another row, spaced three centimeters from the previous one. The sprouts must be kept moist. Within three weeks to a month, they will have rooted and put out leaves. Then, as with tomato plants, the seedlings can be lifted gently out of the seedbed. With dirt still on the roots, they are transplanted into protected beds. In Guabug, the community applies no growth hormones to the shoots or fertilizer to the seedbed.

Fausto Merino went over the arithmetic. A medium-sized seed tuber weighing sixty grams can easily generate six sprouts. Once these are removed, the tuber is returned to storage, where it resprouts again in about three weeks. Given cool conditions, which inhibit dehydration, the procedure can typically be repeated three times, generating a total of eighteen plants, which are rooted in the beds—six sprouts taken from a single tuber three times. Presuming that each of the eighteen seedlings yields fifteen new tubers, the multiplication rate is 18 times 15, or 1 to 270. With a rate like that, the prorated cost per seed tuber generated is next to nothing. No wonder communities had converted en masse to seed production.[10]

In Guabug, the output from sprout multiplication had quickly outrun the capacity of the association's protected beds. So the people there were transplanting directly into potato parcels in frost-protected pockets at lower altitudes. A sprout-generated potato plant soon catches up with one produced from a seed tuber planted whole. Yields are virtually identical. Of course, the process still begins with the virus-free seed stocks supplied by the project. And communities still multiply such stocks at least once in protected beds. In Guabug, for example, the beds were filled with Rosita, a late blight–resistant variety that the association was multiplying for the first time. After that, they would go to sprout multiplication.

When I thought about it, I realized that the Seed Association in Guacoma, the community I had visited the day before, had a diffused-light silo. Sr. Pablito had mentioned sprouts, but I was in a hurry, and the remark's significance had not registered. Before leaving Guabug, we went to the community's upper fields. Flags of different colors identified each of the four potato varieties the association was multiplying. Villagers could compare these new plots to those nearby, planted with less healthy, older seed. We also inspected the ulluco; the association was testing thirteen varieties for frost resistance. We backtracked to the main road, crossed the valley floor, and went up the other side of the valley to the small community of Shobol, situated at 3,400 meters. The Shobol Association had seventeen members with a total of forty hectares, of which five were planted in potatoes and thirteen in an assortment of crops. I saw onions, ulluco, quinoa, and lima beans. The remaining twenty-seven hectares were either in pasture or at rest. The community hardly noticed our arrival. Just about everyone was squeezed into the adobe-brick house that doubles as the community's training center. We entered through the heavy wooden door in front. It had rusty iron hinges, but a fresh coat of green paint. The structure had no windows; for ventilation, there were air spaces near the roof. Inside, one lightbulb dangled from the ceiling. A training session on cheese making was in progress. Merino explained that while the potato is a priority, the Riobamba station's work includes other crops and farming activities, notably milk production, soil conservation, and crop rotations.

Project agronomist Luis Majín had his sleeves rolled up and was answering questions. Women run Shobol's cheese concession. The training had started at dawn with milking. The group had examined the health of the community's fifteen cows and examined each of them for teat infections. They had discussed the importance of good nutrition and hence of improved pastures to milk production. Finally, they had made cheese. Majín claimed that pasteurizing the milk first made for a higher-quality cheese. The women had followed this suggestion and were evaluating the finished product. They seemed excited and satisfied. Majín agreed to come back for additional demonstrations.

We stopped for a break. Shobol's promoter, Sr. Alberto, brought over a steaming pot with a ladle and tin cup. I thought that the pot held some kind of tea, which it did, if *aguardiente* made from boiling water, rum, and cinnamon qualifies as tea.

We did not neglect Shobol's potatoes. And just to make sure I got the picture, we stopped at Santa Isabel and Guabal. In each case, the elements in the story were the same: clean seed, protected beds, diffused-light storage, and sprout multiplication. Community-based participation, training, and on-farm trials had transformed the jargon of "technology transfer" into a dynamic reality.

Between 1995 and 1997, FORTIPAPA distributed 12 metric tons of disease-free seed produced at its Santa Catalina station. This included eleven improved varieties and two native cultivars. In Chimborazo Province, the seed project involved some 1,860 farmers in fourteen communities with thirty protected seedbeds. Yields averaged 29 metric tons per hectare. In 1997, Chimborazo's Seed Growers Association had 160 metric tons of seed tubers in diffused-light silos. With protected beds and sprout technology, communities had reduced the per-unit cost of the seed purchased from Santa Catalina by 90 percent. Farmers, in turn, who bought improved seed saw their yields go up by 20 to 40 percent.[11]

To evaluate its work, FORTIPAPA made a cost-benefit projection. For 1997–2001, its investment in the Chimborazo seed project would come to about $250,000. In the meantime, farmers would plant an estimated 3,400 hectares of quality seed at a net profit of $1.8 million. Over the project's lifetime, from 1990 through 2012, FORTIPAPA calculated its internal rate of return at almost 30 percent. So it would be money well spent.[12]

I admired the way Merino and his team worked with the region's Quechua-speaking communities. They provided respectful, dependable cooperation and expected the same in return. Agronomists like Fausto Merino fight the good fight. But is it not a lost cause? Land-distribution projects, whether Bolivia's violent occupations or Ecuador's piecemeal approach, have fallen far short of expectations. Was it all a big mistake?

Land reform became a *cause cèlébre* in the 1960s. When Fidel Castro seized power in Cuba in 1959, the country's big sugar barons were the first to get the boot; sugar production became a nationalized industry. Fearful that communism would appeal to land-hungry campesinos, the United States–funded Alliance for Progress took up the land-reform gospel. Considering that Bolivian campesinos ran amuck in 1952 without any help from Moscow, Havana, or communists, the fear was not as far-fetched as it may appear in hindsight. The Punta del Este agreement that forged the alliance called for an end to the region's "unjust system of land holding and exploitation." To build a just system based on private property, land distribution was to be coupled with "adequate credit, technical assistance, and market outlets for the products produced."[13] It was a noble thought. From Chile to Colombia, most Latin American countries dabbled in land reform. Such efforts were rarely matched by a systematic plan of action based on credit, training, improved technology, and adequate marketing. Without factoring in such basics, land reform is mostly empty promises. Add these basics in, and the prospects for success go up markedly.[14]

In Japan, Taiwan, and South Korea, land reform accomplished precisely what proved to be so elusive in Latin America. In all three cases, prosperous small farmers formed the backbone of a postwar recovery. Reliable credit, systematic farm-focused extension, and cooperative marketing were the keys to success. That small farmers make for reliable partners in food production is demonstrated every year all across Asia's rice belt. It is not agribusiness that feeds Asia billions but small rice farms, most of which do not exceed a couple of hectares in size.[15]

For Asia to produce enough rice requires both a technical agenda and a social agenda. Latin America's small farmers cannot succeed without both these agendas either. Cut corners, and the result is always too little too late.

PART V
Food and Population

CHAPTER 14

Potato Famine

B etween 1950 and 1990, the size of the world's population doubled, from 2.6 billion to 5.3 billion. The culprit was high growth rates in poor countries, especially in Asia, which accounted for two-thirds of the 2.7 billion additions to the planet's total.[1] Such rapid growth has happened before, but not in Asia. The earlier dramatic surge in the world's population began in northern and western Europe on the eve of the Industrial Revolution; the population explosion in Asia did not begin until the 1950s.

The population of northern and western Europe totaled 62 million in 1750, 81 million in 1800, and 166 million in 1850.[2] Ireland's population, for example, was 4.2 million in 1791 and 8.4 million in 1841; England and Wales had 9 million people in 1801 and almost 18 million in 1851.[3] The pace of Europe's population growth was unprecedented; it did not go unnoticed by contemporaries.

In 1798, Thomas Malthus published the first edition of his *Essay on the Principle of Population*. A clergyman and mathematician, he claimed that "population when unchecked increases in a geometrical ratio; subsistence increases only in arithmetical ratio." By subsistence, Malthus meant the food supply and the rate at which agricultural production goes up. He noted the "immensity of the first power in comparison to the second."[4] Given such unequal powers, population had to be held in check. The ultimate check was "want of food or famine."[5] Malthus did not want famine. As subsequent editions of his *Essay* suggest, he hoped education, moral restraint, and self-interest would put the "hare of population growth to sleep so that the tortoise of food production could overtake her."[6]

In Ireland, the hare did not go to sleep. Between 1845 and 1850 Ireland suffered three years of terrible famine. The 1841 census counted 8.2 million people, the 1851 census just 6.5 million, an absolute drop of 21 percent.[7] Between 1845 and 1850, the death toll reached at least a million people; another seven hundred thousand fled the country.[8] Even these grim figures understate the impact the famine had. For Ireland's population should have grown, not contracted. Taking into account the prefamine birth rate and adjusting for emigration rates, the Registrar General for Ireland's 1851 census estimated that the country's population should have exceeded 9 million. Based on that projection, Ireland's 1841–51 demographic loss was closer to 2.5 million people, or about 28 percent.[9]

Contemporaries blamed Ireland's plight on the potato. A fungus disease called late blight attacked the crop at harvest time in October, 1845. It was worse in 1846; potato production dropped by some 9.5 million metric tons, a loss equivalent to over 600,000 hectares.[10] By the winter of 1847, half the country was on the brink of starvation.[11] To Malthusian critics in London, Ireland's fate seemed preordained, a direct consequence of too many people living too close to subsistence. The true lesson, however, is more complex than a simple food-to-people ratio implies.

Ireland's plight had as much to do with prejudice, free-trade ideology, and land tenure as it did with Malthusian principles. Until Parliament passed the Emancipation Act of 1829, the Penal Codes prohibited Catholics from owning land, voting, teaching school, or entering most professions.[12] Basic education was available only to Protestants and controlled by the Church of England. Ireland's peers, its landed gentry, its great commercial families, and its professional classes belonged almost exclusively to the Protestant establishment. In the towns, a Catholic might end up a shopkeeper, and a few entered medicine or the clergy. But most of Catholic Ireland was impoverished, rural, and illiterate. The highest status an Irish peasant could hope for was that of tenant farmer. Next came cottagers (cottiers), who had little land beyond a potato patch. At the bottom of the heap came day laborers, with no land at all. In economic affairs, the establishment pitted itself against monopoly, preaching free markets and free trade. In social affairs, however, it defended its privileges as part of the natural order. To the better classes, the difference between rich and poor reflected a moral difference; it had little to do with education or economic opportunity.[13]

With hindsight, critics blamed the lazy Irish for relying on a crop so prone to disease.[14] Prior to the famine, however, the potato had fared well enough. Ireland's poor took to the spud because it fit the climate and the social structure. Ireland had long, mild autumns ideal for potatoes. Cool, cloudy weather with heavy rains often bedeviled the wheat and oats. Inclement conditions, however, rarely unsettled the potato; on the contrary, it thrived. The humble spud also accommodated itself easily to rural life. Irish cottages had little furniture, kitchen equipment, or eating utensils. On all scores, the potato fit in. To boil a spud, all a poor family needed was a pot; lacking that, a potato could be roasted on the hearth. The only essential eating utensil was a sharp stick. The skewered spud was held in one hand and peeled with the other. Finally, the potato was easy to grow. For planting wheat and oats, the soil had to be broken up with a plow; when harvested, the hard kernels had to be threshed and milled. With potatoes, all a poor cottager needed was a spade. And potatoes could be cooked fresh, going right from the ground into the pot.[15]

Irish cottagers planted their spuds in raised, trenched beds with lime, sand, and seaweed mixed into the soil. Once planted, the beds did not need weeding. They were left alone until harvest—hence their reputation as lazy beds.[16] Yields ranged between 14 and 18 tons per hectare. Presuming each person in a household of six ate 2.5 kilos a day, the family needed some 5,475 kilos a year (365 x 2.5 x 6), or about 5.4 metric tons. A third of a hectare of lazy beds thus sufficed.[17] By the 1780s, the potato was the country's staple crop. Irish cottagers subsisted on potatoes almost exclusively. Add in a pig, which also ate potatoes, and the household had enough to pay the rent.[18]

Rent was the system's linchpin. In 1841, two-thirds of the Irish population, 5.5 out of 8 million people, depended on agriculture.[19] The country had some eighty thousand landlords, almost all of them Protestants; the rest of the country's rural millions were almost all Catholics, whether tenants or day laborers.[20] Population growth increased the competition for land and hence drove up the rents. To pay the rent and spread the risk, the official tenant subleased the land to others, for whom the land was divvied out in ever smaller parcels. As subdivisions prolifer- ated and rents went up, families ended up even more impoverished.[21] Potato pro- duction kept going up too, as potatoes were the only crop that could keep the poorest afloat. The resulting destitution shocked contemporaries.

Lord George Hill claimed that the four thousand Irish cottagers in his county Donegal parish had among them "one cart, twenty shovels, thirty-two rakes, seven table forks, ninety-three chairs, two-hundred and forty-three stools, three watches, and eight brass candlesticks; they had no coach or any other wheeled vehicle, no clock, and no garden vegetables save potatoes and cabbage."[22] As to how they lived, Mr. N. P. Willis, in his report from county Down, one of Ireland's most prosper- ous, saw only "groveling, despairing, idiotic wretchedness." A typical hovel, he noted in the *Belfast Penny Journal,* had a mud floor set several inches below ground level. Without a bar or raised threshold, filth from the offal heap piled by the entrance drained into the cabin. Ragged children and the family pig traipsed in and out of the cabins, ankle deep in mud and excrement. Willis wondered how people "made in God's image could live in sties like swine, indeed *with* swine, sitting, lying down, cooking, and eating in such filth as all brute animals." Still, he had to admit the children had the "rose of health"; they seemed happy and had engaging smiles, even though "cradled in a dung heap."[23] For this, credit is due the potato. Despite its notorious poverty, on the eve of the famine Ireland had "one of the tallest, healthi- est, and most fertile populations in Europe."[24]

In 1845, Ireland had some 134,000 households with a lease on four-tenths of a hectare or less and 440,000 with two hectares or less.[25] The crux of the matter was how to break the cycle of rents and subdivisions, of potatoes and population. With this in mind, between 1819 and 1844, Parliament commissioned no less than fourteen inquiries into the Irish problem.[26] No solution was forthcoming.

For Ireland's spendthrift, absentee landlords, what mattered were their rents, the higher the better. The country's surplus rural population was actually beneficial, as it kept the rents high and competition keen. Landlords of an improving bent, by contrast, cared less about rents. They wanted to produce export crops, not potatoes, and hoped to convert land from tillage to grazing; for them, improving agriculture typically meant reducing the number of tenants.[27] Irish nationalist Daniel O'Connell had a very different solution. He proposed reforming the land-tenure system to the advantage of the country's tenants. He championed a twenty-year lease, fair rents, compensation for improvements, and a breakup of great estates.[28]

Not all reformers were as radical as O'Connell. John Wiggins argued that both landlords and tenants "were of one and the same family." On his own estate, he shared the cost of improvements, kept his rents moderate, and worked with his tenants to consolidate dispersed plots.[29] William Blacker argued that the "only way to make a pauper population prosperous is to employ it." He accepted tenant subdivisions as a given. What he proposed was boosting rural employment by making Irish agriculture more productive. Using his own estate as an example, he showed that scientific agriculture could be practiced as readily on a small plot as on a large one. The key to success was proper instruction and credit for seed.[30] The schemes Wiggins and Blacker advocated depended on how many landlords converted to a tenant-friendly approach to estate management. Reform of land tenure by Parliament was out of the question. Its members viewed any intervention as an attack on property, and any attack on property undermined self-reliance.[31]

Ireland's poor had few prospects, other than marriage. Matrimony at least meant companionship and family happiness. Nonetheless, to concerned Victorians, imprudent marriage impoverished everyone. "The idea of such a pair of incapables marrying," noted a distraught landowner's wife, "without a home, without employment; it really is a mortal sin, although none of them comprehend its enormity."[32] Contemporaries faulted the improvident potato, which supplied the food for such "incapables"—hence, the spud's bad reputation in genteel social circles.

As Malthus warned, the poor "procreate at a rate out of proportion to the means of subsistence." To prevent this, he argued for "moral restraint," urging couples to delay marriage until they could afford it, a course of action he felt was based soundly on self-interest. Except during a short-term crisis, such as a famine, Malthus opposed poor relief, as it "necessarily destroyed restraints on reproduction."[33] Like

Malthus, England's industrious middle class blamed the poor for their poverty. If the poor avoided work and acted without foresight in England, what could one expect in Ireland? There, reliance on the potato reinforced the irresponsibility and laziness inherent in the Irish race.[34]

In England, farmers produced wheat and oats—crops that required hard, honest labor; godly crops that grew aboveground in the light of day. The slovenly potato, by contrast, grew underground in a lazy bed, rewarding the indolent and hardworking alike. And like the Irish peasant who produced it, the potato was indiscriminately prolific. To its critics, the potato symbolized indulgence: it allowed the ragtag Irish to proliferate; it aided and abetted immorality and sloth.[35]

It is hard to help flawed people one considers unworthy. The squalid Irish were always begging; they had the wrong religion, produced a pernicious crop, and were rebellious to boot. When famine did strike, a critic like Samuel Elly, author of *Potatoes, Pigs, and Politics,* had no doubts about who was to blame. The potato, he complained, was Ireland's undoing; it led to excessive fertility, which, in turn, reduced families to destitution. He viewed the famine as "the means by which a wise and merciful God is preparing to raise the Irishmen to a higher social rank." The means, of course, was to kill off enough of them to improve the lot of the rest. Elly was not alone. The smug view was that Providence "for wise purposes had poured forth the vials of His chastisement upon that ill-fated land." Consequently, why not let the famine take its course?[36] Under such circumstances, is it surprising that Great Britain's newly elected Whig government failed to act with sufficient resolve?

Blight first struck the potato crop in the fall of 1845. The disease did great damage in at least eleven counties. Not all crops failed. Ireland's landed gentry produced wheat, oats, pork, and beef in quantity on their estates, much of it for export to England. The profitable trade continued, even during the worst famine years. In the meantime, poor cottagers, already on the margin, ate through all their food stocks and sold off their animals and most of their clothes to get through the winter.[37]

Sir Robert Peel's Tory government had to meet the crisis; it did so successfully. It set up local committees to distribute food at cost, or for free if necessary; it stockpiled imported cornmeal from the United States. To get money to the destitute, it began public-works projects. Peel's administration, however, lost the June parliamentary elections to Lord John Russell's Whigs. The Whig regime decided that the danger of renewed famine had passed. Consequently, it cut back on famine relief, sold off the imported corn surplus, and closed down the public-works projects. In the meantime, for lack of seed, Irish tenants had planted 100,000 fewer hectares of potatoes in the spring of 1846. When the crop failed again in the fall, this time more completely than the year before, poor tenants faced starvation. With the relief structure dismantled, Ireland faced a great calamity. The Whigs, however, worried more about English taxpayers than they did about the food supply; they fretted that the unworthy poor would take advantage of relief. In defense of free trade, they refused to prohibit food exports from a starving Ireland. Scarcity made food prices rise, but the government delayed importing food. Families were denied relief unless interred in the workhouse; this was precisely the policy Peel had abandoned as impractical and inhumane. By the time the government saw the scheme had failed utterly, it was already January, 1847. A new relief law did not pass Parliament until February 26, 1847. Soups kitchens were not in place until the spring. In May, 1847, eight hundred thousand people were being fed in the kitchens; in August, over three million were.[38]

For many it was already too late. Between the near-total crop failure in the fall of 1846 and August of 1847, tens of thousands of poor Irish tenants died wretchedly. Without the potato, the poor had no food, no pigs, and no way to pay the rent. Families pawned everything they had, down to their clothes and blankets. They were left in rags, which they never changed. At night, families huddled together in their cabins, trying to keep warm. With the people weakened by starvation, ragged and unwashed, contagion spread. The worst maladies were typhus and relapsing fever, spread through contact with lice.

Lice carry in their blood the deadly microorganisms that transmit both diseases. In the misery the famine provoked, the essentials of basic hygiene collapsed; families lived in the same clothes and slept in the same bedding, unchanged for months. In such squalor, lice proliferated, infesting the destitute and their hovels. When a human host crushes or scratches an infected louse, the vermin's blood gets smeared on the skin and the hands or even gets into the eyes, spreading contamination. At death, the lice abandon a cold cadaver, jumping around in search of a warm host. Considering the spotted, itchy rashes and stupor that typhus induces, the vomiting and nosebleeds that come with relapsing fever, and the chronic outbreaks of the "bloody flux" (bacillic dysentery), infection quickly reached epidemic levels. Entire families died in their cabins, their neighbors too weak or too afraid to bury them. Desperate families took to the roads for the larger towns and cities, carrying lice and death with them.[39]

Workhouses and fever hospitals proved pathetically inadequate; the overcrowded facilities made for terrible suffering. Fermoy workhouse was built for 900 inmates but had 1,533 in March, 1847; Killmallock had 1,500 instead of 800. The notorious Kilrush workhouse had as many as 5,000 instead of the 800 for which it was intended. Even those who entered healthy died there. In county Cork, deaths in the workhouse for 1847 came to "the appalling total of 3,329."[40] The dead were heaped up in mass graves. Most people died before they ever made it to a workhouse. The famine hit the hardest in Ireland's western counties, far from Dublin; between 1841 and 1851, seven counties suffered an absolute loss of between 28 and 30 percent of their respective populations.[41]

No one really knew what caused famine fever. A germ theory of disease had yet to be proposed. What worked—keeping a clean house, whitewashing cabins, frequent bathing, clean clothes—worked indirectly. The culprit, or insect vector, was not identified for another sixty years. Anyone who came into contact with the sick ran the risk of picking up lice and getting infected. The toll on the country's courageous physicians, relief workers, and clergymen, Protestant and Catholic alike, was daunting.[42] As many Irish died of typhus and the bloody flux as starved to death. Amidst such affliction came evictions, which made a bad situation more desperate still.

For the terrible toll the famine took in the countryside, Ireland's Catholic clergy blamed the way Parliament organized relief under the Poor Law. As Irish nationalist John Mitchell put it, "the Almighty sent the potato blight, but the English created the famine."[43] Parliament passed the Poor Law Extension Act of 1847 at the famine's height. It allowed families to get relief outside the workhouse but imposed a penalty on anyone who did so. The Gregory Clause, a Tory-sponsored amendment to the Act, refused relief to any family with a tenant holding above a quarter acre. Before such tenants got relief, they had to sign over their tenure rights to their landlords. The choice for many was stark enough: either they starved or they lost their land. When they lost their land, they often lost their cabins too, which were "tumbled." The 1847 act also transferred the cost of relief from the imperial treasury in London to local taxpayers in Ireland; it held Irish landlords responsible for the rates levied on small holdings and cabins.[44] To avoid bankruptcy, landlords began evicting tenants and destroying their cottages. For families left homeless and hungry in the dead of winter, eviction was a virtual death sentence. In the Kilrush district of Country Clare, thousands of families were forcibly expelled by gangs of thugs hired by local landlords.[45] Upwards of half a million people were either forced to surrender their lands and dwellings or were evicted.[46]

In the west of Ireland, the famine did not abate; in the autumn of 1848, disaster struck the potato crop once again. Landlords who had initially sympathized with their tenants now faced ruin themselves. Tenants could not pay their rents, and consequently, landlords could not pay the poor rates Parliament had imposed to cover famine relief. Evictions followed. For its part, the British government refused to make up the difference. During the first two winters of famine, the British government had spent eight million pounds in relief, mostly as loans to Ireland; over the terrible years that followed, it spent just five hundred thousand pounds.[47] During the winter of 1849, the west of Ireland was as desperate as ever; its local relief boards had insufficient funds or were in debt. Nonetheless, an indifferent Parliament, weary of Ireland, passed no additional relief bills. Mortality in the west of Ireland during the winter of 1849 equaled that of 1847. Britain's Chief Poor Law Commissioner, Edward Twistleton, resigned in disgust. "People are dying of starvation," he said, "for want of the advance of a few hundred pounds."[48]

As the famine worsened, flight was the only alternative left for thousands of families. A few landlords helped, tempering "clearances" by paying for fares and provisions. Overwhelmingly, however, emigration was self-financed. Following the 1845 blight, 106,000 people left Ireland; after the crop failure in 1846, 215,000 left. The Irish had a great attachment to their cabins and their potato plots. Families held out as long as they could. The crop failure of 1848 was the final blow. Mass exodus followed. Between 1849 and 1852, over 800,000 people emigrated, most of them, some 84 percent, to the United States.[49] After 1848, the potato blight never hit Ireland again with such virulence. The numbers on relief fell to 833,000 in 1849, to 148,000 in 1850, and to 20,000 in 1851. Nonetheless, the exodus abroad did not stop. Over the next sixty years, an additional 6 million left.[50] Ireland's population in 1911 was just 4.4 million, half of what it was in 1845.[51] Even today, 150 years later, the Irish population, including northern Ireland, is only 5.3 million, compared to over 8 million in 1841.

Ireland was not the only country where the potato failed, but it was the only place where famine so decimated the population. The potato blight first hit the Low Countries and France and then swept north into the British Isles and east into Germany and Poland. Everywhere the blight struck, it devastated the crop, causing great hardship.[52]

The Netherlands had the highest population density per hectare of farmland in Europe; next to Ireland, it was the country most dependent on the potato. Dutch farm families ate through almost four metric tons a year; they found in the potato many of the advantages the Irish did. The potato was a more reliable subsistence crop than grains; it was cheap to produce, easy to cultivate, and fed both the family and farm animals, notably the hogs.[53]

As was true in England, Dutch critics looked down on the potato as "one of the worst kinds of food." Eating too many potatoes led to "insipid, sluggish potato blood," which in turn caused "laziness and indolence in the working classes."[54] The upright Dutch burghers deplored the alcoholism that gin, distilled from potatoes, had allegedly unleashed upon the countryside. The Dutch government, like the British parliament, advocated free trade, and it opposed state-backed social welfare. When blight struck, the country's upper classes saw it as a divine judgement on the potato and its eaters. To meet the crisis, the Dutch government encouraged food imports and local charity but discouraged any intervention in free markets. Fortunately, municipal authorities took the initiative, organizing public works and local relief efforts.[55] Even so, in 1847, for the first time in decades, deaths in the Netherlands exceeded births; during the famine years, the ravages of malnutrition and disease left at least sixty thousand dead. That was still a far cry from the mortality rates in Ireland. The Netherlands had a total population of about 3 million, so as a percentage of its population, the mortality rate was 2 percent.[56] For Ireland, by contrast, mortality estimates are of at least 1 million people, 18 percent of the country's 1841 population of 8.2 million.[57]

France was the continent's largest potato producer; in 1843, per-capita consumption exceeded fifty-four kilos.[58] In France, the potato was a solid citizen. It had taken hold on French farms after 1790. Successive republican governments promoted the potato as a revolutionary crop. Its redoubt was in soups and stews; as such, it soon showed up in almost every meal. In France, potato eating, small holdings, and population growth did not go together. Under the *ancien régime,* land ownership by French peasants was widespread, and it became more so with the revolution. Intent on preventing the fragmentation of their holdings, French peasants kept their families small. In the 1840s, France had the lowest birthrate in Europe.[59] Consequently, the potato in France escaped the harsh attacks meted out by critics in England and the Netherlands.[60] But the poor did not escape. Like their counterparts elsewhere, the better classes in France believed that poverty was due to the moral failings of the pauper: "lack of forethought and thrift, laziness, debauchery, intemperance, and over-large families"—deficiencies that poor relief only encouraged.[61] Laissez-faire individualism prevailed; government intervention, whether in the marketplace or in the guise of social welfare, was anathema. Nonetheless, when the crisis came, French authorities put their ideological scruples aside. Rather than regulate food prices, it authorized eight million francs for public works. Wages were low, but workers and their families also got cheap bread. Local boards also provided direct relief, which included "not only bread but allowances for soap, coal, wood, medicine, and rent." In 1847, over a million people received such assistance.[62] Perhaps vivid memories of the French Revolution made the better classes less doctrinaire. Relief in France was better organized and much more effective than in Ireland. Although mortality rates edged up, the demographic impact came primarily from delayed marriages and a further drop in the birth rate.[63]

In the British Isles, the potato blight also struck with severity in the Scottish highlands. The potato was an entrenched subsistence crop there, but its failure did not create the terrible hardship that occurred in the west of Ireland.[64] Parliament's response in both cases was similar; indeed, Sir Charles Trevelyan was in charge of both Scottish and Irish relief. What then explains the difference?

Compared to their profligate Irish counterparts, highland landlords spent more time on their Scottish estates, and for the most part, such estates were solvent. When Parliament passed the Drainage Act in 1847, which provided loans for local public works, highland landlords quickly applied for the grants. Assistance under the act reached almost half a million pounds. With wages from draining, trenching, and other land improvements, many tenants managed to pay their rents. If tenants fell into arrears, frugal landlords could afford to tide them over. Ultimately, tenants had to liquidate such debts, but not until the 1850s, when the worst famine years were over. For its part, the British government lauded the efforts of highland land-lords. Sir Charles Trevelyan said he was "delighted with the whole conduct of the highland proprietors in the present crisis," compared to Ireland, where "everything both in regard to the people and the proprietors is sickening and disgusting."[65] Landlords did resort to clearances, especially of landless cottagers, and they evicted tenants. Consequently, highlanders, like the Irish, emigrated abroad. In the end, however, the impact on estate structure in the Scottish highlands was not great: a marginal reduction in small tenancies; a net increase in the size of larger, pastoral farms; and greater landlord control over subdivision.[66] In Ireland, by contrast, the famine changed everything.

In 1841, 45 percent of all the holdings in Ireland were under two hectares; in 1851, only 15 percent were.[67] The famine wiped out Ireland's cottager class, the group most dependent on the potato. They died, migrated to the cities, or fled the coun-try.[68] Before the famine, Irishmen married young, had large families, and stayed home. The Irish who survived married late, had small families, and migrated abroad.[69] Ireland's population kept dropping for almost a century.[70] In prefamine Ireland, nationalists like Daniel O'Connell championed home rule within the Brit-ish Empire; after the famine, Irish dissidents embraced rebellion and independence. The explanation for the course the famine took in Ireland, as opposed to elsewhere in Europe, is social, not Malthusian. In Ireland, the famine unfolded within a soci-ety whose social divisions pitted landlord against tenant, Protestant against Catholic, the disenfranchised against the establishment. The famine devastated a society ruled from across the Irish Sea in London. The overlapping divisions embedded in Ireland's social order combined to make dependence on the potato fatal. Nowhere else in Europe did reliance on the potato end in such a disaster. In the Netherlands, local relief did not depend on an isolated, distant government; in France, peasant farmers owned their own land; in Scotland, a solvent gentry could forego tenant rents. Only in Ireland was poor relief designed to transform agriculture, to put insolvent landlords out of business, and to force impoverished tenants off the land.

To Irish landlords bent on reform, agrarian change required a "long, contin-ued, and systematic ejection of small holders and squatting cottagers."[71] When the famine had abated, Lord George Hill concluded that the "Irish had profited much by the Famine" as no teacher "other than Divine Providence" could have induced them to change "their mode of agriculture and their habits of life."[72] The Irish paid a high price for such improvements. Blame the potato, if you must; blame the Irish with their big families. But blame as well how blind self-righteousness can be; blame as well how easily inhumanity cloaks itself in economic theory.

CHAPTER 15

Hungry Ghosts

To say that the Irish starved because the potato crop failed is a half-truth at best. Historians of premodern societies cite bad weather, overdependence on a single crop, population pressure, and peasant idleness for famines.[1] For modern times, ideology must be added to the list. In this respect, Ireland was a harbinger of things to come. London could not stomach taxing the hardworking English to save the lazy Irish. In Ireland, "social engineering" unleashed in the midst of famine was a disaster for tenants but a success for landlords bent on estate consolidation and hence, "improvement."[2] Expelling tenants as a prelude to modernizing agriculture is not unique to Ireland. In modern times, it is standard practice.

In the United States, merchants in the defeated post–Civil War South used credit to extort the region's poor farmers. In 1869, eight Deep South states had a total of only 26 national banks, compared to 829 such banks in just four northern states. Starved for cash, the South's farmers turned to furnishing merchants, using crops for credit. Under the crop-lien system, farmers mortgaged the next year's cotton crop. In exchange, they got credit in kind: an advance of food, clothing, and farming supplies. Such credit was good only at the store that advanced it and kept the ledgers. There was no competition. Compared to the price of goods sold for cash, the price of merchandise advanced on credit got marked up by 40 percent or more. On this inflated base, merchants charged an exorbitant rate of interest. At year's end, farmers had to pay out. They sold their mortgaged cotton crop to the furnishing merchant, who deducted the cost of the goods advanced, plus interest. No neutral third party policed the books. Old ledgers show how the accounting system worked. Items with a cash price of, say, ten cents sold for fourteen cents on credit. Add to this a typical interest charge of 33 percent and the total adds up to nineteen cents, almost double the original price. Once a farmer had signed his first crop lien, he was, in the words of a distraught contemporary, "in helpless peonage," caught in "a modified form of slavery," until he paid out. Given the way ledgers were kept, millions of farmers never paid off their debts; they ended up as tenants and sharecroppers. The Farmers Alliance of the 1880s was the last attempt by rural Americans to defend their self-sufficiency and independence. They lost out to the monopoly power of the railroads, big business, and the banks, which combined against the credit schemes of farmer-backed cooperatives. Land consolidation and migration to the cities followed.[3]

A generation later, Brazil likewise "modernized" its agriculture. Brazil's transformation to an urban society took just fifteen years. In 1965, half of Brazil lived in rural areas; in 1980, only a third did. By comparison, the United States struck a balance between its rural and urban populations in 1920; it took another forty years for the United States to become as urban as Brazil did in just fifteen. Deliberate policy gave impetus to the change. Between 1965 and 1980, Brazil's military government backed a credit scheme that financed mechanization, promoted export agriculture, and encouraged land consolidation. Using 1969 as a base year (100), agriculture's industrial consumption index had risen to 500 in 1980. In the meantime, the amount of land allocated to tenant farming shrank by half; agriculture's share of the labor force dropped from 44 to 30 percent. Throughout Brazil, evictions turned sharecroppers into a rural proletariat housed in slums off the estate's property; families lacked even a small parcel of land for subsistence crops. Brazil's displaced sharecroppers headed en masse for the cities, where they ended up squatters in urban slums. Today, urban Brazil reaps the harvest, counted in abandoned street children, in crackheads, and in urban death squads.[4]

Improving agriculture at the cost of small farmers has a history even more tragic in communist countries. In both the Soviet Union and China, urban cadres imposed collectives on farm families. The consequence was famines and mass starvation on a scale that dwarfs the Irish experience.

In the Soviet Union, Stalin launched his first five-year plan in 1928. In his view, the peasant mentality, fixated on property, religion, and village life, held the country's agriculture back. To change the Soviet Union, the Communist Party had to change the peasants. The remedy devised in Moscow was communal life in "socialist agro-towns." The regime created vast farms that covered thousands of hectares; peasants were to be the "brigades," the "shock troops" who manned the "agricultural fronts" of modern, industrial agriculture. Resistance was fierce; farmers refused to sell grain to the state at the low prices set. The result was state terror unleashed against the so-called rich peasants. The army seized grain stores. Farmers retaliated by burning grain and killing livestock before the state could take it. Villages no longer had enough food to get through the winter. Mass starvation followed. Between 1930 and 1937, eleven million peasants died; five million of the deaths were in the Ukraine, the region expected to feed the country. Soviet agriculture never recovered. In 1934, Stalin relented, allowing households a small plot on which to grow vegetables. Such private plots, treasured by both city folk and rural families, still account for a disproportionate share of the country's vegetable production.[5]

In China, a revolution fought in the name of poor peasants turned into their relentless foe. To be a great country, China had to industrialize. To finance industrialization, Mao was determined to squeeze a surplus from the peasantry. Only the state was allowed to purchase and distribute grain. It bought cheap, sold dear, and used the profits to build up state-owned factories. Because small peasant farmers were "inherently capitalistic," they had to be transformed into loyal cadres. To do this, a sharp break with peasant culture had to be rigidly imposed. In Mao's view, "scattered individualistic production" was the root cause of rural China's poverty and its feudalistic mentality; collective agriculture was the means whereby a reactionary peasantry could be reeducated. During the Party's first five-year plan (1953–57), millions of peasants were forced into collectives. The collectives suppressed the sacred shrines, folk markets, and festivals that shaped peasant culture, replacing them with political rallies and struggle sessions.[6]

Under Mao's reforms, grain harvests started to drop. Average harvests for 1949–58 were below those of 1931–37. Accustomed to working on their own land, peasants lost motivation in collectives. Critics thought Mao was destroying China. Liu Shaoqui, who had seen the disaster wrought by collectives in the Soviet Union, denounced Mao's plans as "false, dangerous, utopian socialism." Mao counterattacked with an antirightist campaign. The 1957 purge silenced Mao's opponents and cleared the way for the Great Leap Forward and the worst famine in China's modern history.[7]

Although not from a peasant background, Mao fancied himself an expert on agriculture. He took the position that collectives, filled with "revolutionary optimism and revolutionary heroism," could use high-density planting and deep plowing to double grain yields.[8] He was not disappointed. Loyal cadres reported what Mao wanted to hear: that the 1958 grain harvest had jumped from 185 million metric tons to 430 million tons. Given such spectacular success, peasants were encouraged to eat as much as they wanted; China started to export grain. In the meantime, the state requisitioned record-breaking quotas for storage. Provincial party leaders who had proclaimed enormous grain surpluses now refused to admit they had lied. They seized communal grain stocks and delivered them for storage. China's peasants now faced winter with grain rations at starvation levels. The next year, 1959, was even worse. The harvest was down 30 million metric tons; notwithstanding, officials reported a bumper crop higher than in 1958. To meet their quotas, party leaders mounted systematic searches for hidden grain. Those who resisted risked being arrested as right-wing opportunists. Thousands of peasants were tortured and beaten to death.[9] By the winter, January–February 1960, millions of China's peasants were starving to death. The old and the young died first; to survive, people ate their children and dead relatives. China starved even though the grain stores were full. The Great Leap Forward had become a great famine, its genesis "rooted in politics, not in nature."[10]

It was Liu Shaoqui who finally saw the terrible truth. Backed by Deng Xiaoping, he dismissed local cadres who had lied and let the people starve. Grain stores were opened for food distribution. In the meantime, Liu Shaoqui modified collective farming, allowing peasants small plots for household consumption. By then, China was in the midst of a great catastrophe. The death toll in just four provinces—Henan, Anhui, Shandong, and Sichuan—came to over thirty million; for China overall, the estimates range between forty and sixty million.[11]

In 1966, Mao countered the challenge to his leadership by unleashing the Cultural Revolution. His Red Guards hunted down those "who had taken the capitalist road." Liu Shaoqui was on the top of the list. Arrested and tortured, he died in 1969 abandoned in prison. Deng Xiaoping, the "number two capitalist roader," was banished to the countryside but somehow escaped death. When Mao died in 1976, it was Deng who ended up in control of China. Almost immediately, he dismantled what remained of the communes. Deng's rural reforms, including the "contract responsibility system," turned China's peasants into small farmers once again.[12]

Malthus worried about the power that population had relative to food production. It never occurred to him that false ideas could be more powerful than objective reality, that terror, delusion, and self-deception posed as great a threat to humanity as starvation. When the potato crop failed in Ireland, London mounted a relief effort, however inadequate. In China, by contrast, disease, crops, and weather cannot be blamed for the famine. No blight destroyed the rice, floods did not wash away the crops, and droughts did not dry up the fields. The country was at peace. The granaries were full. In Cambodia, the Khmer Rouge followed the same path of self-righteous ideology, communes, and terror. For 1970–79, the collective death toll from starvation, civil war, and execution is estimated at between 1.7 and 2.5 million in a country that started the decade with 7 million people.[13]

Malthus's fear that the "hare of population growth" would over take the "tortoise of food production" has a kind of surface validity. Yet, in modern times, a different kind of logic has taken hold. The evil genies of famine and starvation are not unleashed simply by laws of food supply and reproduction. They come from an all-too-familiar place, from ignorance and blind ambition, from greed and want, from that heart of darkness masked by religious presumption, by markets, and by Marxism—by whatever falsehood masquerades as the party line.

CHAPTER 16

Demography

ncrease the food supply and the population grows; decrease it and the population shrinks. More food more people; less food, fewer people. Always. To prove this, put some mice in a large cage; add plenty of food. In short order, the mice proliferate and the cage gets overcrowded. Cut back on the food and the mice starve; the population drops. Hence, the solution to the world's population problem: produce less food, not more.[1] If this is true, the work of agricultural centers like CIP is a waste of time. Feeding a hungry planet is the last thing international organizations ought to promote.

Why did we not think of this before? Actually, Thomas Malthus argued something like this two hundred years ago. He noted that "population has the constant tendency to increase beyond the means of subsistence"—which is to say, more food more people. To Malthus, "scarcity of food" was the most immediate check on population growth—in other words, less food less people.[2] Recent attempts to define the earth's "carrying capacity" likewise include what demographer Kingsley Davis called a "food bias distortion."[3] Despite its terse simplicity, the more-food-more-people paradigm is inaccurate and misleading.

Whose food production is it that adds so dramatically to the world's population? Are agribusiness farmers in the United States to blame? If that is the case, we should expect that most of America's food surplus gets exported to impoverished countries, where the extra food just postpones the inevitable. The facts are otherwise. In 1998, sub-Saharan Africa took less than 2 percent of U.S. agricultural exports. The top customers were Japan, which purchased 17.5 percent of the total, and Canada, with 13.5 percent.[4] Neither country is poor; neither adds much to the world's surplus population. In fact, in both cases, the total fertility rate today is below replacement levels.[5] In Russia, which buys American wheat in lean years to make up for its shortfalls, the total population is actually declining.[6] If more food means more people, Japan ought to be growing much faster. If more food means more people, overfed, overweight America ought to be growing fastest of all.

If anyone is to blame for adding to the world's food supply, it is Asia's small rice farmers. Between 1960 and 2000, the population of China, India, and Indonesia—Asia's most populous nations—doubled, rising from 1.2 to 2.5 billion.[7] Where did the extra food come from? Almost exclusively from each country's domestic production. Rice is an outstanding example. Between 1960–62 and 1997–99, average production for the three countries almost trebled, rising from 125 million metric tons to 374 million tons. This was much faster than population growth.[8] If more food creates a population surplus, then Asia's small rice farmers must be stopped. But who is to stop them? How likely is it that Asia's politicians will downsize their populations by cutting back on rice production? Not very. One-worlders used to claim that "development is the best contraception." Today, hard-nosed realists imply that hunger, although regrettable, is a much better check. They complain that "alleviating poverty just spurs population growth," and blame the potato for famine in Ireland. Those who argue Third World populations should be "left alone to find a natural balance" do not live there.[9] They live well-fed lives somewhere else.

But let's grant the argument that the main factor driving population growth is food. Is starvation a good way to teach profligate poor countries a lesson? Would starvation be good news for rich countries? Such thinking is muddleheaded and exceedingly dangerous. Starvation is one thing in a hapless place like Somalia but quite another in technologically sophisticated India, Pakistan, or China. Such countries could exact a terrible toll on affluent adversaries who sat on top of food reserves while they starved. Biological warfare, computer viruses, and nuclear terror are no longer rich-country monopolies. Even if the starving stoically accepted their fate, a malnourished, dying population is a fearsome breeding ground for infectious disease. Who knows what kind of lethal virus nature might unleash? Some one hundred million international travelers enter and exit the United States each year.[10] No place is too far away for a virus anymore.

For too many rats in a cage, less food may be the solution. But that is not the solution for humankind on planet Earth. For us, life is not a simple experiment whose parameters are set by food supply and reproduction rates. Human beings hold weddings, save up for dowries, and worry about cross-cousin marriages. They grow, process, and distribute food in complex systems of exchange. For humankind, producing, preparing, and eating food, along with courting, marrying, and raising families, are cultural activities. Rats eat and procreate, but they do not write books about it. The planet can have too many rats, and it can end up with too many people. Even so, analyzing how the surplus comes about in each case requires fundamentally different approaches and assumptions. For rats, experiments with controlled food inputs is fine. For human beings, action takes place in a cultural context shaped by group memberships, customs, beliefs, and expectations. To ignore such basic facts about the human condition is to profoundly misread demographic history.[11]

To get more people, either fertility rates must go up or mortality rates must go down; whatever impact food supply has is secondary and indirect.[12] Consider a preindustrial population in which the age of a woman's first marriage is fixed, as is the percentage of women who ever marry; add to this high fixed birthrates countered by high fixed death rates. Under such circumstances, the population will not grow. How much food households produce is irrelevant. The more-food-more-people paradigm implies that households want to produce as much food as possible. In premodern societies, however, what households produce is necessarily limited by how much they can store and sell. In medieval Europe, the benefits from a bumper grain crop did not last long. Even if households stuffed their barns and attics to the rafters, even if they filled every available barrel with beer, "it is doubtful they could store enough to get through two consecutive bad harvests." In such a situation, the size of the population is not the key factor. In 1300, northern Europe's population was less than half of what it was in 1750. Even so, widespread crop failures in 1315, 1316, and 1318 unleashed a terrible famine. To understand what happened, historians point to climatic change, to primitive exchange systems, and to taxation exacted in kind; population pressure is not the explanation.[13]

When Malthus published his *Essay on the Principle of Population* in 1798, northwest Europe was in the midst of an unprecedented demographic expansion. What he did not know was the result of that expansion. We do know. Between 1750 and 1850, northwest Europe's population doubled. Over the next century it doubled again. Compared to its 1750 population of 60 million, northwest Europe had a population of 205 million in 1950, almost a four-fold increase.[14] Over the same period, the West's share of the world's population went from about a fifth to over a third.[15] England, which had 5.8 million people in 1750, had 42 million in 1950, a seven-fold increase.[16] Despite such massive demographic gains, the population did not starve. Quite the contrary. Living standards, income, and life expectancy all went up, the opposite of what Malthus predicted. Northwest Europe, despite its demographic growth, ended up as one of the planet's most prosperous places. Ireland's fate, it turns out, was the exception, not the rule. What happened?

What social scientists call the "demographic transition"—a dramatic change in how populations maintain a balance—is a good description of the historical shift that took place.[17] Prior to 1750, a crude death rate estimated at about 40 per 1000 kept the population of northwest Europe from growing rapidly. Then mortality dropped. Since birthrates stayed high, the population started to grow at an unprecedented rate. At the end of the transition period, however, fertility had dropped to replacement levels and population growth stopped. Low death rates and birth-rates had replaced high death rates and birthrates.[18] England is a good example.

In 1750, on the eve of Europe's population explosion, England had high rates of infant mortality. On average, almost a quarter of the male children born died in the first year; half were dead by age twenty-eight. Eighty years later, in the 1830s, the mortality rate for male infants had dropped to 160 per 1000 born, or to 16 percent. And those that survived lived longer; half were still alive at age forty-four. In the meantime, marriage patterns were changing. The percentage of people who never married dropped from around 27 percent in 1650 to an average of around 10 percent in 1800; the age at first marriage also fell, from 26.8 years to 23.8 years. Given high total fertility rates—on average, over seven children per couple—England's population started to grow at an alarming rate. More people got married earlier, and more children survived to form additional households.[19]

Not only did England avoid a Malthusian catastrophe, however; it prospered. The Industrial Revolution created jobs, productivity expanded on farms, coal and the steam engine reshaped the country's energy system, and higher wages improved the standard of living.[20] Using 1750 as a base year, England's gross domestic product (GDP) was up 60 percent in 1810, 150 percent in 1830, and 350 percent in 1850; by then, output per capita had doubled.[21]

When a high birthrate is combined with falling rates of infant mortality, a large net population gain occurs. In England, the gap began to narrow in the 1890s. For success in a more technologically sophisticated world, children had to be educated. In 1851, only 8 percent of school-age children in England were enrolled in primary schools; in 1891, almost 60 percent were. In the following decades, the complexities of urban living and the investment raising children required changed cultural views of family life and ideal family size. The result was an extraordinary change in fertility, a "quiet revolution" that led to a much lower birthrate. England's population, for example, had stabilized by 1950, with low infant mortality and much smaller families. Today, all the countries of northwest Europe have total fertility rates at or below replacement.[22]

Western Europe's population may soon start declining. Policy makers no longer fret like Malthus did that numbers will exceed subsistence limits. Instead, they worry about aging populations, overburdened social-security systems, and health care for the elderly.[23]

A demographic transition took place all across northwest Europe, but each country took a slightly different path. In northern France, for example, the total fertility rate from the 1740s to the 1770s is estimated at between eight and nine births per married woman, much higher than it was in England. Shortly thereafter, infant mortality started to fall in France, from about 28 percent in the 1780s and 1790s to 18 percent in the 1820s and to 15 percent in the 1840s. Despite the drop, France's population did not grow as fast as England's did. In 1750, France had twenty-five million people; a century later, it had thirty-six million, an increase of just 45 percent. By contrast, England's population almost trebled during the same period. The difference had little to do with the food supply. In France, inheritance laws divided property equally among heirs; to prevent fragmentation of their holdings, peasant households consciously limited family size.[24] Even before France started to industrialize and peasants left agriculture for the cities, the birthrate had dropped. The French Revolution reduced adherence to religious norms, created new paths to social status, and strengthened individualism.[25] As a result, infant mortality and birthrates in France fell in tandem, without the big gaps and consequent population gains that took place in England.[26]

As a description of past events, the demographic transition tells us what happened; it does not tell us why. Only a case-by-case historical analysis can do that. As a description of the future, the demographic transition tells us what to expect in the long run; it does not predict how long it will take each country to get there or what path each will take.

Between 1750 and 1950, the West dramatically increased its share of the world's population. Now the tables are turned. Between 1950 and 2000, the West's share dropped from 36 percent to 24 percent, thus restoring the pre-1750 regional balance. What took the West two centuries to accomplish was reversed in just fifty years. Most of that population growth has occurred in Asia. Of the 3.5 billion people added to the earth's total between 1950 and 2000, Asia accounted for over 2 billion, or 60 percent. Europe's demographic revolution added millions; Asia's has added billions, and at a much faster rate. In Asia, the demographic transition has taken twists and turns quite different from the path taken in northwest Europe.[27]

Asia's population in 1950 was 1.4 billion, which was already enormous by world standards.[28] Consequently, a sharp drop in the death rate without a commensurate fall in fertility was bound to add enormous numbers to that already large total. What few demographers foresaw was how fast infant mortality would drop and thus how fast the numbers would go up.[29] In England and France, it took a century for infant mortality to drop by half, from a high of roughly 28 percent in 1750 to around 15 percent in 1850.[30] In Asia, the rates fell further and faster. India is a good example. For 1918–20, infant mortality averaged 230 deaths per 1000 children born, or 23 percent; for 1943–45, it had dropped to 16 percent; and for 1978–80, it was down to 12 percent. This was an overall drop of 50 percent in just fifty years.[31] And in India, the pace of change was comparatively slow. Infant mortality in the Philippines dropped from 18 percent in the late 1920s to 10 percent in the mid-1950s and was down to just 5 percent in the late 1970s.[32] In Malaysia, infant mortality fell from 15 percent in the late 1930s to 8 percent in the mid-1950s to only 3 percent in the late 1970s.[33]

Improvements in public sanitation, community health, and child care—especially infant vaccinations against infectious diseases—made the rapid decline possible. Sri Lanka is the classic case. Between 1945–46 and 1949–50, an antimalaria program, combined with inoculation campaigns, antibiotics, and improved public health, reduced the country's infant mortality rate by almost half in just five years, from 14 percent to 8 percent.[34] In northwest Europe, by contrast, most advances in preventive medicine and public health had little impact until after 1850, and even then, the advances were much more gradual.[35]

In Asia, infant mortality dropped much faster than cultural norms about family size. The consequence was a massive upsurge in population. In part, this reflected rules about family life. When the demographic transition began in northwest Europe, new rural households were expected to be self-sufficient, a norm that discouraged early marriage. In much of rural Asia, by contrast, extended families are the rule. A young couple often joins the husband's family. The burden of early marriage and childbearing is thus spread across a larger social unit. Add to this the work children do in rural households, limited educational opportunities, and the security a large family provides, and the result, at least in the short run, is high fertility.[36]

In 1958, the United Nations predicted "as almost a matter of certainty" that the world's population would total between 6 and 7 billion by the century's end. Even that estimate presumed a drop in fertility rates; otherwise, the experts jacked up the forecast to 7.4 billion.[37] Fortunately, the world's rate of population growth dropped by more than a third, from 2.2 percent a year in 1965 to 1.4 percent in 1999.[38] The world's population reached almost exactly 6 billion in 2000, on the low side of earlier estimates. In the meantime, fertility rates in many countries have dropped sharply.

In just thirty-five years, from 1950 to 1985, the total fertility rates in Taiwan, South Korea, and Singapore dropped from highs of between five and seven children per woman to below replacement.[39] So rapid drops in fertility can happen, even from high initial levels. Of course, the aforesaid cases are Asia's economic success stories. Can fertility decline even when families are poor? The experience of Kerala, one of India's most densely populated states, demonstrates that it can.

With some 35 million people, Kerala has a larger population than most countries, including Peru, Venezuela, and Canada. One of India's poorest states, its per-capita income is below the country's average, while its unemployment rate is much higher. In the 1950s, Kerala had a crude birth rate of 44 per 1000; forty years later, in 1991, the rate had dropped to 18, or by almost 60 percent. The state's total fertility rate is 1.8 children per woman, much below the overall rate for India, which for the year 2000 is estimated at 3.1; it is even below the U.S. fertility rate of 2.1 children per woman. A low-cost, labor-intensive approach to social investment has been the key to Kerala's success. Since the 1960s, the state government has steadfastly supported rural education, community health, land reform, and adequate food for all. The results are impressive by any standard. Almost every village in the state has a primary school, a health dispensary, and all-weather roads. Infant mortality is just 1.6 percent. The female literacy rate in Kerala is 86 percent, the highest in India. More people read daily newspapers in Kerala than in any other state. Life expectancy for men and women is seventy-one and seventy-four years respectively, which is comparable to developed countries.[40]

Kerala is not alone. In Bangladesh, a family-planning campaign helped reduce the total fertility rate by half, from seven children per woman in 1975 to 3.3 children in 2000. A further reduction, however, is unlikely without a greater emphasis on female education. In the Indian state of Tamil Nadu, the total fertility rate declined from four children in 1971 to 2.2 in 1991. Like Kerala, the state had low infant mortality, a well-organized family-planning program, and high female literacy rates. Today, women marry later in Tamil Nadu and more of them are in the labor force.[41] Overall, the combination of basic education and community-focused health, including birth control, has a good track record in a diverse set of countries. Today, Taiwan, Sri Lanka, Thailand, South Korea, and China all have total fertility rates at or below replacement.[42]

Fertility rates are already below replacement level in China. India and Indonesia are not far behind; by 2010, their fertility rates are expected to drop to 2.6 and 2.3 children respectively, compared to average rates of 6 children in 1960.[43] The drop in Asia's total fertility rate was not due to food scarcity, a consequence of the less-food-less-people paradigm. On the contrary, food security, especially with respect to rice production, has been an integral part of success. That success comes just in time.

Few facts of the modern world are as startling as the cumulative growth in the numbers of humankind. In 1650, the world's population numbered 600 million; a century later, it was up an additional 160 million. In 1850, the planet had perhaps 1.2 billion people. That figure had doubled to 2.5 billion by 1950 and had doubled again by 1990 to 5.2 billion. Between 1960 and 2000 the world added some 400 million people every five years, which is twice the population increase of 1650 to 1750.[44]

Population growth on such a scale should have spelled disaster for poor countries. By and large, it has not. Between 1980 and 1990, GDP growth in low-income countries expanded by an average of 6.6 percent a year, compared to 3.1 percent for high-income countries. For 1990–98, the rates were 7.3 and 2.1 respectively. On a per-capita basis, the GDP figures were 3.9 percent a year for low-income countries and 2.8 percent a year in wealthy countries.[45] Between 1980 and 1998, China's GDP expanded by more than 10 percent a year, a feat unequaled by any nation in the postwar period.[46] Using the years 1979–81 as a base period, food production per capita for 1991–93 was up 23 percent in India and 39 percent in China.[47] For other low-income countries, the overall food-production index for 1995–97 was up 23 percent over 1989–91 and up almost 50 percent over 1979–81.[48] In the meantime, food prices fell. Rice led the way with a 40-percent drop.[49] The explanation, of course, has much to do with technology. The International Rice Research Institute (IRRI) initiated what analysts later called the "green revolution." In fact, some demographers argue that population pressure, far from pulling a society down, stimulates innovation.[50]

It is hard to deny the strong correlation between the pace of technological change and the buildup in the world's population. Since 1800, the world has changed its energy systems, how it communicates, and how it produces food. Is the computer itself not a response to more people, organizations, and countries processing, storing, and retrieving ever greater quantities of information? Correlation, of course, is not cause. If more people matters most, then nineteenth-century China should have monopolized technological change; instead, it lost out to small countries like England and France.

Just because the rapid population growth of the past did not overtake the world's resource base does not mean it will not do so in the future. True, demographers do not expect the world's population to double again. For most regions of the world, fertility rates should be at or near replacement by 2025; by then, the world's population will be some 7.9 billion.[51] Momentum may keep the numbers going up until midcentury.[52] How much these numbers grow depends on assumptions about the age of marriage, the spacing of children, and family size—factors crucial to how long it takes a population to stop growing—and reflects assumptions about what people want and expect from life.[53] A stable population presumes a cultural transformation driven by changes in infant mortality and life expectancy.

In 1950, experts doubted that medical science could bring down Third World death rates. As it turned out, the cost of public-health strategies was much lower than expected. The driving force behind subsequent population growth was declining infant mortality, not higher birthrates.[54] The result is a world in which the old calculus of early death and epidemic disease does not apply, a world that middle-class America has long taken for granted. In a world where the young die, having just one child runs against the odds. In a world where almost all children survive, a one-child family is a sensible option. Educational opportunities and lifestyle factors also come into play. As a society modernizes, marriage and child rearing must compete with schooling, careers, and a couple's standard of living.

Consider Japan. The proportion of unmarried women ages twenty-five to twenty-nine increased from 18 percent in 1970 to 48 percent in 1995. Japan's fertility rate of 1.4 is currently one of the lowest in the world. Soon, Japan's population will start to fall. And it is getting older.[55] In 2005, more than 18 percent of Japan's population will be over sixty-five years of age, compared to 12 percent in the United

States.[56] Why don't Japanese women get married earlier and have more children? In part, because they find the single lifestyle more attractive. Marriage means giving up a career and staying home to cook for one's husband and raise the kids. To understand Japan's fertility rate, one has to analyze the workplace and the home and how both of these filter individual choices. The food supply is irrelevant.

The promise behind the demographic revolution is that children will not die. Infant mortality in the United States, western Europe, and Japan is less than 1 percent. For children between one and four years of age, it is .002 percent or less. In developed countries, the death of a child or a young adult is a great tragedy; we wonder who is to blame. This world of low mortality and a long life is the greatest revolution of modern times; everyone has a right to enter it, not just the rich. Kerala shows that families in poor countries can shift from high to low fertility. It does not cost much: a decent education, access to community health, family planning, and enough to eat. Is this too much to ask? Rich and poor countries alike stand to gain from stabilizing the world's population as quickly as possible. In the meantime, as the momentum slows, producing enough food is both a moral imperative and a practical necessity. Relying on starvation and disease to balance the planet's population returns us to Thomas Hobbes's nasty world, where life is brutish and short.

PART VI
Potato Diseases and Pests

CHAPTER 17

Late Blight

"It's like a dragon spewing flames, burning everything in its path." That is how CIP pathologist Edward French described late blight when I met him in 1988.[1] I did not believe it. Years later, researching the Irish potato famine, I finally met French's dragon.

Irish folk tradition remembered "the heavy stench of decay, the blackened stalks, the potato pits sagging with rot." Families went to bed "with the fields green as holly"; when they awoke, "the lumpers were as black as soot." They watched helplessly as a "swath of brownish-black decay spread from field to field."[2] A Catholic priest who went from Cork to Dublin on July 27, 1846, reported the potato crop "had bloomed in all the luxuriance of an abundant harvest." When he returned a week later, all that remained was "one wide waste of putrefying vegetation."[3] Worst of all was the sheer terror of watching one's livelihood destroyed, the dread of the suffering ahead, the insecurity of not knowing when the monster would return.

A disease that could lay to waste a crop as green, prolific, and dependable as the potato was a terrifying spectacle. Contemporaries compared its impact to a plague; the *Dublin Evening Post,* for example, dubbed the pestilence "Potato Cholera."[4] The malady's underlying cause was uncertain and controversial. In the 1840s, medical science presumed an organism had to "lose vigor" before infection occurred. To critics, the potato's susceptibility to blight reinforced their belief that the plant had hereditary defects. Only later, after Pasteur's experiments of the 1870s, was it widely accepted that germs cause disease and that infection can randomly strike down a perfectly healthy organism. Hence, in retrospect, the blight had nothing to do with the potato's moral character.[5] But it did reflect the vulnerability that genetic uniformity engenders. Potatoes reproduce asexually; each tuber is a clone. Consequently, all plants of the same variety are identical genetically. Once late blight hit the Lumper, Ireland's most popular variety, it struck it down without mercy.

Observers at the time recognized that a rank-smelling fungus accompanied potato blight; most of them, however, presumed it developed after the plant died, not before. The French biologist Anton de Bary was the first to describe the malady's causal agent, a fungus he dubbed *Phytophtora infestans,* which roughly translates as "plant-destroying infestation." His work gave birth to a new scientific specialty, plant pathology.[6]

In the 1840s, Irish farmers sometimes protected wheat seeds from bunt disease by dipping them in a solution of copper sulfate mixed with water. At the height of the Irish famine, David Moore, curator of Dublin's botanical gardens, tried a similar treatment on the potato. He was on the right track. Even so, a proper formula was not devised until the 1880s. French botanist Alexandre Millardet noticed that vineyards sprayed with an inorganic mixture of copper sulfate, hydrated lime, and water did not get downy mildew. This confection, subsequently called "Bordeaux mixture," was demonstrated to be effective against late blight. By 1900, spraying Ireland's potato crop was common practice.[7]

The fungus strain that attacked the Irish potato, probably A-1, gradually lost its virulence.[8] Field sanitation, healthy seed, and systemic fungicides allowed farmers, the potato, and late blight to coexist. Until the early 1990s, late blight seemed under control, dismissed even by the experts as "merely a costly nuisance."[9]

How severe a late-blight attack is depends on multiple factors. By all accounts, Ireland's potato crop failed under an ominous combination of mists, rainstorms, and unseasonable gales.[10] In fact, weather determines how quickly the fungus multiplies and how far it spreads. CIP pathologist Greg Forbes describes late blight as a swimming spore, a night-growing fungus. "It needs moist, wet conditions to proliferate," he told me, "and such conditions typically prevail at night."[11] A late-blight specialist, Forbes worked at Ecuador's Santa Catalina potato-research station. I was in Ecuador learning about potato production problems, and late blight was at the top of the list. Forbes explained that the fungus thrives with cloudy, damp weather combined with mild daytime temperatures of between fifteen and twenty-five degrees centigrade (fifty-nine to seventy-seven degrees Fahrenheit). Hot, dry weather and direct sunlight can kill off the fungus; colder weather slows down multiplication.

The fungus replicates itself asexually by producing sporangia. Proliferation starts when the airborne spores alight on potato plants in a wet field. Spores require moisture both to germinate and to penetrate plant tissue. Rain droplets and wind spread the spores to healthy plants nearby or even to distant fields five to ten miles away.[12] A single leaf lesion can generate as many as three hundred thousand sporangia per day. Consequently, outbreaks can be explosive, as the cycle of penetration, colonization, and sporulation unfolds rapidly. All parts of the potato plant, including its leaves, stems, and tubers, are susceptible to infection. Rainfall can wash sporangia from the leaves to the soil and hence down to the tubers, which soften and rot once infected.[13]

A hard frost will kill off the clonally produced type A-1 spores. In temperate climates, A-1 cannot survive the winter unless it holds out unfrozen inside discarded tubers—hence the importance of field sanitation.[14] To control the fungus, farmers should destroy any diseased tubers left in their fields rather than heap them up in cull piles. They should uproot volunteer plants from the previous season, lest they carry infection. Rotating from potatoes to grains or legumes likewise helps eliminate the pathogen. Finally, farmers should make sure that the seed they plant is not infected. Such defensive practices keep the fungus out of fields much longer. If the A-1 malady does strike, a systemic fungicide absorbed by the plant can reduce sporalation and contain the disease before it spreads.[15] In the meantime, farmers can hill up soil around the plants to protect the tubers below from infection. Before harvesting, they should cut back the plant's foliage so that it dies out completely before digging tubers. Otherwise, the disease can spread from the fresh vines to the harvested spuds. The infection may go unnoticed at first, but rot will set in during storage.

Wherever potatoes are produced, a late-blight outbreak is possible. What varies are the malady's frequency and severity, factors determined largely by climate. In zones with a hard winter frost, the fungus has to restart every year. Where frost never occurs, as in Ecuador's Carchi Province, late-blight spores are in the air all the time. Fortunately, the cool mountain temperatures slow down the disease. In the Andes, the higher a potato field's elevation, the less the late-blight risk. So the fungus is more threatening in misty Carchi Province than on Bolivia's semiarid altiplano. In the United States, summers in North Dakota's Red River Valley are often too hot and the humidity too low for late blight; outbreaks are infrequent. Michigan, in contrast, can turn into a late-blight hot spot. When late blight occurs, farmers have to start spraying or risk losing the crop.[16]

Late blight has made the potato the world's most chemically dependent crop. CIP puts the global fungicide bill for the tuber's protection at over $2 billion a year, of which some $600 million is spent by farmers in developing countries.[17] How often do farmers apply fungicides?

Applications begin once the plant emerges and the canopy starts to fill up the space between rows; they often continue until the foliage dies back. In Ecuador, Doña Marta, from Carchi Province, said she sprayed a dozen times, which is not unusual for highland farmers.[18] Bolivian pathologist Oscar Navia, who worked with PROINPA, said farmers spray a susceptible variety like Waych'a at least eight times. When they plant varieties with some resistance, they can cut back to two applications, save money, and double yields.[19] In the United States and Canada, farmers spray damp fields every four to five days. On average, during the 1998 season, potato farmers in Michigan applied fungicides ten to twelve times.[20]

There is no fixed rule on frequency. How often a farmer needs to spray depends on local weather conditions and the variety planted. Each season's battle conditions can be fundamentally different. The challenge is to equip farmers to face an invisible and unpredictable adversary.

Greg Forbes felt Carchi's farmers should spray only when needed, not on a rigid schedule. "What farmers lack," he said, "is a practical way to decide on the probabilities." The formula includes how much it has rained, whether the rain came at night or during the day, and how much sunlight there has been. Such forecasts are not just for Andean farmers. In Nova Scotia, Canada's top spud-producing province, farmers get "blightcasts" on the radio or from the Internet, with warnings updated twice weekly for each of three districts in the fertile Annapolis Valley.[21] In the United States, blightcasts, fact sheets, bulletins, and alerts are likewise commonplace. Oregon's extension service calculates "severity values" based on temperature and humidity. When these exceed a predetermined threshold, farmers are told to apply fungicides forthwith.[22] For the 2000 growing season, Michigan had nine late-blight stations issuing severity values every twenty-four hours.[23]

Farmers needed all the help they could get. In the 1990s, late blight recouped its reputation as a dragon-like destroyer. In the United States and western Europe, virulent A-2-type strains have put the potato crop at risk again. The threat is worldwide. In the past, even after late blight had established itself, farmers could expect a systemic agent like metalaxyl to halt its spread. No longer. The A-2 fungus is resistant. In Ecuador and Peru, A-1 genotypes have mutated and are likewise resistant. Against such virulent adversaries, farmers use protectants sprayed prophylactically. A protectant must be on the plant before the fungus attacks; it creates a chemical barrier on the foliage that inhibits sporulation, but it does not kill the fungus.[24] Stop spraying, and the disease spreads.

Pathogens that build up resistance to antibiotics are the scourge of modern medicine. Today, a wonder drug like penicillin is useless against the tuberculosis bacillus it was originally designed to counteract. New super strains resist almost all the antibiotics in the medical arsenal.[25] The same is true of the drugs used to suppress malaria. Quinine gave way to chloroquine-based drugs, which gave way to mefloquine, which is now losing its effectiveness.[26] Overreliance on antibiotics forces pathogens to mutate; new, more virulent strains result. The same principle applies in plant pathology and, in this instance, to late blight. Dousing the fungus with powerful, systemic fungicides was asking for trouble.

Scientists used to think that all known strains of potato late blight belonged to a population now called A-1. Even though the A-1 pathogen produces both male and female structures, it was thought to reproduce exclusively by asexual sporulation. In the early 1950s, World Food Prize laureate John Niederhauser identified Mexico's Tuluca valley as a zone of late-blight diversity. He discovered there a second mating type, A-2. Together, the two types can sexually reproduce, creating genetically unique, thick-walled oospores rather than clonally propagated, genetically uniform sporangia. Such oospores can winter over in the soil, even without a potato host. Fortunately, the two populations have yet to intermingle on a large scale. Even so, the new strains of both A-1 and A-2 are more virulent than the ones they have replaced. They may not sexually reproduce, but they are still resistant to an old standby like metalaxyl.[27]

In the 1980s, the A-2 mating type showed up in the Netherlands. Since then, scientists have identified A-2 populations in many countries. The intrusive strains are becoming dominant, pushing out the less aggressive types. In the United States, for example, a deadly A-2 strain called US-8 was found in New York state in 1992. Three years later, it had established itself in twenty-three states. Cornell University pathologist William Fry, who tracks the new strains, characterized the speed of the pathogen's migration as "mind-boggling in its magnitude."[28] Infected tuber seed helped to spread the fungus.[29]

The new pathogen has done enormous damage. In 1994, farmers in New York state had crop losses estimated at over $100 million.[30] In 1995, the pathogen affected sixty-six thousand hectares in the Pacific Northwest, adding some $33 million to production costs.[31] Even Idaho's $2.6 billion processing industry was in jeopardy.[32] Prior to 1995, only one case of US-8 had been reported in the state. The next year, it had infected thirty fields. Today, almost every state and province of the United States and Canada reports A-2 late blight that is resistant to metalaxyl. Not only is A-2 dominant, but the most lethal strains of it, US-8 and US-11, are found most frequently. According to on-farm surveys, in 1999 growers in Idaho, Maine, Washington, and North Dakota's Red River Valley used fungicides on 95 percent of their potato acreage—1.9 million kilos (active ingredients) on 289,000 hectares.[33]

The future will not be secure until the new late-blight menace is countered. The world's small farmers produce some 37 percent of the global potato crop; it adds an extra $17 billion annually to their income.[34] What would they do without it? As one CIP scientist put it, "we must conquer the fungus before we lose the potato." For developing countries, CIP puts losses from the older, less aggressive A-1 types at 15 percent of the total harvest; in sub-Saharan Africa, where many farmers cannot afford pesticides, the damage is twice as severe.[35] With more lethal A-2 pathogens loose, the potential for disaster is greater still. To meet this threat, CIP put together a Global Late Blight Initiative (GILB). By linking research, extension organizations, and farmers, the project hopes to contain the late-blight menace.

In the United States, no commercial cultivar planted in the 2000 season had late-blight resistance of any significance. Unfortunately, the prima donna of potato processing, Russet-Burbank, is notorious for its susceptibility. The U.S. approach is to find a better fungicide; Acrobat MZ, Curzate M-8, and Tattoo C are at the top of the list.[36] Agribusiness can afford the protection costs, but small farmers in developing countries cannot. Consequently, a key component in GILB's strategy is to breed potatoes with both A-1 and A-2 resistance and get them to farmers. CIP has worked on late blight for years. Fungicides are expensive, posing both environmental and health hazards. Recall that in Ecuador's Carchi Province, farmers mixed fungicides with insecticides, a practice that heightens toxicity. Countering the new late-blight threat with more lethal chemicals could make a bad situation even worse. With a more resistant potato, by contrast, farmers get built-in protection for just the cost of the seed. That is what CIP did with A-1 late blight, and it paid off.[37] During 1987–93, CIP cultivars released to Third World farmers had, as a matter of course, a significant level of A-1 resistance.[38]

The results added up quickly. Farmers could reduce the quantity and potency of the fungicides used. Compared to what is needed to protect a susceptible cultivar, the savings can be as much as seven-tenths of a kilo of active ingredients per hectare per week.[39] Canchán, a CIP cultivar released in Peru in 1990, reduced fungicide use against A-1 late blight by half, saving farmers $74 a hectare.[40] For East Africa, CIP estimates the return to farmers on A-1—resistant cultivars at $10 million a year, a benefit reaped both in higher yields and reduced costs.[41] Unfortunately, such varieties are likely to succumb to more lethal A-1 or A-2 genotypes. But at least CIP was on the right track.

Specific genes in the potato plant called "R" genes match up with and resist particular strains of the late-blight pathogen. When R genes are incorporated into a new variety, as with Canchán, the resistance level is impressive, especially when combined with a judicious use of fungicides. Once the pathogen mutates, however, a formerly resistant plant can suddenly become vulnerable again. In fact, vertical resistance increases the selective pressure on the late-blight pathogen, which speeds up mutation. So a decade ago, CIP opted for "horizontal resistance," a low-key approach that puts the pathogen under less selective pressure. The objective is not to make the potato plant immune but to allow for a mutual coexistence beneficial to the potato. Horizontal resistance is durable because it combines multiple minor genes, or "r"-type resistance genes.[42] "Late blight is like a thief with a ring of keys," explains CIP breeder Juan Landeo. "If the door has just one lock, even a very strong one, it doesn't take long for the thief to find the right key. But if the door has a lot of locks that work in combination, it's going to take much longer." With r-gene resistance, the fungus can still multiply, but not fast enough to damage the plant significantly or reduce yields.[43]

Such at least is the goal. To get there, the r genes must first be found. CIP's search started with *Solanum phureja*, a domesticated but lesser-known Andean species.[44] Farmers plant phureja types in warmer valleys; they do well, despite higher humidity and late blight. The common tuberosum potato, by contrast, has little late-blight resistance of its own. With the r-gene search in mind, CIP increased its biotechnology budget even before the GILB took shape. The objective was to find genetic markers that indicated where the resistance genes in phureja types were located. Once these traits are "tagged," scientists can transfer them to commercial varieties. With molecular genetics, it's much easier to verify that resistance has been passed on to progeny. Scientists no longer have to infect plants with the pathogen and then monitor the resistance. Instead, they can check in the lab to make sure the tagged markers show up on strands of chromosomes.[45] Of course, breeders will still rely on field trials to make sure they were right, but such trials will be fewer and less frequent.

In 1995, CIP already had a set of cultivars with promising horizontal resistance. Since then, it has bolstered this with r genes from more diverse genetic sources, hoping to make such resistance durable.[46] With A-1 late blight, the so-called resistance can be location specific. Consider Pimpernel, a Dutch variety that seemed to have A-1 resistance. When tried out in Peru, the hardy import did no better than Yungay, a local, susceptible variety.[47] So what looked like genetic resistance proved ephemeral. Once out of its niche in the Netherlands, Pimpernel's A-1 resistance collapsed. With this lesson in mind, GILB cultivars that seem to have A-2 resistance must run the gauntlet of late-blight hot spots.[48]

Promising A-2 clones face Standard International Field Trials (SIFT) in Africa, Asia, and Latin America. Since no single variety suits the many farming systems, climates, and consumer tastes involved, CIP works on three potato sets simultaneously: yellow-fleshed andigena types for South America, white-fleshed European-style tuberosums for East Africa, and crosses between the two for tropical Asia.[49] The question is, can CIP and its GILB partners develop, test, and circulate new cultivars faster than mutant late-blight strains can ravage the old ones? Can they conquer the fungus before we lose the potato?

In 1997, El Niño rainstorms and an aggressive A-1 genotype dubbed EC-1 battered Peru's coastal potato farms. Under this lethal combination, Canchán fell prey to the disease; not even metalaxyl could save it. "We watched potatoes all around our test sites in Peru wither and die," said Wanda Collins, CIP's director for research. But there was good news too. CIP saw the r-type lines it had banked on do much better. They "grew up sound and healthy, despite the rains and wet weather." Since its new potato lines have passed the El Niño test, CIP hopes they can withstand attacks "by all existing forms of late blight."[50] Only time will tell.

A potato that can take on the new strains will advance the cause, but it will not

Fig. 8. Farmer examines late-blight spores. Courtesy CIP and R. Nelson, from CIP, *International Potato Center Annual Report 1999*, p. 10.

win the day. Resistance is not immunity; even with r genes, no one knows for sure how durable a new cultivar's resistance will be. As with old-fashioned A-1 strains, the key is to train farmers to fight the disease in their own potato fields. They have much to learn. Farmers do not know the difference between fungicide brands, what an active ingrediant is, or what makes a systemic fungicide different from a protectant. Late-blight spores are invisible; the disease can be present before farmers can spot the symptoms. And there is no fixed formula that tells farmers how much fungicide to use or what kind. It depends on a cultivar's resistance level, the health of the seed planted, and the weather. In the end, success requires both farmer training and biotechnology. To organize the ground war, CIP is setting up on-farm, late-blight training schools in Asia (China and Bangladesh); South America (Peru, Ecuador, Bolivia); and Africa (Uganda and Ethiopia). The three-year project is backed by the International Fund for Agricultural Development. Farmers will learn how the fungus spreads, how it multiplies, and the factors that favor its rapid proliferation. They will learn how to track the weather and estimate the danger of an attack. Working with SIFT, farmers will evaluate and select resistant cultivars. Local trainers will be armed with field guides, maps of late-blight hot spots, and portable microscopes.[51]

In Peru and Bolivia, extension workers, farmers, and NGOs teamed up in 1997–98 to start the project. Farmers conducted on-farm experiments to see firsthand how the disease spreads. First they looked at spores under small, hand-sized microscopes. (See fig. 8.) Then they dipped one healthy potato leaf in water with spores and put another healthy leaf in clean water. Both leaves went into plastic containers for observation. Farmers thus discovered the mechanism of infection. Wet conditions alone did not cause late blight; spores did. Farmers also tested new CIP lines selected for their r-type resistance. In Peru, communities have already chosen the variety Chata Rota for seed multiplication.[52]

One hundred and fifty years after Ireland's tragic famine, the culprit was at last to be fought properly—not with ideology and chemical warfare, but with on-site training and resistant plants.

CHAPTER 18

Tuber Moth

"What future does the potato have," Aziz Lagnaoui asked, "if we can't curb its dependence on insecticides?" A tall, intense Moroccan in his thirties, Lagnaoui is an entomologist trained at the University of Minnesota in the United States. His specialty is integrated pest management (IPM)—how to fight insects without overkill. Lagnaoui headed up CIP's regional office for the Middle East and North Africa; headquarters are in the Tunisian capital, Tunis. I arrived in June, just in time for the potato harvest. Lagnaoui met me at the airport. The next day, we teamed up with agronomists from Tunisia's National Institute for Agronomic Research (INRAT). In a hectic ten days, we covered the country's chief potato-producing zones, including the Cap Bon Peninsula, which accounts for over half the potatoes produced in Tunisia.

In the United States, insecticide use is entrenched. In 1999, farmers surveyed in Idaho, Washington, Maine, and North Dakota's Red River Valley said they used insecticides on 94 percent of their potato acreage—some 920,000 kilos (active ingredients) on 287,000 hectares.[1] By comparison, Tunisia's potatoes escaped almost unscathed. In 1990, only 14 percent of the country's farmers used insecticides on their potato crop. They had other ways to ward off the lepidopterous tuber moth (*Phthorimaea opercuella*). This was not always the case. Tunisia's potato was recovering from insecticide abuse. How this recovery happened is the story that brought me to CIP's Tunis office.

The tuber moth infests tubers both in the ground, prior to the harvest, and later, when they are placed in storage. To protect the crop, Tunisian farmers used chlorinated hydrocarbons (DDT) and organophosphates (parathion), even on potatoes stored for consumption. Such use was fraught with difficulties. Contaminated runoff from irrigated fields is hazardous to the aquatic environment and groundwater. Insecticide residues had shown up on potatoes sold in the markets, which posed the risk of pesticide poisoning. Indiscriminate use, storage, and disposal endangered farm families and their workers.[2] To counter such threats, CIP and INRAT began a joint offensive in 1979.

During the first phase of CIP's work with INRAT, researchers tried out safer, less toxic alternatives. Trials showed that synthetic pyrethroids, whose chemical toxicity breaks down more rapidly, were just as effective. In 1984, the Tunisian government restricted DDT and parathion imports; INRAT began recommending the pyrethroid substitute deltamethrin (DECIS) to farmers.[3] Problem solved? Not quite.

Breeders at CIP expected to find varieties with tuber-moth resistance in the center's vast collection of potato germ plasm. If they could backcross resistance genes into commercial varieties, farmers might avoid insecticides altogether. That was twenty years ago; the search goes on. Fortunately, CIP had other ways to defend the potato. Pest management relies on field practices to reduce pest levels. Insecticides come into play only as a last resort. CIP's conviction that farmers can manage pests has withstood the tests both of time and of scientific rigor. Today, entomologists acknowledge that heavy insecticide use puts insect populations under enormous selective pressure; as a result, they mutate and develop resistance. As Lagnaoui explained, a powerful insecticide may kill off 99 percent of the tuber moths in a field, but the 1 percent that survive exposure will have some level of resistance.[4] Slowly but surely a new, more resistant population evolves. As the pest makes a comeback, farmers switch to new chemicals, which force the pest to mutate, which in turn requires an even more lethal insecticide. The result is what critics call a "pesticide treadmill."[5]

Pest management is not just good public relations. In the long run, it may be the only practical way to protect crops like the potato. The drawback is the tedious research, fieldwork, and on-farm extension required to validate controls. It is much easier to recommend a new insecticide. Consequently, converting to pest management is no mean achievement.

Like the potato itself, the common tuber moth comes from South America. Back in its native Andean habitat, the potato must contend with four different tuber-moth species. In Tunisia, fortunately, it faces but one. Each species has its own population dynamics, peculiar habits, and disease proclivities; what works with Tunisia's common tuber moth does not necessarily transfer to variants like *Scrobipalpopsis solanivora*, the Guatemalan pest.[6] When the CIP-INRAT team took on Tunisia's tuber moth, they first tracked the pest's incidence in farmers' fields, thus determining when the infestations peaked. Calculations done elsewhere did not apply, as the timing is location specific. To estimate pest levels, CIP-INRAT made inexpensive traps out of gallon-sized plastic containers. (See fig. 9.) A large opening is cut near the top so the insects can fly in, and soapy water is added in the

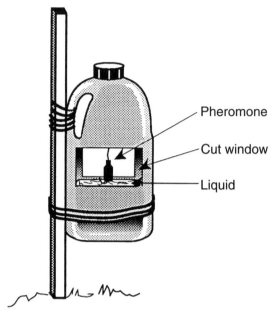

Fig. 9. Plastic container made into a pheromone trap. Courtesy CIP, from Walker and Crissman, eds., *Case Studies of the Economic Impact of CIP-Related Technologies,* p. 6.

bottom. What attracts the male tuber moth to the trap—and only the male—is the bait, a sex pheromone that has the female's scent. The pheromone is suspended from the container's top inside a thimble-sized rubber cap. Male moths swarm to the bait, flying around it until they drop into the soapy water below. The precise pheromone formulation was perfected at CIP. In 1988, CIP reported that its lab had distributed thirty thousand samples to more than twenty-five countries.[7]

Researchers put the traps in potato fields, counting how many tuber moths they caught. In pest management, the first line of defense is escape—timing crops to avoid an insect onslaught. Tunisia was no exception. Monitoring showed that the number of tuber moths goes up sharply by mid-June, when Tunisia's weather turns hot and dry. A farmer who keeps potatoes in the ground later than that is asking for trouble.

Adult moths are grayish brown and about ten millimeters long (.39 inches). To get at the tubers, female moths tunnel through the cracks in the soil; they lay between 100 and 150 eggs each in depressions on the skin's surface, especially around the eyes. If the moths cannot reach the tubers, they deposit their eggs on the plant's foliage, or failing that, on soil debris. The larvae hatch 3 to 5 days later and bore into any available tubers. After feeding for between 10 and 15 days, they form cocoons. The metamorphosis is completed within 5 to 10 days, and a new adult generation emerges. As the weather turns drier, the modal reproductive cycle drops to just 15 days. Tunisia's most popular potato variety, the Dutch tuber called Spunta, matures in about 120 days. So there is ample time for the pest's cycle to accelerate. Proliferation continues throughout the harvest period and into storage.[8]

Having worked out the pest's dynamics, the CIP-INRAT team tested control components. Planting early so as to harvest by early June is essential. Even so, the tuber moth still stalks potato fields. The problem is how to keep them away from the tubers. In Tunisia, almost all farmers irrigate their potatoes. The trick is to keep on irrigating up to harvest time. This keeps the soil around the potato plants moist, so it does not crack in the heat. Since tuber moths cannot bore through the muddy soil, their path is blocked. Hilling up soil around the plants also keeps the pest at bay.[9] Finally, farmers can reduce tuber-moth buildup by rotating to other crops—except, of course, to solanaceous potato relatives like peppers, tomatoes, and eggplants, crops that aid and abet tuber-moth proliferation.

Once on-farm trials had validated the key components, or "cultural practices" involved, it was time for large-scale implementation.[10] INRAT trained extension agents, who then trained farmers; they set up demonstration plots and designed simple brochures. So successful was the campaign against the tuber moth that CIP subsequently estimated the project's impact. Using farm surveys from each of Tunisia's potato-producing districts, the study compared the pest-management practices farmers used in 1986 and 1990. The percentage of farmers who sprayed their fields with chemical insecticides dropped from 46 to 14 percent. Those relying on irrigation as a control doubled from 40 to 80 percent. Almost all the farmers interviewed recognized that an early June harvest was the best way to avoid tuber-moth damage. The weather, however, does not always cooperate. Crops get in late, which in turn delays the harvest. Moreover, Tunisian farmers are accustomed to leaving their spuds in the ground until the skins harden. The practice improves a spud's storage quality and reduces spoilage, but it delays the harvest and increases tuber-moth exposure. Nonetheless, almost half said they harvested early. Finally, some 15 percent hilled up soil around the potato plants; hardly anyone did so in 1986. As to the economic impact, farmers saved $165,000 annually on pesticides. Even more important, the field practices outlined above so reduced the number of infected tubers stored that losses were cut by half. In 1991 that generated an on-farm storage benefit of over $2 million.[11]

Tunisia needs all the potatoes it can produce. Between 1960 and 2000, its population more than doubled, rising from 4.1 to 9.6 million. The change to smaller families, however, has taken hold. Since 1985, the country's total fertility rate has dropped by almost half, from an average of 4.5 children to an estimated 2.3 in 2000, far below the rates for the rest of North Africa.[12] Potato production more than kept pace with the population increase. For 1968–70, the harvest averaged 63,000 metric tons; for 1997–99, it was 315,000 metric tons. At the same time, average yields nearly doubled, from 7.1 to 13.7 metric tons per hectare.[13] Part of the explanation is the high priority CIP and INRAT gave to improving seed quality.

Tunisia's main-season potato crop is planted from late January into early March and harvested from May into June. The ever-popular Spunta is a large, white-skinned, oblong tuber with a smooth skin, shallow eyes, and yellow flesh; it accounts for between 60 and 70 percent of the total crop. Its advantage is its short dormancy period. Farmers can harvest Spunta in June and then recycle the seed into the late-season crop, which they plant in August/September for a November/December harvest. Most of Tunisia's farmers start off the main-season crop with new seed. For this, Tunisia imports between 16,000 and 20,000 metric tons of seed tubers annually; most of the seed comes from the Netherlands, and most of it is Spunta.

As we bounced our way from one production site to another, Mohammed Fahem, director of Tunisia's seed program, explained all of this patiently.[14] Powerfully built, with thinning black hair, he had an infectious, can-do enthusiasm. He needed it to coordinate a seed program. Tunisia targets the late-season August crop, when imported seed is not available. In the past, farmers simply replanted what they had held over and stored. Such stocks can be infested with the tuber moth or with diseases that reduce yields. Farmers are much better off with virus-free seed purchased from the national program. Although most of that seed is Spunta, Fahem noted that the variety has its drawbacks. "Spunta is too watery," he told me. "It varies greatly in size, damages easily, and stores poorly." Gradually, the national program is working in new varieties like Kondor. Kondor is a red-skinned, medium-sized spud; its uniformity and oval shape make it ideal for processing. Its flesh has the same taste and texture as Spunta. Also in stock was Mondial, a white-skinned type with one Spunta parent. Since it took a couple of weeks longer to mature than Spunta, it was best suited to the mild climate of the Cap Bon Peninsula. In 1996, Tunisia's seed-multiplication program had taken some 240,000 cuttings from virus-free mother plants; these cuttings were then transferred to outdoor, netted greenhouses. The small, seedling tubers thus generated went to a select group of farmers for an additional, main-season multiplication. Tunisia's national program purchased this certified seed at harvest time in June. The stocks went into cold storage until August, at which time the seed was sold to farmers for the late-season crop.

I accompanied the CIP-INRAT group on its June inspection rounds. Headed by Lagnaoui and Fahem respectively, the group stopped at storage depots and then backtracked to potato fields. With seed potatoes, farmers cut back the plant's foliage and pile it on top of the rows; they do this at least two weeks in advance so the skins can harden. Potatoes destined directly for market do not require such special treatment. Cap Bon's sandy soil was already dry and powdery. The potatoes came out of the ground clean, smooth, and attractive; washing them was unnecessary. They were sorted by size directly into plastic crates. June days are long, and the sun is intense. Farmers harvest from dawn until about 11 A.M. After that, it's too hot. Around 4 P.M., when it begins to cool off, they can finish up. Uncollected potatoes dug in the morning should be covered with straw or foliage. Otherwise, the strong sunlight and heat can turn them green. They cannot be left out on the ground overnight either. Nighttime is when tuber moths are most active and lay their eggs.

Tunisia's potato farmers rarely sell the entire harvest. They usually put part into some kind of rustic, bulk storage. In this way, they can sell off gradually, avoiding gluts and fetching higher prices. For potatoes to be stored successfully, they must be sheltered from the sun. Farmers select a shady spot, pile up their potatoes, and then cover them with a thick layer of straw.[15] The hard part, however, is keeping the tuber moth away. Farmers must first sort through their potatoes, discarding any with obvious signs of tuber-moth damage. Even so, the tiny holes left by intrusive tuber-moth larvae can be hard to spot, not to mention the microscopic eggs deposited in the hidden recesses of a potato's eye. No matter how carefully farmers sort, tuber moths will end up in their piles. So farmers dip each potato in a solution to which the pyrethroid insecticide DECIS has been added, or they spray the pile with it.

Farmers complain that the DECIS they buy locally is diluted. According to Lagnaoui, DECIS is in fact less effective than it used to be, not because it is less potent but because the tuber-moth population is more resistant. Fortunately, remedies equally effective yet safer to all concerned are in the works.[16]

The CIP-INRAT drive against the tuber moth had unexpected results. Researchers noticed that the pinkish tuber-moth larvae sometimes turned a sickly white and died. They sent samples back to CIP for analysis. Results showed that the larva had contracted a granulosis virus (GV) lethal only to the tuber moth. A crash program followed. The CIP-INRAT team worked out methods to multiply infected larvae, mash them up, and produce an organic powder to spread over stored potatoes. The emerging tuber-moth larvae contract the virus as they crawl over the tubers and ingest the powder. Starting out with small, experimental quantities, INRAT's lab in Tunis produced about five hundred metric tons of GV in 1996. The objective is to replace chemical insecticides like DECIS with GV, which is harmless to humans, even when consumed in quantity. In 1996, testing had reached its final stage: large-scale, on-farm trials. So far, GV protects stored potatoes as well as the best insecticides.

It was now almost 2 P.M. and time for lunch, or should I say dinner? As in most of the Mediterranean world, the afternoon brought the day's main meal. We headed for a small restaurant with shaded, outside tables. First came the salads—cucumbers, hard-boiled eggs, and tomatoes with a hot chili sauce; then the french fries, pasta, fried eggplant, and fava beans boiled in tomatoes; and for the meat dish, grilled lamb chops. We finished with watermelon and green tea. No wine. Most Tunisians are Muslim, and the Koran forbids alcohol. For alcohol, one has to go to the more secular Tunis.

Tunisia's history makes it part of Africa and part of Europe. When Tunis was a treasured province of the Roman world, its peaceful villas produced wheat, wine, olive oil, and figs to ship back to Rome. In the Cap Bon Peninsula, olive groves and fig trees still hold out against the potato; the north coast is still famous for its wheat. The Roman mosaics collected in Tunisia's Bardo Museum are the most beautiful ever found. Those taken from old Roman villas are famed for their portrayal of agriculture, fishing, and family life.[17]

Prior to imperial times, ancient Tunisia was home to Carthage, Rome's great enemy. And later, after its Islamic metamorphosis, it turned foe once again, this time against a Christian Rome. There were no "dark ages" in the Maghreb countries of North Africa. Islam built upon the foundations of the Roman world, reinforcing its urban sophistication, scholarship, and trade.[18] Moorish architecture and influence spread up the Iberian Peninsula. What seems so Spanish about old Spain—its enclosed balconies, its señoritas hidden behind mantillas, the narrow covered streets, the heavy wooden doors and carved balustrades—is as much Moorish as Spanish. The Medina of old Tunis is a maze of courtyards—a walled-in city of cool, covered passageways, with shops, cafes, and Mosques, its cramped neighborhoods filled with chatter. At night, when the sun goes down, everyone goes out, women included, much like in Grenada in southern Spain, or Cartagena in Colombia. Tunisian men embrace and kiss in the street, like the Italians—or the Argentines. The eddies of a common Mediterranean culture still persist, despite the diaspora of its people and the contentious gods they have embraced.

CHAPTER 19

Natural Enemies

A standard practice in the insect wars is to enlist the support of a pest's natural enemies. A good example is the brown plant hopper in Indonesia, a pest that attacks the country's rice. Experience shows that the plant hopper's predators, including an assortment of spiders, parasites, and pathogens, can hold it in check. Farmers in Indonesia have learned to sample and count the number of hoppers per rice plant. When the numbers exceed a predefined threshold, they call in the insecticides as a last resort. When they do spray, they use chemicals that target the plant hopper specifically, leaving spiders and other beneficial insects unscathed.[1]

If pest management can work so well in rice, why all the fuss about the potato tuber moth? Because Indonesia's rice is a special case. Rice has been planted in quantity in Asia for thousands of years, ample time for an insect balance to evolve. Once Indonesian farmers curtailed insecticide use, the equilibrium between plant hopper and predator reasserted itself. In North Africa, by contrast, the potato is a relative newcomer. True, the potato got there before its insect nemesis showed up. Eventually, however, the tuber moth caught up with the potato. When it did, it found plenty of spuds around to infest and few natural enemies to keep it in check. In potato-producing Tunisia and Egypt, the tuber moth holds sway—unless checked by insecticides. Or so it seemed. In its Andean homeland, the tuber moth has enemies aplenty. In Peru, for example, a parasitic wasp lays its eggs in the tuber moth's larvae, breaking the pest's reproductive cycle.[2] In North Africa, the tuber moth has no such lethal insect opponents. That is why the granulosis pathogen discovered in Tunisia is so important; it turned the tables on the tuber moth. But nowhere has the ground war changed so decisively as in Egypt. Over the past five years, Egypt's Plant Protection Institute and CIP have developed, tested, and released biological controls that can replace chemical insecticides.

Egypt has over sixty-eight million inhabitants, almost as many people as in the rest of North Africa.[3] The country receives no appreciable rainfall; its cities, industries, and farms are all nourished by the same source, the Nile River. Most of the arable land, all of which is irrigated, lies in the fertile Nile delta, between Cairo and Alexandria. The country's urban population, however, continues to mount. Cairo has a population of over fifteen million; Alexandria has five million; and Tanta, a dusty, unremarkable place in the middle of the delta, has two million. Competition for space is thus intense. The delta is where most of Egypt's food is grown, especially its potatoes. And it is where most of the country's farmers live. For despite urbanization, more than a third of Egypt's economically active population still works in agriculture.[4]

After wheat and rice, the potato is Egypt's top food crop. Harvests have risen at an annual rate of 5 percent for the past thirty years, one of the highest rates for potatoes anywhere in the world. During 1979–81, Egyptian farmers harvested 1.1 million metric tons of potatoes. For 1997–99, production was almost 2 million metric tons with yields of 17 metric tons per hectare.[5] Egyptian sources put per-capita potato consumption at thirty-two kilos; a generation ago, it was only eight kilos.[6] Most families eat their potatoes freshly boiled, enhanced with olive oil and tomatoes.

In Egypt's delta, the winter is short and frosts rare; it is the summer's extreme heat that limits the potato's growing season. Farmers plant the year's first potato crop in early January; they harvest it by the end of April or early May. Given irrigation, Egyptian farmers can squeeze in two, sometimes even three potato crops a year. Regardless of the season, the tuber moth is the worst pest, both in the field and in storage. Land is cultivated intensively, with successive crops on the same or adjacent plots. The result is tuber-moth buildup. According to Ramzy El-Bedewy, who headed up CIP's delta research station at Kafr El-Zayat, local farmers sprayed the potato crop six to eight times. "They used a wide range of chemical insecticides," El-Bedewy said, "with little concern for themselves or consumers." To make matters worse, a decade of heavy pesticide use had generated a more resistant tuber-moth population. "The standard chemicals, whether used in the field or in storage, no longer worked effectively," El-Bedewy said. "Farmers reported frequent control failures."[7]

The pest-management tactics worked out in Tunisia made it to Egypt in the nick of time. Both Egypt and Tunisia have about 2.8 million hectares of arable land, but Egypt has seven times the population.[8] That puts tremendous pressure on farmers to double and triple crop. The land is used constantly, and so are agriculture's chemical warriors. The pesticide escalation that resulted posed a greater hazard in Egypt's densely populated delta than in Tunisia. Even though Egypt started its research on the granulosis virus (GV) late, it soon caught up. In 1996, farmers had taken to biological controls en masse. According to El-Bedewy, they had little choice. Egypt's Ministry of Health regularly tests potatoes earmarked for consumption; if they contain a detectable level of pesticides, they are destroyed. In 1995, some 19,000 metric tons were so identified. Compared to a harvest of some 2 million metric tons, that is not much, but it was enough to scare consumers away from potatoes, which left an unsold glut in local markets. Prices plummeted. Moreover, Egypt exports about 400,000 metric tons of its early spring potato crop to Europe. A reputation for pesticide overdose put that lucrative market in jeopardy. For all these reasons, pest management is now the party line.[9]

Egypt has North Africa's most ambitious biological-control program. Tunisia showed the steps farmers could take to reduce insecticide use in the field. Egyptian farmers followed suit, planting early, hilling around plants, keeping soils moist, and rotating out of solanaceous crops. It was not enough. Tuber moths start to peak in March, long before farmers can harvest the potato crop. There is no escape. Even early-maturing varieties need at least ninety-five days to mature. Like it or not, farmers had to spray their potato crop at strategic times. The question was, with what? Since chemical insecticides had a bad reputation, Egypt turned to biologically based alternatives. Using techniques worked out in Tunisia as a model, Egypt's Plant Protection Research Institute in Cairo opted for GV and mass production.

I went to Egypt with Aziz Lagnaoui; Ramzy El-Bedewy met us at the airport. The hectic schedule of El-Bedewy, a stocky, hardworking man in his fifties, did not mean he neglected his role as host. Since I had never been in Egypt before, he headed immediately for Giza and the pyramids. For dinner, he insisted on an inconspicuous restaurant, but one crammed with patrons. It was the best meal I had in Egypt: stuffed grape leaves, olives, potatoes boiled with tomatoes, tender slices of roasted lamb, stuffed peppers, eggplants, okra, and stuffed zucchini. We finished off with steaming hot tea.

The next day, Lagnaoui, El-Bedewy, and I went to the Institute to meet its director, Dr. Galal M. Moawad.[10] A hefty, effusive man, he is a staunch pest-management advocate. "The tide has turned against insecticides in Egypt," he said. "Years of heavy use destroyed any semblance of balance between insects and their predators. Farmers plant several rounds of the same crop on the same land. With potatoes, it hardly matters now how much farmers spray, given the tuber moth's level of insecticide resistance." Egypt has lots of crops to protect and lots of insects to contend with. The Institute's work on the tuber moth is its most advanced, but is by no means the only control program. Additional crops on the Institute's hit list include cotton, sugarcane, fava beans, and citrus; targeted insects include cutworms, white flies, aphids, leaf miners, and fruit flies. Problems are interrelated. When farmers depend on insecticides, they may win a battle, but they lose the war. Even if they defeat the tuber moth, they still face leaf miners and white flies later on. The outbreaks intensify as the ranks of useful insects get decimated. Given the intensity of the tuber-moth war, the potato crop was an obvious place to draw the line.

Dr. Moawad took us to the GV production unit. The first step in making GV is to raise healthy tuber moths. This is done inside plastic jars covered with circular pieces of black construction paper. The adult moths lay their eggs on the paper. When covered with eggs, the black discs are removed from the jars and placed in a square plastic container filled with tubers coated with the GV virus. It takes about three weeks for the eggs to hatch and another three weeks for the larvae to eat into the tubers and reach full size. When the larvae emerge, they crawl across and feed on the infected tubers, thus picking up the virus. Once infested, the tubers are dumped out on top of a screen and subjected to intense light. In ten to fifteen minutes, the heat brings the tuber-moth larvae to the surface; they crawl out of the tubers and fall down on the screen. The infected larvae are then gathered up and macerated in water. To make a powder, a solidifying agent, usually talcum (magnesium silicate), is added. The mixture then goes into shallow pans to dry out. When solid, it is eased out of the pans, crushed with a roller, and packed in one-kilo plastic bags.[11]

GV powder is used to coat potatoes for storage; as a spray, GV can be applied directly to the crop in the field. Just eighty infected larvae suffice to make a kilo of GV powder or 150 liters of GV spray. The powder can also be mixed with water and converted into a spray. In 1996, the Institute reared a million infected larvae, the equivalent of twelve metric tons of GV-based powder or 1.9 million liters of GV-based spray. It takes about three kilos of GV powder to cover a metric ton of potatoes for storage. Put the powder in a sack, add potatoes, and shake. When the potatoes are coated with the white GV powder, they are ready to store.[12]

When Tunisia's national program first tested GV powder, it had to work out the proportions: how many infected larvae per kilo and how many kilos per metric ton of potatoes. To do that, it had to know how much virus a typical larva contracted; in short, it had to sample and count. Thus, much detailed research stands behind the now-standard GV formula. Even so, GV does not kill on contact like an insecticide. When farmers spray their fields, it takes a couple of days before they see the results. To use GV, farmers have to be sure it works. Confidence comes from repeated on-farm trials and demonstrations.

GV is not the only biologically based option. A microorganism of the soil, *Bacillus thuringiensis*, commonly known as Bt, has also been enlisted against the tuber moth. It may seem odd that a soil bacterium qualifies as a "natural" enemy. The search for allies, whether for our own protection or that of our plants, can make strange bedfellows. Penicillin was isolated from a soil bacterium. From streptomycin to tetracycline, dirt has given medical science the greater part of the drugs it uses against infectious diseases.[13] The Bt bacterium is common in soils world-wide and comes in multiple strains. As it sporolates, Bt produces a protein crystal that is toxic to the larvae of certain insect types, notably of the butterfly order, *Lepidoptera*, to which the tuber moth belongs. Stored in the spores, the toxic crystals are released when ingested by the insect.[14] The stumbling block is discovering which Bt strain kills off which insects. To tell one strain from another requires a DNA analysis. Then each strain has to be tested against a battery of insects. In Egypt, the Institute isolated a Bt strain lethal to the tuber moth. It has since multiplied the spores, which it freezes for later use. The Institute's Bt formula is marketed as PROTECTO and comes in both powder and liquid form. In 1996, Egypt's Ministry of Agriculture banned the use of fenitrothion, a toxic insecticide made from organophosphorus compounds. It recommended Bt as a replacement.[15]

Egypt recently added over $2 million to its pest-management budget. Included are four new research labs and a special biological control unit at Cairo University. To advance its Bt research, the Institute will team up with the Agriculture Genetic Engineering Research Institute, which is located nearby. El-Bedewy did not consider my orientation complete without a look at Genetic Engineering. To enter the building, we had to register, go through security, and then clip on identity cards. By transferring the Bt bacterium to a potato plant, Genetic Engineering hopes to create a new plant with insecticidal properties. We met with the Director General, Magby A. Madkour, who received us graciously—tea, coffee, and plates of pastries. He explained how Bt gets into a potato plant.[16] By way of demonstration, we went to one of the labs.

Genetic Engineering starts with a virus-free potato plantlet produced in a sterile medium. Technicians clip a small leaf from the plant; then they insert the Bt bacterium into an abrasion, using an agrobacterium vector.[17] The leaf is subsequently rooted in a test tube. Gradually, the Bt bacteria starts to multiply inside the leaf's cellular tissue. From this "mother" plant, technicians take additional cuttings, which they transfer to a greenhouse for tuber production. "If the Bt level is high enough and holds up under tuber multiplication," noted Madkour, "then we can get a resistant plant." The final test is to expose the Bt-enhanced tubers to larvae during storage. As the larvae ingest the tubers, the Bt toxin should be released and the larvae should die off. Each leaf generates a new plant line and must be tested separately; the problem is to find a plant with a high enough Bt level, which means evaluating thousands of plants. "Total immunity against the tuber moth is unlikely," Madkour said, "but the resistance should be very high, sufficient to reduce the damage by 90 to 95 percent." Adding the Bt "gene" to plants so that they produce their own insecticide is not an Egyptian innovation. Agrochemical companies like Monsanto have worked on the technology for years, with notable success. Most of the cotton produced in the United States, for example, is "Bt enriched."

That is bad news for the boll weevil but good news for the environment. There is no pesticide runoff, no groundwater contamination. That seems too good to be true. And in some respects it is. Bt's downside is the all-too-familiar treadmill syndrome. Putting Bt into a plant exposes the target insects that feed on it to the bacteria. That puts pressure on the pest to mutate. Farmers in South Carolina already report Bt resistance in boll weevils and fear a "Bt-resistant super bug will evolve."[18] Perhaps a different Bt strain can be enlisted against the weevil. Still, as Aziz Lagnaoui says, that simply delays the day of reckoning. "Just as insects develop cross resistance to an entire insecticide class," he said, "chances are, they will develop cross resistance to different Bt strains."

GV is not immune either. Lagnaoui has already generated resistant tuber moths under laboratory conditions. "It's easy," he said. "Feed a tuber-moth population on GV-infected potatoes. Most of the larvae will die. The few that survive, however, can still lay hundreds of eggs. When this population hatches, a somewhat larger number will survive the GV treatment. With each cycle, the percentage of resistant tuber moths goes up until they reach a majority." Of course, in nature, the situation is dynamic. New tuber moths enter a potato field, and others leave; the population is not so artificially confined. So it takes much longer for resistance to build up. "But it will," Lagnaoui insisted, "given reliance on a single biological control used repeatedly."[19]

To confuse the pest's immune response, CIP and its partners test an assortment of controls that combine GV and Bt with neem extract, which is derived from the leaves of the tropical neem tree. In Lagnaoui's view, "managing resistance" is as important as managing pests. Even with biological controls, farmers should use them only as needed. That means monitoring pest levels as opposed to spraying on a fixed schedule. And they should alternate between types of controls, which reduces selective pressure from a single source. The objective is coexistence and balance, not the tuber moth's destruction. Egypt takes such selective control seriously. In 1995, the Ministry of Agriculture imported 250,000 pheromone capsules for tuber-moth monitoring by farmers.[20]

Ramzy El-Bedewy was getting tired of Cairo and anxious to get back to Kafr El-Zayat and its farmers. "I wear two hats," he told me, "one for CIP and one for Egypt. I am like a doctor. Farmers call at night, worried about their potatoes. I make house calls, going from one farm to another. Really, I hardly have time for a social life."[21] El-Bedewy was supposed to run CIP's research station; he was not expected to be a potato troubleshooter. On the other hand, farmers want advice. And they value what El-Bedewy has to say. It is hard to say no.

The next morning we left Cairo for the delta. The four-lane thoroughfare between Cairo and the port of Alexandria is Egypt's single most important road. The heaviest traffic is along the ninety-kilometer stretch from Cairo to Tanta, where much of the country's heavy industry has located. It was 6:30 A.M. The paths that straddle the main road had filled up with people walking or bicycling to work. At outdoor stands, folks lined up for a morning snack of fava beans and flat bread. Past Tanta, a tidy, lush countryside opens up. It was July. Corn, cotton, and sunflowers can take the dry heat of an Egyptian summer; by contrast, the onions and garlic had turned brown, ready for harvest. We arrived at the Kafr El-Zayat research station. El-Bedewy pointed out his neighbor, a backyard brick factory, with its straw and sand piled up in front. Egypt builds with brick; there are no forests. The recipe for bricks is to mix water, straw, sand, and cement and bake in a kiln with a smokestack. From the station, I counted the smokestacks I could see. I stopped at one hundred.

"Farmers trust us," El-Bedewy said. "That is why they agreed to try biological controls in the first place." Such alternatives require farmers to change both their practices and their expectations. With both Bt and GV, farmers spray only when they catch at least twenty tuber moths in a pheromone trap over three days, or if they find tuber moths on 5 percent of their potato plants.[22] A biological spray kills off insects gradually, not on contact. Farmers do not like that. They want to see the results right away. The station sponsored on-farm trials so farmers could see that biological control worked, that it was safer, and that it cost less. "Farmers are now convinced," El-Bedewy said "They are the ones pushing for it, asking for it."

In Egypt, potatoes are harvested by the end of April. Those not sold immediately go into storage, where the battle against the tuber moth goes on unabated. "Farmers store potatoes in any protected space they can find," El-Bedewy noted, "including in the house, even in the bedroom. So when they use insecticides, they create a terrible health hazard." Keeping the tuber moth from eating the stored crop matters as much as protecting them in the field. Maybe more so. Farmers sell off their stored potatoes gradually, as prices go up. It is like having money in the bank. To lose the account to the tuber moth is a blow to a family's budget. Getting farmers to try out biological controls for storage was a final, risky test. If our on-farm visits were any indication, it was a test already passed.

Most farmers stash away potatoes in some type of rustic storage shed, or *nawalla*.[23] Makeshift models are fashioned from bamboo poles, cornstalk husks, and tree branches; the more substantial ones use adobe bricks. Sometimes farmers store them in shaded piles covered with a thick layer of straw. Wherever we went, the story was the same: "I never believed this powder would work. But having tried it out myself, I can say the results are better than with insecticides." Our last stop was at what El-Bedewy called a "traditional" *nawalla*, which was built entirely of adobe bricks, including the roof. The adobe does not erode, as it rarely rains in the delta. For ventilation, the *nawalla* had diamond-shaped holes studding the walls in random fashion from top to bottom. In spite of the sun's intensity, it seemed cool enough under the shade of two ancient mango trees. The farmer arrived. A tall, rail-thin man in a tattered robe, Ragib greeted us warmly, clasping each outstretched hand between both of his. He seemed to have an inner serenity that drew people to him. He kissed those he knew, including El-Bedewy. He sent his son to pick apples for us as gifts; then he opened his *nawalla*. It was dark inside, but deliciously cool and well ventilated. The farmer asked El-Bedewy about ways to improve his *nawalla*. Suggestions included wire mesh put over the air holes to keep out rats and an exterior coat of whitewash to deflect the sunlight.

The potatoes were piled in rows, each covered with a half-meter of straw. Like everyone else, I reached down through the straw to pull out a couple tubers. They were surprising cool and without any sign of tuber moth. Each row used a different kind of control, including GV; PROTECTO, which is the Bt made in Egypt; and DIPEL 2X, a Bt formulation made in the United States. Ragib did not bother with a control group, a pile treated with the customary organophosphate-based insecticides. He is so negative about insecticides he refuses to have them anywhere near his potatoes. A pheromone trap hung from the ceiling. One trap suffices for about twenty square meters of storage space. Ragib checks them every three days. If he finds more than twenty adult tuber moths, the potatoes need another treatment.

A technical discussion followed. Finally, Aziz Lagnaoui stopped and turned to the farmer. "I don't care what he says," pointing to Ramzy El-Bedewy. "I want to know what you think." Ragib hugged him and held his hand. "Even if this way saves no money at all," he said, "it is still better, better for eating, better for our health." Back outside, we ate our apples.

"Farmers know a good result when they see it," El-Bedewy remarked. "It does not have to appear in a scientific journal first."[24] He explained that delta farmers had accepted biological control almost five years ago. It had taken that long for production and official policy to catch up with farmer demand. Next year, finally, supply would not be a problem. From what I could tell, the GV story was certainly a good one, so I asked El-Bedewy what factors he felt made for the success. "The Kafr El-Zayat station is in the middle of a dense farming district," he said. "We have demonstration plots and field days; farmers can see what we are doing and why. We train people in pest management. What matters most though is the right attitude. In the end, work on the tuber moth is not for CIP, or for science; it is for Egyptian farmers."

CHAPTER 20

Freezing a French Fry

Most of the world still boils its potatoes. But the frozen french fry, the flagship of fast food, waits in the wings, even in Egypt. Lest I miss the potato in its modern garb, Ramzy El-Bedewy and Aziz Lagnaoui arranged a stop at the Farm Fry corporation, a state-of-the-art frozen french fry facility.[1]

The factory is forty kilometers outside Cairo, past the international airport at Heliopolis. The airport used to be in the desert; today it is surrounded by busy streets, shops, and apartment buildings. Past Cairo's suburbs, a new world of satellite cities, recreation areas, factories, and big farms opens up. As everywhere in Egypt, such reclamation depends on water pumped from the Nile.

The Egyptian government wants new projects to locate outside the crowded capital—which is to say, in the desert. Such settlement is heavily subsidized and bunched up along the four-lane highway that links Cairo to Ismailia and the Suez Canal. The government put in the roads, the power lines, and the water pipes, not to mention the mass-transit rail links to Cairo and the subway. Factories in such reclaimed areas pay no taxes during the first decade of operation. Farm Fry is in an industrial zone; its neighbors include oil refineries, specialty steel mills, furniture factories, and ceramics factories.

Farm Fry was scrupulously clean, spacious, and uncluttered. The business is a joint Dutch-Egyptian enterprise. We stopped at the manager's office for coffee; the manager deferred questions until after our inspection. Like all the employees, we had to change into the Farm Fry uniform, which consisted of white lab coats, hair nets, and caps. We began in the potato reception area.

The factory processes some 32,000 metric tons of potatoes a year, or about 150 metric tons a day.[2] It operates twenty-four hours a day, six days a week. It closes only on Friday, the Islamic holy day. Potatoes arrive by the truckload, usually in burlap bags; most consignments, about 90 percent, come from large, company-managed potato farms located in reclaimed areas. Company production goes exclusively to the factory; hence, such production does not compete with the delta's small farmers, whose fresh potatoes go almost exclusively to urban markets. The favored varieties are Diamond, King Edward, and a red-skinned cultivar, Cardinal. Before a truck is unloaded, samples are checked for sugar content and dry-matter content. For good fries, Farm Fry wants its potatoes to be low in sugar but high in starchy dry matter. Shipments are also examined randomly for tuber size. All the tubers from a sample bag are laid out end to end in rows on a meter-long measuring table. Eight to ten potatoes per meter is the rule of thumb for a good french fry. If too many rows fall short, the shipment is deemed below french-fry quality and demoted to potato-chip status.

Once a load is approved, the tubers are emptied onto a conveyor belt. At the end of the belt is a vibrating, chain-link grate used for sorting. The smallest tubers fall through the holes and get consigned to chip making. The tubers that pass muster are conveyed to the washer and from there to the pressure cooker for blanching: a rapid steam-heat treatment at 180 degrees centigrade (356 degrees Fahrenheit). Under pressure at high temperature, the potatoes' skins swell, but the flesh is left firm. When the tubers get to the rotating rasper, the skins can be scraped off easily. Thus peeled, the potatoes are washed again and then inspected by a gauntlet of workers looking for signs of insect damage or diseased tubers. According to our Farm Fry technician, tuber-moth losses at this stage were twice as high as usual.

The tubers now face a common destiny, the french-fry cutter. Regardless of size, they all get sliced to the same thickness. The fries stay on the conveyer belt, and the waste drops below. Such waste has potential as animal feed or organic manure. The Farm Fry practice now is to discard it.

The freshly cut potatoes are next precooked and prefried. First comes a steam-heat blanching at 90 degrees centigrade (194 degrees Fahrenheit) for four minutes, followed by twelve minutes at 60 degrees centigrade (140 degrees Fahrenheit); then comes frying in palm oil for a minute at 180 degrees centigrade (356 degrees Fahrenheit). Prior to freezing, the hot fries must be cooled down in special cold units by steps: from 180 degrees centigrade to 40 degrees (104 degrees Fahrenheit), then down to about 5 (50 degrees Fahrenheit), and finally to -1 degree centigrade (30 degrees Fahrenheit). Then comes deep-freezing at -13 degrees centigrade (8 degrees Fahrenheit) in a unit with an air temperature of -32 degrees centigrade (-26 degrees Fahrenheit). Once frozen, the fries pass onto a vibrating grate that separates out short or broken fries from the rest. This "waste" product is packaged separately and sold at a reduced price.

In the air-conditioned packaging department, machines separate the frozen fries into uniform piles of approximately 2.5 kilos. Each pile is sealed in a plastic bag and boxed in cardboard cartons just big enough for four bags. The cardboard boxes then pass under a metal detector. A labeler imprints the date the fries were packaged, the product's expiration date, and the total number of grams per package. The 2.5-kilo bags go to restaurants and fast-food chains. Farm Fry also packs by the kilo and half kilo, amounts geared to households. The french fry is not the only potato product the factory makes; it also makes potato wedges, a thick frozen chip for frying, and a thin one for potato chips.

The final step—tasting the finished product. It was now past noon; we were hungry enough to taste our way through several kilos of fries and chips.

We returned to the manager's office for more details.

About 60 percent of Farm Fry's production is exported abroad; 40 percent is for the local market, which includes McDonald's. A kilo of frozen fries costs about 3.5 Egyptian pounds, or one U.S. dollar, a reasonable price for the country's middle class. Farm Fry has about 35 percent of the frozen-potato market in Egypt.

The waste from sorting, peeling, and cutting typically runs at about 20 percent; this year it is twice that. The problem is the tuber moth. In the past, Farm Fry used as much insecticide as it wanted: three foliar sprays during the growing season and one directly on harvested potatoes. But this year, regulators from Egypt's Ministry of Health cracked down. Consequently, the percentage of Farm Fry potatoes with tuber-moth damage went up dramatically.

As tubers are cut and washed, the water, which is used repeatedly, ends up with a high starch content. The starchy water could be recycled into agricultural irrigation, or the starch could be separated out and dried. Farm Fry, however, wants to check out the local starch market before committing itself to production. Currently, all the water used in the industrial zone goes to a common treatment plant before being discarded.

Before leaving, Ramzy El-Bedewy wanted a look at the company's cold-storage unit. To make sure it has potatoes all year, Farm Fry stockpiles for the off-season. In storage, tubers are kept at 8.5 degrees centigrade (47 degrees Fahrenheit). If the temperature gets much cooler, the sugar content in the tubers starts to go up, and processing quality plummets. The facility had eight storage bins with a total capacity of twelve thousand metric tons. Computerized sensors track the temperature in the bins. When a bin is ready to be processed, the air-conditioning goes off, and fans suck in warmer air from the outside. The ideal temperature of a potato for processing is 18 degrees centigrade (64 degrees Fahrenheit).[3]

I asked El-Bedewy why Farm Fry did not fight the tuber moth with pest management and biological controls. In fact, he told me, the Egyptian operation had proposed such a switch, but the home office in the Netherlands had overruled them. Then came the insecticide crackdown. The company called El-Bedewy for help, but it was too late. Anyway, no one had enough Bt to take care of Farm Fry's one thousand hectares. "Why did they call you?" I asked El-Bedewy. He laughed. "Almost anyone in Egypt who works on potatoes has been to Kafr El-Zayat," he said, "including most of the company's Egyptian technicians."[4]

Fighting tuber moths and freezing french fries both involve potatoes, but technically the two are worlds apart. With the tuber moth, the knowledge needed is site specific. Agronomists had to track the pest's seasonal incidence and identify its natural enemies; they had to understand the production technology in use, the timing of harvests, and storage needs. Such information shapes the pest-management strategy recommended to farmers. When elements in the equation change, such as potato-planting dates, climatic conditions, or the pest's population dynamics, then the solution will also change. Controls suited to Tunisia's Cap Bon Peninsula, for example, are less effective in Egypt's warmer Nile delta. When Tunisia's farmers plant early, harvest on time, and keep the soil moist, they escape the worst of the tuber moth; they can often eliminate spraying altogether. For Egypt's farmers, however, there is no escape. Even with irrigation, hilling, and a timely harvest, sooner or later they must turn to insecticides or to biologically safe alternatives like GV or Bt spray. In each case, the dynamics of the tuber moth are similar but not identical.

Compared to pest management, it's much easier to freeze a french fry. All that is needed is the right equipment. To deep-freeze in Egypt, Farm Fry did not design new technology; its engineers simply imported duplicates of what the company used back home. By contrast, to fight Egypt's tuber moth, Farm Fry is better off with advice from Ramzy El-Bedewy than from the home office in Amsterdam.

PART VII
Potato Seed Systems

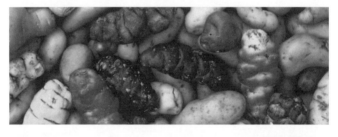

CHAPTER 21

Mother Plants

"My grandparents had plenty of fresh land," said Ugandan potato farmer Steven Tindimubona. "They could rest the soil for years."[1] No longer. Tindimubona lives in Kabale District, in the temperate, East African highlands. Rapid population growth has led to small fragmented plots, continuous cropping, and depleted soils. The highlands stretch from Rwanda and Burundi north through Uganda and west into the Congo. The cool nights and dependable rainfall are just right for the potato. Terraced plots cover every patch of land on the mountainsides.

In Kampala, I stayed at the Speke Hotel, a grand old structure badly damaged during Uganda's decades of civil strife. Now that the hotel is back in the hands of its former Indian proprietors, the hardwood floors, balconies, and graceful staircases are being cautiously restored. Rooms are simple, but spacious, clean, and quiet. An overhead fan made sleep possible, at least with a mosquito net. Years ago, working with Crossroads Africa, I had learned to pack a mosquito net, a lesson I had not forgotten. Uganda was once home to a prosperous class of Indian merchants and shopkeepers. They came with the British during the heyday of the Empire. During the 1970s, Idi Amin expelled them by the thousands and confiscated their property. Uganda has since had a change of heart; the Indian community is returning slowly, rebuilding its temples, and reclaiming some of its property.[2] Kampala is set in low-lying hills an hour's distance from Lake Victoria and the Entebbe airport. The city still showed the scars of past strife. Almost every street still had a bombed-out building encircled by a sheet-metal fence. Still, a sense of optimism had taken hold. Ordinary people wanted to put the past behind, rebuild their country, and rebuild their lives.

I joined Haile Kidane-Mariam in Kampala. An Ethiopian, Kidane-Mariam studied agronomy in the United States at the University of Wisconsin. He had worked with CIP out of its Nairobi office in neighboring Kenya for over twelve years. A breeder by training, Kidane-Mariam coordinated an ambitious seed-production program, which involved Uganda, Kenya, and Ethiopia. We left Kampala for Kabale at dawn. On the city's outskirts, the markets were already awake and busy. The banana is a favorite here, eaten in a steamed, mashed dish called *matooke.* Shoppers stocked up, selecting huge bunches from the truckloads heaped up along the road. There were bags of rice from Uganda's wetlands and sorghum, a grain crop domesticated by Africa's ancient farmers.[3] Then came piles of fresh yams, taro root, sweet potatoes, manioc, and of course potatoes. The spuds were separated by color into stacks of white, pink, and red.

From Kampala the road west skirts Lake Victoria and its marshes. The Nile rises here, draining Lake Victoria, swampy Lake Kyoga, and Lake Albert to the north. As far as Masaka, some one hundred kilometers southwest of Kampala, the rolling hills are lush with banana plantations. Then the road veers sharply west, gaining altitude, going from 1,000 meters (3,280 feet) at the lake to almost 2,000 meters (7,500 feet) at Kabale City. The dry season had set in around Kampala, making the place dusty and giving the flowers a spent look. But the closer we got to Kabale, the more rugged the mountains and the greener the countryside. Clouds gathered; a steady, misty rain began. Haile Kidane-Mariam had warned me it would be cool in Kabale. But after the heat of Kampala, it was surprising that going four hundred kilometers west could make so much difference. I put on my coat. It rains most months in Kabale; there is no dry season comparable to that on the savannas, and temperatures vary little. Farmhouses are built with the rain and cool nights in mind, although it never frosts. Rectangular in shape and constructed from adobe brick, they have heavy wooden doors and corrugated tin roofs. Kabale reminded me of potato-producing Carchi Province, in Ecuador, which I had visited a few months before. Like Ecuador's Andes, Uganda's mountains provide a perfect niche for the potato.

Kabale's temperate to subtropical climate allows farmers to keep an unusual assortment of crops under cultivation. Legumes range from cool-weather peas and green beans to warm-weather groundnuts and yard-long beans. Per-capita legume consumption in neighboring Rwanda is put at fifty kilos, the highest in the world.[4] Vegetables include tropical squashes like susuti, plus cabbages, cauliflower, peppers, greens, and tomatoes.[5] The most popular cereal crops are corn and sorghum. Of course, there are roots and tubers: sweet potatoes and taro root, false banana, yams, and potatoes. Flowers that show up weeks or months apart in my garden back home bloomed in tandem in Kabale: the irises and dahlias, the roses and daylilies, not to mention the daisies, fuchsias, and impatiens. With the Rift Valley to the east, mountains to the west, and Lake Victoria to the south, Uganda seemed a good candidate for the legendary Garden that was humankind's first home.[6]

By late afternoon we reached Kabale City, the district capital, and checked into our hotel. We had tea in the garden, avoiding the tables set up under the jack tree; its fruits are as big as watermelons and almost as heavy. After dinner, we gathered around the huge fireplace in the hotel's lobby, talking potatoes with Kidane-Mariam's friends and counterparts.

I met Mohammed Yusuf, program manager for Africare, a U.S.–based NGO. Africare is a partner in the African Highland Initiative, a consortium that combines the work of national programs, international research centers like CIP, and NGOs. According to Yusuf, the priorities in Kabale include soil conservation, erosion control, and on-farm training.[7] A civil engineer trained in the United States, Yusuf described himself as an ethnic Somali originally from Ethiopia. He is now a U.S. citizen. A tall, distinguished young man in his thirties, Yusuf explained that the Kabale District has no new areas for farming left. Plots are small; few families have enough land to leave fallow, so it is in use continually. "A first step," he said, "is to encourage crop rotations that build up the soil's fertility."

The potato is Kabale's most important cash crop. According to Kidane-Mariam, the district's farmers plant some 25,000 hectares annually; in good years, they harvest between 150,000 and 200,000 metric tons, a third of Uganda's total.[8] The reason for the "more or less" figures is that farmers in Uganda think in terms of parcels and the number of bags from each, and not in terms of hectares or metric tons. The main complaints farmers have about the potato are its susceptibility to bacterial wilt and late blight, plus the degeneration of seed stocks. The problems are intertwined. Berga Lemaga, a CIP pathologist stationed in Kabale, is a bacterial-wilt expert. A short, quiet young man in his mid-thirties, and an Ethiopian, Lemaga explained why bacterial wilt presents such a challenge to Kabale's farmers.[9]

Caused by the bacterium *Pseudomonas solanacearum,* which is carried both in the soil and by infected tubers, bacterial wilt is a destructive and widespread potato disease. Race 3, the highland type, prefers a cool climate. The most robust disease-management tool is typically genetic resistance. But in the case of bacterial wilt, progress has been slow. Resistance involves multiple genes, and these genes react differently in different environments. As to chemical controls, they are expensive, and their impact is temporary. Even after the disease is eradicated aboveground, the bacterium can survive underground for long periods. Consequently, to fight bacterial wilt successfully requires an integrated approach; no single measure is sufficient by itself.[10]

Working with farmers, Lemaga and his counterparts from Uganda's National Agrarian Research Organization (NARO) designed a set of affordable, user-friendly recommendations. Race 3, notes Lemaga, rarely gets into a healthy root; it infects roots damaged during weeding, or it infests unhealthy plants. "To get a healthy plant, farmers need wilt-tolerant varieties and clean seed," noted Lemaga. "We tell them to sow whole tubers rather than cut-up pieces," he said, "and to hill up around the tubers." Once the potatoes are in, the rule is minimal cultivation, as hoeing can damage the roots and thus promote bacterial infection. Farmers are told to uproot and destroy any wilted plants. Lemaga's team visited potato plots and talked with farmers every fifteen days. On forty-six farms where the new technology was applied, the yield increase over four seasons averaged almost 50 percent. "The onset of bacterial wilt was delayed," Lemaga said. "Wilt incidence was reduced, and the number of disease-free plants was much higher."[11]

The challenge now is to get the disease out of the soil entirely.[12] To accomplish that, Lemaga's team is evaluating rotation sequences that reduce the amount of bacterium buildup; strategies to improve the soil's organic content are also essential. The best cropping patterns so far combine corn, wheat, and beans. "We started with fields that had a 90- to 100-percent bacterial-wilt infection rate," said Lemaga. "Just one corn crop reduced the rate to 50 percent; successive bean crops brought it down to just 40 percent." After such treatments, potato yields on test plots at the Kalengyere station went up dramatically, from 3.2 to 11 metric tons per hectare. With low-fertility soils, additional measures are required. A leguminous green-manure crop, for example, adds nitrogen and organic matter to the soil. The impact such efforts have will be short-lived, however, if farmers reinfect the soil. That happens when they plant seed potatoes that carry the bacterium.[13]

In cooperation with NARO, NGOs like Africare, and local seed growers, CIP is determined to break the cycle of disease-ridden seed and low yields. "Breeding better varieties has little impact when seed stocks are insufficient and degenerate rapidly," Haile Kidane-Mariam noted. The strategy is to produce clean seed in quantity for three or four popular varieties, in what Kidane-Mariam describes as a "farmer-based, flush-out system."[14] That means a regular influx of virus-free seed stocks each year, multiplication by small-scale seed growers, and then direct sale to farmers. Seed stocks are thus liquidated or "flushed out" each year, which reduces disease buildup. The first multiplication round starts with tissue culture at the Maguga Quarantine Station in Kenya, outside Nairobi. From there, the elite stock produced goes to Kabale District's Kalengyere Research Station for multiplication by stem cuttings from mother plants. Finally, local seed growers multiply out the disease-free seed and sell the stocks directly to farmers.

CIP initiated the farmer-based seed project in 1996. Target countries included Uganda, Kenya, and Ethiopia, all of them participants in the African Highland Initiative. Together, they produce 1.3 million metric tons of potatoes a year on approximately 215,000 hectares.[15] The first step is to decide which varieties to multiply. "We could not multiply all the favorites in three countries," explained Kidane-Mariam. "So we selected those that had the best disease resistance." For Uganda, that included Victoria, Kabale, and Cruza, improved CIP varieties already released and familiar to farmers. Kabale and Cruza are resistant to type A-1 late blight; Victoria combines moderate A-1 resistance with some tolerance for bacterial wilt. So by selecting carefully, seed multiplication advances the cause of disease control.

My orientation complete, it was time for a good night's sleep—this time with plenty of blankets and no mosquito net. Kabale is too high up for the protozoan parasite *Plasmodium falciparum* and its host, the *Anopheles gambiae* mosquito. For once, I could rest without fear of coming down with the shakes. Most Africans cannot. At lower altitudes, malaria is endemic and debilitating. The malady "damages the brain and other vital organs; it causes respiratory distress, kidney failure, and bleeding disorders." Almost everyone living in the lowlands gets infected repeatedly. Globally, twice as many people die from malaria as from AIDS. In sub-Saharan Africa, half of all childhood deaths are malaria related. Unfortunately, "in malaria research, money is not a problem because, basically, there is no money." The entire infectious-disease budget for the World Health Organization is about equal to that of a major teaching hospital in the United States.[16]

In the morning, following a breakfast that included fried potatoes, I met Steven Tindimubona, head of Kabale District's recently founded Seed Growers Association.[17] A bony, articulate man in his forties, Tindimubona explained that members purchased their disease-free planting stock from the Kalengyere Research Station, multiplied it once, and then sold it to farmers. A typical member produces about fifty bags weighing one hundred kilos, or about five metric tons. Seed growers plant an entire tuber, pull out or "rogue" any infected plants, and rotate crops. "Unless the soil is clean," he said, "we will not plant in the same parcel twice in a row." Growers usually put the seed in diffused-light storage until it breaks dormancy and sprouts. In Kabale, the two main potato seasons are March to May and September to December. According to Tindimubona, there is a lot of demand for clean seed. When farmers buy seed in local markets, varieties are mixed together indiscriminately; there is lots of disease, and much of the seed has yet to sprout. "We could multiply twice as much seed as we currently do," he told me. For that to happen, however, the Kalengyere station will have to increase its output, which is currently at about thirty metric tons a year, or about three hundred bags.

Tindimubona wanted me to meet some local seed growers, so we set off in a four-wheel-drive jeep. It had rained all night, however, and the rain had continued into the morning. The dirt roads had turned to mud and were now impassible. We could not get to a farm located any distance from a paved road. So we stuck to the tarmac and visited Mrs. Rubereti.

Dressed in a colorful, full-length skirt, with a silk scarf tied around her head in traditional style, she greeted us warmly and invited us inside. A self-confident woman of perhaps thirty, she was in charge of her household's potato production.

Her husband, who treated her respectfully, was in his fifties. They had a solid house surrounded by flowers. On the wall inside was a campaign poster for Uganda's president, Y. Kaguta Museveni, with his party's slogan across the top: "Peace, Unity, Democracy and (in big letters) MODERNIZATION." The living room had a fireplace, a small coffee table, a broken television, and stuffed chairs. The walls were painted in aqua, rose, and cream. The floor had linoleum.

Mrs. Rubereti spoke from experience and with authority.[18] She said she bought fifteen bags of seed last season from the the Kalengyere station, paying 30,000 shillings, or about thirty dollars, for an eighty-kilo bag. The previous year, she had attended training sessions with ten other growers. "I put clean seed on clean soil," she said, "so I have little trouble with bacterial wilt." She rogues her fields, uprooting both diseased plants and any volunteers that resprout after the harvest. To protect against late blight, she sprays a fungicide twice a week for a month. After the harvest, she rotates to crops like beans, sweet potatoes, garlic, and cabbage, but never to tomatoes or eggplants, "which are from the same family as the potato." Mrs. Rubereti said she does not go back to potatoes for at least four seasons (two years) and that she reserves her best land for seed multiplication. Last year, she planted fifteen bags of seed and harvested sixty-five bags, which she sold presprouted to her neighbors, to farmers from outside Kabale, and to NGOs with potato projects in other districts. She charged 40,000 shillings, or about forty dollars, for a one hundred–kilo bag. It takes fifteen bags of seed to plant an acre; if she could get enough seed, Mrs. Rubereti said, she could plant three times as much.

Although she currently had no seed in stock, Mrs. Rubereti let us take a look at her diffused-light silo. We left through the front door of the house and went around to a protected courtyard in back. She said she had room for forty-five bags weighing one hundred kilos each. Square in shape, the storage area had five shelves about a meter wide lining the walls on three sides; a fourth set of shelves was constructed from floor to ceiling in the center. Each shelf was made from woven bamboo slats and covered with straw; the holes were large enough to let the air circulate between shelves. For air, the roof had an open overhang. Two walls were constructed of bamboo poles spaced so that indirect, or diffused light entered the warehouse.

The seed Mrs. Rubereti multiplied came from the Kalengyere Research Station. A farmer field day was scheduled there the next morning, which was one of the reasons Haile Kidane-Mariam had come all the way from Nairobi. We planned to leave early the next morning for Kalengyere. Since it was dusk when we got back to the hotel, we did not sit around after dinner. It takes at least two hours to reach the station from Kabale City.

Located at an elevation of 2,450 meters (over 8,000 feet), Kalengyere has a cool climate and is well suited to both potato multiplication and storage. Soils are still free of bacterial wilt. We went up to the station with Mr. Sentaro, a local seed grower.

He was partial to Victoria, an early-maturing variety that could be harvested in seventy to ninety days.[19] I liked it too, mainly for its delicate, rose-colored skin. The dirt road to the station followed along the edge of a swamp and then started up the mountainside. Women were out in the fields, clearing the land, weeding, and harvesting. "Producing food is linked to household chores, so it is women's work," Mr. Sentaro explained, "but if it is a cash crop like potatoes, you can be sure the men are involved." As we gained altitude, we could look behind us down into the valleys and terraces below. The neatly divided plots wound their way up the valley, reminiscent of an Asian landscape painting. We arrived at Kalengyere. Emblazoned on a colorful banner over the station's main gate was the day's motto: "The Potato for Food Security and Poverty Elimination in Uganda."

It was late morning; by now, scores of farmers had converged on the station. The objective for the day was to show farmers how the station multiplied seed stock from mother plants. Francis Alacho, the station's seed technologist, urged farmers to exchange ideas. "Don't let the technicians do all the talking," he said. "They don't know everything." We divided into three groups. Some started at the station's diffused-light storage facilities or went to look at the field trials. I went with the first and largest group for a lesson in potato arithmetic.[20]

A handsome young man with an engaging manner, Alacho began with an accounting problem. "Kalengyere pays the Maguga Station in Kenya 260,000 shillings (about $260) for an eighty-kilo bag of elite seed stock.[21] We sell the same quantity to seed growers for just 30,000 shillings (about $30) a bag. How do we manage?" Alacho explained that the secret was in multiplication. The Kalengyere station generates mother plants from the seed it gets from Maguga. From each of these mother plants come multiple cuttings. (See fig. 10.) He took us over to a long line of such plants for a demonstration. Deftly clipping off the top piece of a stem, he snipped off all but a couple leaves. Next, he dipped the stem in a rooting hormone and gently edged it into a sandy-soil medium. Alacho said that once a mother plant's foliage has branched out, some twenty-five cuttings can be taken at a time. The plant is then fertilized and quickly recuperates. A typical mother plant can be cut back four times in two months. The material has to be kept clean; technicians check cuttings periodically to make sure they are virus free.

Technicians take the cuttings to the station's rooting shed, planting them at a density of about a hundred per square meter. Cuttings are watered daily for two weeks. If left so closely packed together in boxes, the miniature plants will produce marble-sized minitubers. Usually, once the cuttings have rooted, they are transplanted and spaced so that larger, egg-sized tubers result.

Consider the calculations. Begin with 100 elite tubers from Maguga. These tubers go to the Kalengyere station, where they are planted a meter apart, thereby generating 100 virus-free mother plants. From these plants, technicians take successive

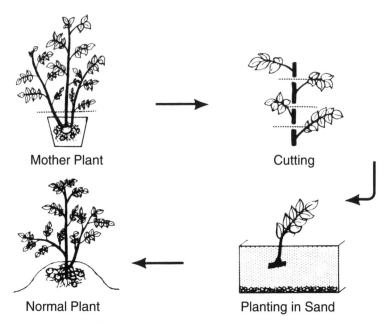

Mother Plant

Cutting

Normal Plant

Planting in Sand

Fig. 10. Rapid multiplication from a mother plant. Courtesy CIP, from CIP, "Stem Cuttings," slide 24, p. 10.

cuttings—as explained above—which they insert in nursery beds at a density of 100 per square meter. On average, over the two-month lifetime of a mother plant, 100 cuttings can be taken from each. In the case at hand, this generates 10,000 cuttings (100 mother plants x 100 cuttings per plant). Once the cuttings root, they are separated and transplanted at a density of 20 x 50 centimeters. So closely spaced, each plant yields on average 5 small, egg-sized seedling tubers, weighing perhaps sixty grams apiece. Total production thus adds up to 50,000 virus-free tubers (10,000 transplanted cuttings x 5 seedling tubers per plant) for a multiplication rate of 500 to 1 (50,000 seedling tubers from 100 mother plants).

The Kalengyere station does not convert all the elite tubers from Maguga to mother plants. Most go directly to the field for traditional multiplication, which yields about 10 seedling tubers for each one planted. What matters here is the flexibility the station has to control costs. As needed, it can shift to mother plants and rapid multiplication. "Last year," Alacho said, "the station sold three hundred bags of basic seed; we could have sold twice as much." That is no exaggeration. NGOs like Africare got some of the seed, but most of it was bought by members of Uganda's Seed Producers Association, such as Mrs. Rubereti.

It had started to rain again. Kidane-Mariam wanted to get back to Kabale before the road turned into a muddy morass. So we inspected the warehouse and skipped lunch. Of the varieties multiplied at the station, my favorites were Kabale and Cruza-148. Kabale is a white-skinned andigena with distinctive purple-pink eyes. Cruza-148 originally came from Mexico. It has a purple ring inside its otherwise creamy white flesh; short dormancy and late-blight resistance make Cruza-148 a favorite with farmers.[22] As we looked at the varieties, Kidane-Mariam complained that some of the seed was too big.[23] "Farmers do not like getting big heavy tubers in their seed bags," he said. "Seed is sold by weight; since farmers plant their tubers whole, a lot of big ones mean fewer plants overall." A good seedling tuber is egg sized, with multiple eyes for sprouting. The advantage of planting a whole tuber is lots of eyes and sprouts. Cutting them up means fewer eyes per piece and a tendency to apical dominance—that is, to a single strong shoot, which lowers yields. The threshold on small is thus quite small. What matters is not size but the number of shoots generated. A tuber the size of a strawberry does as well as one the size of a grapefruit.

We started back for Kabale. Producing seed of multiple varieties for three different countries was a daunting task. Was CIP's celebrated flush-out system worth the investment? Apparently so.

CIP and its national research partners tracked ten pilot seed growers from each target country during the project's second phase, the spring season of 1997. Growers multiplied virus-free stock purchased from their respective national potato programs. The results showed that the project's impact was mounting fast. In Uganda, growers in Kabale District planted 128 bags of virus-free seed. Of the 540 bags harvested, they sold 414 bags to ninety-two farmers in sixty-three different locations; Ethiopia's pilot growers in Shashemene District planted 100 bags of disease-free seed and harvested 852 bags, they sold most of this to sixty-three farmers; in Kenya, the seed growers from Burguret planted 36 bags, harvested 271 bags, and sold this to eighty-two farmers.[24]

What kind of yield advantage did farmers get who invested in better seed? In Uganda's Kabale District, multiplication rates increased from three to six; in Ethiopia, yields per hectare went up from eight to twenty-five metric tons; in Kenya, farmers said they had a multiplication rate of seven.[25] As a result, farmers made much more on the sale of their crop. This can create what Haile Kidane-Mariam calls a "sustainable cycle," in which more farmers demand quality seed and specialized seed growers expand production.

CHAPTER 22

Tissue Culture

C IP considers itself a research institution, not a development agency. The party line is that seed production is best left to national programs; CIP can explain the steps involved, but it is not responsible for implementation. In East Africa, however, CIP cannot get off the hook so easily. When CIP's work in the region was evaluated in 1995, its failure to work on seed production was roundly criticized.[1] After all, better varieties are mere technicalities unless enough seed gets to farmers. CIP's response was the flush-out system described in the previous chapter. But we have left out an important step. To keep the process going, CIP must generate enough elite seed to meet the demand for mother plants—not just at the Kalengyere station but at multiplication sites in Kenya and Ethiopia as well.

Production starts at Kenya's Agricultural Research Institute (KARI), which has opened its Plant Quarantine Station in Maguga to CIP.[2] This includes virus-testing facilities, a tissue-culture laboratory, and screened-in greenhouses. Joint sponsorship allows the virus-free seed multiplied at the station to carry a Kenyan phytosanitary certificate, which makes it valid for distribution throughout East Africa.

One small station multiplies enough elite seed each year to supply three countries with mother plants. How is this possible? By way of example, consider Victoria, an improved CIP variety popular with Uganda's farmers. The first step is to send a sample back to CIP's tissue lab in Lima. There, CIP scientists generate a virus-free sprout, which they ship back in vitro to the Maguga Quarantine Station. From this sprout, KARI scientists can take about 5 nodal cuttings, which they place in a medium in test tubes. A month later, the nodes have sprouted and new cuttings are taken. (See fig. 11.) In this fashion, multiplication is continuous, as each node generates a biologically new plant.

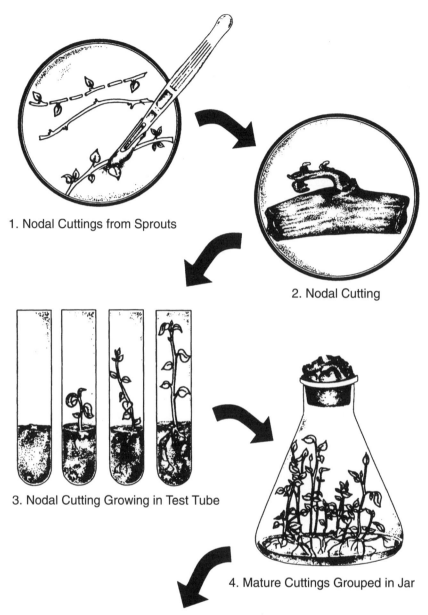

1. Nodal Cuttings from Sprouts

2. Nodal Cutting

3. Nodal Cutting Growing in Test Tube

4. Mature Cuttings Grouped in Jar

5. Cuttings Transfered to Potting Soil

Fig. 11. Rapid multiplication from tissue culture. Drawing by Jeremy Teaford, based on CIP, *Tissue Culture Propagation of Potato,* slide 34, p. 12.

The arithmetic is straightforward. At the end of the first month's cycle (1 x 5), technicians have 5 virus-free nodal cuttings from the original in-vitro Victoria plantlet. At the end of the second cycle (5 x 5), there are 25 in-vitro cuttings: those initial 5 now cut into 5 additional pieces. For the third cycle (25 x 5), technicians take 5 cuttings from each of the 25 in-vitro Victoria plants, which generates a total of 125 new plants. Cycle four (125 x 5) yields 625 in-vitro plants, cycle five (625 x 5) 3,125 plants, and if they continued taking nodal cuttings, cycle six (3,125 x 5) would yield 15,625 plantlets. By now, however, technicians have started to let the cuttings mature, grouping them ten to a jar. When they reach a height of approximately ten centimeters, cuttings are transplanted at high density to one of the station's three screened-in greenhouses, each of which holds about 6,000 plants.

The soil in the greenhouse is sterilized to kill off soilborne pathogens such as bacterial wilt and nematodes. At every step, from in-vitro cuttings to greenhouse transplants and the final harvest, material is sampled and tested to make sure it is virus free.

On average, a virus-free plant produces about 5 small seedling tubers. Thus, the total seed stock from a greenhouse comes to about 30,000 tubers (6,000 plants x 5 tubers each). Overall, each greenhouse produces one crop a year, for a total production of about 90,000 virus-free tubers (30,000 tubers x 3 greenhouses). This is the elite stock that goes to national program research stations in Uganda, Ethiopia, and Kenya. The stations use it to generate mother plants, or they multiply the seed directly in the field. Once multiplied, the seed is sold locally for farmer-based seed production.

Considering that the Maguga station multiplies different varieties for each country, managing three complete cycles a year is a great accomplishment.

CHAPTER 23

The Potato Revolution

"**B**otanical potato seed is a wild thing," said CIP plant physiologist Noël Pallais. "Trying to domesticate it is devilishly tricky." That was back in 1988, the first time I visited CIP.[1] Potatoes are typically propagated from other tubers. Just cut them up, or plant them whole; the progeny will be genetically identical to the tubers from which they came. But there is another way. Much like its close relative the tomato, the potato also produces a fruit with seeds. (See fig. 12.) These seeds can be harvested, processed, and planted. Breeders have crossed potato plants like this for years. Pallais was determined to make true seed a farmer-friendly way to grow potatoes. The challenges were many.[2] In 1988, despite a decade of work, true seed was still not competitive commercially. When harvested, a parcel of CIP's botanical seed yielded a mixed batch of tubers that had different shapes, sizes, and colors. Only poor farmers desperate for cheap carbohydrates were likely to find CIP's "true potato seed" (TPS) acceptable. And even if they did like it, who would mass-produce the tiny, delicate seeds?

Nonetheless, the logic behind TPS is compelling. CIP argues that the poor quality of seed tubers is often the main obstacle to higher yields—hence, CIP's sponsorship of rapid-multiplication schemes, from Bolivia's protected beds to Ecuador's tuber sprouts and Uganda's flush-out system. The tuber seed so multiplied, however, has to be stored for varying lengths of time. In Egypt and Tunisia, potatoes put into rustic storage must brave the tuber moth. In India, where potatoes are typically a winter crop, seed must be kept in cold storage over the monsoon season, an expensive proposition.

A moment's reflection shows that the modern potato's Achilles heel is the unwieldy system needed to produce tuber seed in bulk. And farmers need a lot of it, too, typically from two to three metric tons per hectare.[3] In a first-rate system, seed

Fig. 12. A handful of TPS harvested from seed berries can produce a hectare of potatoes. Courtesy CIP Archives, from CIP, *Potatoes for the Developing World*, pp. 11, 92.

is supposed to be inspected and certified as disease free. Stocks should be plentiful, cheap, and available when needed. Despite decades of CIP preaching, seed systems in developing countries fall far short of the norm.

"What makes TPS so attractive," Pallais explained, "is its revolutionary potential." With just fifty grams of these tiny seeds, farmers can plant a hectare of potatoes. That means no more storage silos, no more bulky tubers to lug to the field, and no more disease-ridden, insect-infested tubers to endanger each year's crop. For tuber seed can spread bacterial wilt, late blight, and a host of potato-specific virus infections, not to mention weevils and tuber moth. By contrast, botanical seed like TPS is virtually disease free.[4]

By the time Pallais finished his sermon on TPS, he had a convert. During my many visits to CIP, I had never met anyone who could match Pallais's passion for the potato. His original career plan was intentionally lowkey. A Nicaraguan, Pallais was studying at the University of Notre Dame in the United States in 1978, the year the Sandinista revolt toppled the Samoza regime and politicized the country. "When that happened," Pallais told me, "I decided I wanted a career without an ideology, without a right or a left. That's how I ended up in agriculture; that's how I ended up with the potato."[5] When it comes to TPS, however, Pallais is on the vanguard, in revolt against entrenched, old-fashioned seed production. How did the modest potato end up enmeshed in revolution? It's worth considering.

The potato has two strategies for survival, a conservative one based on tubers and a radical one based on seed. Since ancient times, Andean farmers selected potato plants based on the tubers produced; they paid little attention to the seed. The result was a domesticated plant that was good at setting tubers but that had wild and erratic seed. With crops like lima beans, Andean farmers did just the opposite. They selected plants with the size, taste, and vigor of the botanical seed in mind. After all, it was the seed they ate, and it was the seed that farmers planted. That's true of Andean grain crops too, from corn to quinoa, and it's true of nuts. Indirectly, it is also true of squashes, berries, and most fruits. People may not eat the seed, but they eat the seed's food supply, the pulpy flesh that surrounds it. By saving attractive, large-sized squashes, berries, and fruits, ancient farmers indirectly selected good seed producers. By comparison, the potato is an anomaly. It is the tuber that people eat, and it is the tuber farmers plant. That left the botanical seed to fend for itself.

Of course, Andean farmers knew the potato plant had true seed. In its native habitat, seed balls the size of cherry tomatoes form prolifically on most plants. When ripe, the fruit falls to the ground. The next season, after the rains set in, volunteer seedlings come up like weeds. Andigena potatoes are self-fertile; so most of such seedlings would be "selfed" from the flowers of the same plant. But outcrossing also occurs, and that generates natural hybrids. This takes place spontaneously because Andean farmers plant multiple potato varieties, even within the same patch. In addition, wild species, which are outcrossers, often grow in close proximity to domesticated species. And do not forget the volunteers from last year's crop lurking about in nearby fields. Hence, in the Andes, different varieties—even different potato species—can cross with each other. Any curious farmer who saved the tubers from such offspring would be creating, technically speaking, a new variety. Such fortuitous selection interacted with a multifaceted environment to create the Andean world's enormous potato diversity.[6]

Ancient Andean farmers, however, used botanical seed deliberately, not just accidently. They certainly do so today. Communities wash, macerate, dry, and store true seed, using it to regenerate old varieties and create new ones. Chances are, Andean farmers have used true seed for generations, perhaps since the dawn of agriculture.[7]

In their adopted European environment, potato plants do not usually bear fruits. Even so, this happened often enough to be noticed early on. The selfed seedlings from these first imports had a great range of variation. It was not long before "skilled cultivators purposely planted the seed from naturally formed berries, selecting the plants with the characteristics they sought." In 1707, John Mortimore, writing on the *Whole Art of Husbandry,* observed that the potato could be propagated by seed as well as by tuber. By then, the practice was already well established.[8] Modern plant breeders improve the potato by crossing varieties and saving the seed. To make the cross, the stamens in the tiny flowers of the female parent must be emasculated by hand to prevent self-fertilization. Then pollen has to be carefully collected from the male parent. The next step is to hand pollinate the flowers of each female. The result is two or three berries the size of large cherry tomatoes. Each typically has forty to fifty seeds, whose total weight does not come to half a gram. Such exacting techniques yield enough seed for breeding purposes but not for commercial production.

The final step is to evaluate the progeny. That means planting the seeds and tracking the traits selected for, such as disease resistance and tuber quality. To do this, a germination rate of even 50 percent is quite sufficient. In the above example, if each of three berries had fifty seeds, a total of seventy-five seeds would germinate (150 seeds x .5 rate of germination). For commercial production, as opposed to small-scale breeding projects, a germination rate that low is unacceptable. "To get good seed," Pallais said, "we need early emergence and vigorous germination, at least 90 to 95 percent; we need seed that will get up and grow." To push up the rate, CIP scientists suggested treating TPS with growth hormones. But Pallais would not have it. "When you do that, you are just fooling yourself," he said. "TPS has to act like real seed."[9] In short, CIP breeders needed to find good parental plants. For large-scale TPS production, a good plant sets lots of berries packed full of seeds.

The seed has to germinate rapidly at high rates, and it has to grow vigorously.[10] In 1988, CIP breeders had a pool of good parents. But problems remained. TPS took from six to eighteen months to break dormancy, and it did not do so uniformly.[11] And as previously discussed, the progeny produced a mongrel spud assortment. When I left CIP after a three-week stint, my faith in TPS had wavered; I accepted it in theory but not in practice.

That changed in Egypt, but not before I had strayed from the true-potato path. I veered off to rice and the green revolution.[12] I did not get back to the potato until October, 1995. I started out with a six-week field trip to Indonesia. The story there was that the TPS experiments of the late 1980s had largely fizzled. I went to Egypt the next year to cover the tuber moth. I had mostly forgotten about TPS. On my last day at the Kafr El-Zayat station, CIP scientist Ramzy El-Bedewy showed me samples from the station's last potato harvest.[13] Crates of local varieties were on one side, crates of TPS-generated tubers on the other. I was astonished. I could not tell a TPS-generated tuber from a conventional one. The skin was just as smooth, the tubers just as large, the lots just as uniform as those raised from tuber seed. The yield from TPS exceeded the yield of local varieties by 20 percent. I could not believe it.

"A decade ago," El-Bedewy explained, "the poor quality of TPS plants was a serious drawback. They took too long to mature; the tubers had deep eyes, which made them hard to peel; and their dry-matter content was too high for the Egyptian market." All that has changed.[14] The germination rate and homogeneity of TPS tubers is now very good. At CIP, Noël Pallais found a simple solution to breaking TPS dormancy.[15] Moreover, breeders have taken some of the wildness out of botanical seed. As a result, TPS potatoes are as uniform on the outside as any commercial crop; they have smooth skins and early maturity. TPS seed is plentiful, cheap, and easy to produce. Nowhere has TPS made greater inroads than in India. Its farmers, extension workers, and scientists took a technology confined to CIP enthusiasts and converted it into a practical alternative for farmers. In 1996, India produced 680 kilos of TPS, enough to plant thirty-four thousand hectares of potatoes.[16] That was more TPS than any other country in the world. India was on the forefront of the potato revolution.

CHAPTER 24

True Potato Seed

"I did not know what true potato seed was," said Dr. S. K. Bardhan Roy. That was in 1992. Roy had just joined West Bengal's Directorate of Agriculture, assigned to run Anandapore Farm, a research station a hundred kilometers south of Calcutta. Four years later, Anandapore Farm produced eleven kilos of true potato seed, or TPS. For the 1996–97 season, Roy expected to process over fifty kilos.

India's great rivers flow through West Bengal. The Hugli, which passes through Calcutta, empties into the Bay of Bengal, forming one of the many mouths of the Ganges. Most of the state's farmland, some three million out of five million hectares, is lowland floodplain. During the summer monsoon, the rivers flood slowly and inexorably; the rice paddies fill up with water. Potatoes come in the winter, after rice. Much of the state is a delta. When the rivers flood, they leave rich silt on the land as the waters gradually recede. Not only is the soil fertile, but the rivers flush out salts and leave new nutrients behind.

It was December, and the morning was cool. I met Dr. Roy at 6 A.M. at the Harry Guest House in a middle-class Calcutta suburb. I had arrived in Delhi two weeks before; TPS was the main item on the agenda. Dr. Roy and I headed for the station to catch a local train to Midnapore. Calcutta is run-down, but its narrow city streets and trams still give it a certain charm. In Delhi, despite the broad, tree-lined avenues, the press of crowds and motored vehicles have turned the city into a nightmare. Every driver blows his horn constantly. The lingua franca of Indian thoroughfares is "Horn Please" or "Blow Horn," which is stenciled on the back of almost every vehicle. And everyone complies. The result is an incessant din that gets even worse at night. After long drives on crowded highways, I always had a headache. So I was relieved that we were going through Calcutta before the morning rush hour set in, for only one bridge spans the Hugli River, and the train station is on the other side. Once at the station, we had tea, bought newspapers, and got in line for tickets. Dr. Roy booked first class. This guaranteed us a seat but nothing else. We entered a dilapidated coach with wooden benches and no amenities.

For 1997–99, India's potato production averaged almost 22 million metric tons, more than for all of South America.[1] A generation ago, the potato was a minor crop. In 1961–63, for example, India grew just 2.8 million tons a year.[2] The subsequent increase is due to expanded acreage in Uttar Pradesh, West Bengal, and Bihar, all big states in the Ganges valley. Farmers in these three states grow over 80 percent of the country's potatoes.[3] Developments in the cropping system, in the organization of seed production, and in storage helped make this possible.

The high-yielding rice and wheat types introduced into India during the 1960s matured much faster than traditional ones. For rice, farmers can now get a crop in just 90 to 100 days, compared to what used to take 150 days or more. So farmers have time for another crop. The potato fits in well. Compared to grains, it draws on different soil nutrients, attracts different pests, and is subject to different diseases. Consequently, a grain to potato rotation does not exhaust the soil or risk pest and disease buildup. Moreover, the potato adds a cash crop during what used to be the slack season. In Haryana and Uttar Pradesh, states in the upper Ganges valley, farmers plant the crop in October and harvest from late January into February. In West Bengal, in the lower part of the valley, the weather is too hot for potatoes until November; farmers harvest into March.[4]

In the 1950s, disease-free seed tubers could only be produced during the summer, up in the cool Himalayan hill country. The seed was subsequently recycled to the Gangetic Plain for the winter crop. The impact on seed supply, however, was minimal. Only a few varieties did well under both long-day (summer) and short-day (winter) conditions. Moreover, the amount of land set aside for multiplication was insufficient and the transport problems formidable. So as the demand for seed increased, India's Central Potato Research Institute (CPRI) had to look for other options. Its research showed that aphid populations, which carry potato viruses, are dormant on the Gangetic Plain during the winter. That being the case, it could safely relocate seed production to the plains, precisely where the demand for seed was increasing. Subsequently, the institute greatly increased its seed production capacity. It grew basic seed at selected research stations and then sold it to state governments and private farmers for further multiplication. At the end of the chain came certified seed, which was sold to farmers.[5]

With time for an extra rotation, and with better seed available, potato production mounted. Despite the institute's hard work, however, barely 10 percent of the potato stocks farmers plant is certified seed—although the 10 percent of the 1990s is six to seven times greater than the 10 percent of the 1960s. Most is still farmers' seed, recycled from crop to crop. In monsoonal India, potatoes kept in rustic stores over the long summer rainy season are in sad shape by the fall. To keep seed firm and healthy, farmers turned to cold storage during the 1970s. A cold-store warehouse keeps seed stocks under refrigeration at a constant temperature. These multistoried, cement-block buildings, which number in the thousands, are found all over northern India, wherever potatoes are produced in quantity.[6]

How good is the planting material kept in a refrigerated warehouse? Much better than that kept in rustic storage but still not as good as it could be. Farmers face long delays, as harried truckers line up to make potato deposits and withdrawals. Power outages are frequent, often because warehouse managers cut corners to save on electricity. Consequently, the temperatures inside are unstable, which compromises storage quality. It is an imperfect, expensive system at best. The price of tuber seed, which includes transport, storage, and handling charges, accounts for between 40 and 60 percent of the cost of producing potatoes. Recent surveys in Asia still cite seed problems, including storage, as serious obstacles to production.[7] That being the case, agricultural officials like Roy see TPS as a way of extricating farmers from a cumbersome system.

From the train, which had large, open windows, we saw the undulating flatlands stretched out before us. The irregularly-shaped rice paddies fit together like pieces of a jigsaw puzzle. It was winter, and most of the rice had already been harvested. The main job left was to gather up the straw. Since it was the off-season for rice, villagers had less to do; even so, the fields were full of people. Up at dawn, they drove their cows, water buffalo, and goats to the paddies to feed on leftover rice stubble. Ox carts with huge wheels carried loads of straw, threshed rice, plastic irrigation pumps, and families. It was morning, time to wash in the streams and canals. In the marshes, villagers had started seedbeds for the next rice crop, a hopeful sign of a rainy season yet to come.

The sun climbed above the tree line. In the train, passengers took off their scarves and woolen hats. The night before, the temperature had dropped to six degrees centigrade (forty-three degrees Fahrenheit), very cold by local standards. Compared to Uttar Pradesh, which occasionally gets a winter frost, West Bengal's potato country is more tropical, a place for banana trees, coconuts, guava, and papaya. In the Midnapore district, enclaves of the jungle are still intact. Great shade trees, nature's air conditioners, surround the villages. According to Dr. Roy, marauding elephants still raid the sugarcane fields and banana groves. Fortunately, they do not like potatoes. As we closed in on the town of Midnapore, the district capital, bright-green potato patches appeared, sandwiched between spent rice paddies. We arrived at the station and then headed for the local guest house. We got the best room for ten dollars per night. Birds chirped in the shade trees, and the air was sweet and fresh; it was a welcome change from a week of Delhi's traffic.

After lunch, we went in Dr. Roy's jeep to Anandapore Farm, located about an hour's drive from Midnapore. The countryside had little motor traffic and no honking. Most villagers walked between settlements or to their fields; some had bicycles or an ox cart. The Communist Party has run West Bengal for years. In the 1960s, when they were struggling for power, they encouraged strikes and worker protests; now in charge, discipline is the golden rule. Their policies have done much good, especially in the countryside. To protect the forest, the state government subsidizes both propane gas and cooking oil. To judge from the evidence on the roadside, however, dried cow dung is much more popular and certainly cheaper, subsidies or not. The dung is mixed with a little straw, soil, or sand. Villagers dry the dung on any flat surface, including pressed up against the adobe brick of their homes or the cement-block walls of public buildings. The dung makes a pungent cooking fuel.

Dr. Roy considered West Bengal's land-reform program a reasonable success.[8] It divides farms by category and sets a limit on their size. For village agriculture, for example, farms cannot exceed 6 hectares. "The practice of breaking up a farm," said Dr. Roy, "and registering each separate part in the name of a different family member, a common but fraudulent stratagem in other states, is not permitted in West Bengal." That certainly seemed true in the Midnapore district. Of some 1.1 million holdings, over 900,000 were a hectare or less in size, only 18 were larger than 20 hectares. The total area held by the smallest farms came to 450,000 hectares; the largest 18 farms had less than 1000 hectares in all.[9] Land reform is not the whole story. The state has invested heavily in education and in rural extension. Dr. Roy said literacy rates were high in Midnapore district, even among tribals.

West Bengal devotes over 300,000 hectares a year to potatoes. After Uttar Pradesh, it is the country's most important potato-producing state, and Midnapore is one of the state's most important potato districts. Yields in West Bengal average twenty-three metric tons per hectare, the highest of India's top producers.[10] In Midnapore, most farmers plant potatoes from late October into early November and then harvest in February. Winter weather is cool at night, which is favorable for potatoes, and comfortable during the day. At Anandapore Farm, the average high in late December is between twenty-five and twenty-seven degrees centigrade (seventy-seven to eighty degrees Fahrenheit); the average lows are from five to seven degrees centigrade (forty-one to forty-five degrees Fahrenheit); it never frosts. Between February 15 and March 15, however, the thermometer starts to climb; average highs go from twenty-nine to thirty-six degrees centigrade (eighty-four to ninety-seven degrees Fahrenheit), and the nighttime lows go from seven to fourteen degrees centigrade (forty-five to fifty-seven degrees Fahrenheit).[11] By March, aphids are a problem, and the potatoes must come out. All told, farmers rarely have much more than one hundred days to squeeze in a potato crop. It can be fewer still. If a late monsoon delays the rice crop, the potatoes are delayed in turn. Hence, early-maturing varieties are imperative. Anandapore Farm covers about

80 hectares. Besides the farm's work with TPS, it has cashew trees and a mango orchard. The cashews, of which the station has grafted many types, have wide, waxy green leaves much like magnolia trees. It was marigold season; almost every farm, including the Anandapore research station, had bright plots of their bold, orange-yellow flowers. In India's villages, strings of marigolds are customary offerings at temples and popular decorations at every festival. West Bengal is famous for its devotion to the goddess Kali.

Trained in rice production, Dr. Roy worked for several years as an intern at the International Rice Research Institute (IRRI) in the Philippines. When he returned to West Bengal, he spent four years as project leader for India's Rice Development Program. Without prior potato experience, how did Dr. Roy, a rice specialist, manage to produce eleven kilos of TPS on his first try? Roy explained that he drew on the proven TPS technology that India's potato program (CPRI), CIP, and local farmers had already worked out. That technology included high-quality parental lines, techniques for inexpensive, large-scale seed production, and an interim step between botanical seed and a commercial crop, the minituber.

For a good TPS crop, start with the right parents. The female parent Dr. Roy used, MF-1, was developed by Mahesh Upadhya and Khyal Thakur, breeders at CIP's regional office in New Delhi. They did this in collaboration with Indian potato farmer Satyendra Bhargava and his son Arvind. Disease-free planting stock for both TPS parents came from the institute's research station in Uttar Pradesh. Because MF-1 is self-sterile, its flowers do not require emasculation. For male parents, Roy uses TPS-13, a variety from CIP.[12]

I had met Dr. Thakur back in Delhi. A determined, rail-thin man in his forties with jet-black hair, he had spent fifteen years selecting TPS parents.[13] Thakur explained that for high yields, the best crosses were tuberosum females with andigena males, a combination that also widens the seed's genetic base and confers more disease resistance. Andigena types flower prolifically for a longer period than tuberosum types, which makes for good pollen production. "It took us a long time to get the right parental lines," he said. "We needed crosses that gave us ten to fifteen flowers, that conferred resistance to late blight, and produced vigorous seed with a germination rate of at least 90 percent; we wanted a compact plant with strong stalks that set numerous fruits with lots of seed." In addition, plants needed to set uniform tubers early, tubers that had the same color, shape, and flesh texture. And they wanted a yield potential of at least thirty metric tons.

The Ganges valley in winter has a perfect climate for andigena varieties. With short winter days, cool nights, and moderate daytime highs, it is a virtual duplicate of the Andes in summer. The main problem is the tuberosum females, which need longer days to flower. This was solved with sodium lighting. The artificial light tricks tuberosum plants into thinking the days are longer.

Satyendra Bhargava pioneered the technique. The first potatoes I saw in India grew on his forty hectares of land in Sadabad, Uttar Pradesh, called Mahendra Farm. In fact, MF-1 stands for "Mahendra Farm-1." A seed producer, Bhargava had worked with potatoes for forty-five years. He learned to grow them in the 1950s in the United States. As a young man, he had participated in a farmer-to-farmer exchange in North Dakota sponsored by the Rockefeller Foundation. When he got back, he planted his first crop, just half an acre. Since then, he has specialized almost exclusively in potatoes, primarily as a seed producer. Over the years, the Bhargava family has provided a home for students from the International Farmers Exchange Program; they have hosted scores of Indian researchers and CIP scientists. Now almost eighty, Bhargava had lost none of his zest for life, or the potato.[14] Besides traditional seed tubers, Mahendra Farm specializes in TPS. Bhargava said he loves to experiment. He has developed sugar-free varieties that are excellent for chipping and varieties with A-1 late-blight resistance. His TPS material is very early, setting tubers in seventy-five to eighty days. He grows five sterile female parents for each male parent; the males are planted in a separate plot. The sodium lights come on at dusk and stay on for about five hours. They are used from planting time until the berries are ready to harvest. Each light is sufficient to cover one hundred square meters. Bhargava irrigates his potato crop from a tube well; he found the water source himself, for he is renowned as a dowser, a skill he picked up in North Dakota. Apparently, such ability is innate. The old Dakota farmer who taught Bhargava could not pass the talent on to his own son. Nor has Bhargava managed to teach his son Arvind the skill.

Planting potatoes year after year ought to exhaust the soil. To enrich the organic content of his plots, Bhargava practices "soil building." He plants a green-manure crop each year. Composed of nitrogen-fixing sesbania and millet, this mixture is sown in July, when the monsoon season starts. In forty-five days, both are waist high and ready to be incorporated. I spent the night at Mahendra Farm and soon found out why Bhargava is known for his hospitality. For the afternoon meal, we started with trays of cashew nuts, almonds, raisins, and pistachios. Then came rice pulao, deep-fried potato chips, fresh yogurt, mango relish, squares of cottage cheese, and kheer. Kheer is made by soaking fresh shredded carrots in milk, sugar, and cardamon, with just a pinch of saffron. We had boiled potatoes and cauliflower refried in curry, plus green peas and hot soup. There was no meat, and I did not miss it. For dessert, there was a sweet made from flour and condensed milk, which were rolled into a ball, fried, and served with sugar syrup. Breakfast was hard-boiled eggs, chapatis, and tea; baked sesame seed with pistachio; toasted lentils with graham flower; and jellybee, a deep-fried batter soaked in honey. For fruit, we had guavas, bananas, and oranges, the latter native to north India.

Much of the credit for Dr. Roy's TPS parent, MF-1, and his sodium lighting goes to Mahendra Farm. But the list is far from complete. How did Dr. Roy manage to mass-pollinate hundreds of plants? He follows practical procedures worked out at the Potato Institute's research station in Modipuram. At flowering time, Roy cuts flowers from the male plant, TPS-13, and puts them in a plastic, zip-lock bag overnight to dry. The next day, he crushes the flowers and strains out the pollen, which is used for direct, hand pollination of the female parent, MF-1.[15] Recall that MF-1 is self-incompatible, so emasculation of the anthers in its flowers is unnecessary. Once a plant is pollinated, berries start to form within six to seven days; in forty-five days, they are ready to pick. Dr. Roy told me he got between eighteen and twenty berries on each plant; each berry had at least 450 seeds.[16] Such abundant seed production was no accident. CIP's breeding program had selected for such traits for years. To extract, wash, and dry the seed, Roy uses simple, manual equipment engineered by P. C. Pande at the Potato Institute. When the time comes to break the seed's dormancy, Roy keeps it at a temperature of at least 40 degrees centigrade(104 degrees Fahrenheit) for four months, a foolproof method worked out by Noël Pallais at CIP, Lima.[17]

What does Dr. Roy do with the TPS produced at Anandapore Farm? It does not go to farmers, at least not initially. In West Bengal, as in the rest of India, TPS is rarely used to plant a commercial crop directly; it is still one season away from a more farmer-friendly form, the minituber, which *is* directly produced from TPS. To produce minitubers, TPS is planted densely in small plots. This is not as laborious as it seems. Indian scientists devised a meter-square wooden frame that makes planting the tiny seeds easy.[18] The frame has four vertical strips, each with twenty-five pegs. Inverting the frame and pressing it into the ground makes a pattern of one hundred holes per square meter. A couple of seeds go in each hole. Once the seeds germinate, they are thinned to one plant per hole. These one hundred plants will produce tiny potatoes—TPS minitubers—that vary in size, from that of a pea to that of a large marble. When sown, a minituber produces a plant that is as large and productive as the highest quality seed tubers. For when it comes to high yields, what matters most is a tuber's health and age, not its size. In 1995, Anandapore Farm produced six metric tons of minitubers from one kilo of TPS. A kilo of minitubers costs no more than a kilo of conventional, certified seed, but it goes much further. That gives the minituber a tremendous advantage. A cost-benefit study calculated that with minitubers, farmers doubled their rate of return.[19]

The potato arithmetic involved is as simple as it is elegant. (See fig. 13.) One gram of TPS contains at least one thousand seeds. That suffices to plant at least five of the square-meter sized plots noted above. Each plot yields approximately four kilos of seed, that is, 350 to 400 minitubers, each weighing from eight to ten grams.[20] It is these TPS minitubers that farmers in West Bengal substitute for traditional seed tubers. To plant a hectare of potatoes, farmers use about four hun-

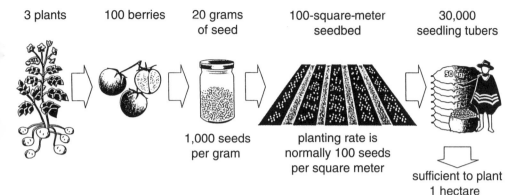

3 plants	100 berries	20 grams of seed	100-square-meter seedbed	30,000 seedling tubers

1,000 seeds per gram

planting rate is normally 100 seeds per square meter

sufficient to plant 1 hectare

Fig. 13. TPS arithmetic for minitubers. This illustration traces the process required to generate enough seedling tubers to plant 1 hectare of potatoes from just 20 grams of true potato seed (TPS). The arithmetic involved is as simple as it is elegant. Potato plants produce berries that contain true seeds. One gram of TPS contains 1,000 tiny seeds. Twenty grams of TPS is sufficient to plant a seedbed 100 meters square. This 100-square-meter seedbed produces 30,000 minitubers that weigh approximately 500 kilograms, enough to plant 1 hectare. That is a far cry from the 2.5 metric tons of conventional seed tubers required per hectare. Courtesy CIP, from CIP, *International Potato Center Annual Report 1996*, p. 26.

dred kilos of TPS minitubers, an amount that twenty grams of TPS can produce. That is a far cry from the 2.5 metric tons of conventional seed tubers needed per hectare.[21]

Local farmers, whose holdings rarely exceed half a hectare, were skeptical; they did not believe a marble-sized tuber could be such a prolific potato producer. To convince them, West Bengal's extension service distributed TPS minitubers to some 150 farmers in five districts of the state. The result: the potato yields from minitubers came to 32 metric tons per hectare, much higher than the 24 tons averaged by Kufri Jyoti, the local favorite.[22]

According to the extension service's post-harvest surveys, over 90 percent of the farmers who tried TPS minitubers wanted to keep planting them. Farmers concluded it cost them much less to plant minitubers and that the yields they got were higher. Their experience confirmed that a crop of TPS minitubers had more resistance to late blight, a trait reinforced by the diverse genetic makeup of TPS minituber progeny. With TPS, each seed produced, and each plant generated from such a seed, is genetically distinct. So when late blight tries to spread from plant to plant, it has to overcome the genetic barriers built into each.[23] Because of this genetic variability, anyone can spot a field of TPS-generated plants, for the height, bushiness, and leaf configuration of each are different. Nonetheless, the size, color, and shape of the tubers developing underneath the plants are uniform. Only farmers who harvest too early complain about tuber size. For the potatoes to reach maturity—and hence, full size and uniformity—farmers cannot skimp on time. Minitubers need from ninety to one hundred days, which is a couple weeks longer than the competition. When left that long, however, the TPS potatoes harvested are comparable in size to the popular Kufri Jyoti.

A TPS minituber is only one multiplication old, which makes it much younger than either certified seed or recycled farmers' seed. Once the minituber-generated crop is mature, farmers sell most of it commercially but hold back enough seed stock to replant. The tubers so retained are now full sized, but they are still younger and more vigorous and have greater disease resistance than their conventional rivals. Thus, farmers who plant minitubers get dependable, first-generation, high-quality seed at a reasonable price.

The tour of the TPS plots over, we headed back for the main office. Villagers had set up chairs for us under shade trees adorned with marigolds and roses. We had roasted greengram snacks. Village women make a dough by crushing the beans and adding flour. The dough is rolled out, cut, and toasted. We had guava juice but no alcohol. In the Indian states of Haryana, Andra Pradesh, and Gujarat, alcohol cannot be bought or sold. When men drink, they become belligerent, aggressive, and antisocial, a danger to their wives and families. Alcohol is considered more dangerous to communities and harmful to individuals than marijuana or opium. Experience shows that drugs calm people down rather than rile them up. Marijuana and opium are sold in licensed stores at less cost than alcohol.[24]

We made out way back to Midnapore. It was evening; the entire town was out walking, men and women, children, families, even groups of single girls. Everyone was neatly dressed; they certainly did not seem oppressed. According to Dr. Roy, the grip of the caste system, and the expensive burden of dowries, is much less in West Bengal than in the states of old Mughal India, such as Punjab and Uttar Pradesh. In the markets, women are often the shopkeepers; in rural areas, they walk or ride their bicycles everywhere. "In West Bengal," said Dr. Roy, "villages are not controlled by big landlords. Most families have land, and they do the physical labor themselves. We are a people that love drama, art, and poetry." The Indian poet Tagore, who won the nobel prize for literature, wrote in Bengali.

The next day was set aside to visit potato farms, but we got off to a late start. The jeep's fuel line plugged up, and we did not want to waste the morning getting it fixed. So we hired a taxi and headed for Chandrakona, a town famous for potato marketing. Dr. Roy explained that the Midnapore district used to be exclusively a rice monoculture; farmers grew two rice crops a year. Short-duration rice and irrigation had changed all that.[25] Farmers can now fit in a third crop during the winter, and not just potatoes. They export all manner of vegetables and flowers to Calcutta, including tomatoes, chile peppers, mustard seed, and marigolds. Roads and transportation links today are much better than in the 1970s. Still, it was farmers who worked out these rotations, and they did it without scientific research as a backup. Of course, they need irrigation, as it rains little from November through March, the winter dry season.[26] For irrigation, farmers collect excess monsoonal runoff in natural depressions and later pump it to their fields. Most of the stored water is gone by March. In the lowlands, they can tap into rivers and streams, which dry gradually, or the water table, which is just below the surface.

Secondary roads get little traffic, so villagers use the pavement to dry their rice. We were expected to drive around it. We entered potato country; it seemed that wherever farmers had a source of water, they planted potatoes. We passed cold-storage warehouses, some of them under construction. They rise up from the flatness of surrounding rice paddies, looking like abandoned, windowless factories. They are the largest structures in the countryside, bigger even than the temples. Dr. Roy said total storage capacity in the district was three hundred thousand metric tons; production is twice that. In every town square, potatoes from the cold stores were being spread out and separated. Prices were high now, as the new crop had yet to be harvested; anyway, room had to be made for the next cycle.

In Midnapore, upland mesas and lowland floodplains abut each other. Farmers start potatoes in their upland plots, as these dry out first. Later, they plant in the lowlands, where the paddies stay soggy longer. If farmers sow tubers into wet ground, they are likely to rot. Even from a distance, it was easy to tell an upland plot from a lowland one; upland plants had a head start, so they were taller, with a well-developed canopy. To get to the lowlands from the uplands was a simple matter of scurrying down an embankment. For all practical purposes, upland and lowlands plots were cheek by jowl. As much as possible, roads follow the natural contours of the mesas. When a road crosses the floodplain, it is built on top of levees some four meters high, which keeps it above water when the fields flood. We stopped to inspect the potatoes. Dr. Roy knew which farmers in the vicinity planted TPS minitubers. We met Mr. Ghosh and his son working on their diesel-fueled water pump. They had tapped into an underground channel about three meters below the surface and were connecting up their plastic irrigation pipes. As Dr. Roy noted, even though the plots are small, farmers like Ghosh are conscientious and progressive. "They are making a lot of money on potatoes," he said. "The farm-gate price is currently 6 rupees per kilo. The city price is 7.5 rupees per kilo [35 rupees = one U.S. dollar]. So the potato fetches a good price." Midnapore's potatoes go to Calcutta and to states farther south, such as Orissa and Andra Pradesh.

Shade trees cooled Mr. Ghosh's wattle-and-daub compound. A thick layer of rice straw protected the roof; it had a narrow, thatched porch in front. Farmers also use rice stubble as animal fodder. Three great piles were heaped on top of bamboo platforms. A young farmer in his early thirties, Mr. Ghosh was enthusiastic about his potatoes. I could see why; they were a beautiful dark blue-green, tall and healthy. He told me this was the second year he had planted minitubers. "Last year," he said, "I planted fifty-five kilos of minitubers and kept all the tubers I harvested for seed. With this, I planted several plots, almost a hectare."[27] Just to be safe, he has planted some Kufri Jyoti, the old favorite. Mr. Ghosh said he planted the crop at the beginning of November; he expected to harvest it at the end of February. He plants potatoes only once a year. I asked him about the size of the potatoes he got from TPS minitubers. He said that he was very satisfied, that size was not a problem. According to Mr. Ghosh, nine out of ten farmers who had planted TPS minitubers were happy with the results. So far this season, he has not used pesticides of any kind. Dr. Roy and Mr. Ghosh discussed how much fertilizer was best. Like most farmers, Ghosh is using too much. Dr. Roy told him he could reduce the amount substantially and his yields would not drop; in fact, they might go up.

We headed for our last stop, the Garhbeta Block Office. In India, each state government organizes agricultural extension. In West Bengal, the Directorate of Agriculture took a labor-intensive approach to its work. In Midnapore, almost every village had an office. The district had seven subdistricts, each of which was subdivided into blocks—in this case for a total of fifty-four. The Garhbeta office was one of seven block headquarters in the Chandrakona subdistrict. It was small, but friendly and efficient. Mr. Choudhuri, the block leader, insisted we have tea. To make Indian tea, roughly equal amounts of milk and water are brought to a boil and poured into a pot. Only then are tea leaves and sugar added. When the brew is strong enough, the leaves are removed. It is delicious. Mr. Choudhuri was in his early twenties, dressed in dungarees, a colorful short-sleeved shirt, and sandals. He made his rounds to villages by motor scooter.

The Garhbeta block collects practical, basic data.[28] Choudhuri showed me his ledgers. Each farmer was listed along with the total area devoted to potatoes, the dates of planting and harvesting, and the yields obtained. The ledger noted how many times a farmer irrigated his crop, the dates he applied fertilizer, and when he hilled up around his plants. There was information on pest and disease incidence and on the number and composition of the chemical sprays applied and the comments each farmer had on the problems faced that season. Farmers in this block have planted TPS minitubers for three years. Most farmers still plant Kufri Jyoti, but minitubers are spreading. The minituber yield last year exceeded the local average by over seven metric tons; most of the harvest was sold commercially. For the typical farmer, that meant an extra 14,000 rupees (four hundred dollars) for each hectare of TPS minitubers planted.

The Garhbeta block is taking the next step, showing farmers they can produce their own minitubers directly from TPS seed, rather than buying them from Anandapore Farm. Choudhuri explained that the block's method is to plant TPS densely in a seedbed: "After the seed germinates and the seedlings take root, farmers transplant them into plots. Farmers already do this with rice. We think it is easier than using a frame with a hundred holes and then thinning, which is wasteful and time-consuming."

Farther north, in the states of Uttar Pradesh and Haryana, farmers have already taken Choudhuri's "next step." Four years ago, Jagbir Singh Mann and his wife, Sarla Mann, potato farmers in Haryana state, got five grams of TPS. From this, they grew minitubers. Now they have 15 hectares of potatoes descended from the original minituber crop. And they are not alone. India's farmers currently have over 10,000 hectares of TPS-generated minitubers and potatoes under production. Given the multiplier effect, that total is expected to mount quickly. When I asked Mann if the marketing quality of his minituber-generated potato crop was good enough, he replied, "If it wasn't, I wouldn't have so much of it." He is now growing minitubers to sell to other farmers. For the 1996–97 crop, he sowed some four hundred grams of TPS in his seed bed, transplanting enough to cover 3.5 hectares.[29]

A decade ago, critics thought mass-production of TPS seed was not feasible. They were wrong. A decade ago, critics thought TPS-generated potatoes would always vary too much in shape, size, and color to be marketed commercially. Wrong again. The recipe for success was the give and take between research at CIP and the problem-solving approach of India's national programs, which were backed up in turn by extension. Without such cooperation, the critics would be the winners, West Bengal's small potato farmers the losers.

A recent impact study concluded that the net return to farmers who invested in a TPS-based system came to $415 per hectare. For 2000, that meant an extra $16 million in household income for India's potato farmers.[30]

CHAPTER 25

Super Seed

Mahesh Upadhya did not think the potato revolution had gone far enough: "With the minituber, potato production still depends on multiplication and storage. Farmers should get a commercial crop directly from TPS; they do not need minitubers at all." Upadhya had worked on the TPS parental lines that had done so well in India. He was simply the best breeder of TPS populations in the world. That is why CIP had brought him to its headquarters in Peru.

Upadhya was in his sixties and impatient. He had worked on TPS lines for over twenty years. In Upadhya's view, the potato's contribution to food production was hostage to expensive multiplication projects. "Conventional seed multiplication requires a huge investment in land, storage, transportation, and management. India has Asia's best seed system, but supplies only a fraction of what is needed. The rest is recycled farmer seed. As disease problems mount, it degenerates and yields fall."[1] He had a point. Some fifteen viruses depend on the potato to survive. They use the potato as a host; all of them can be transmitted by tuber seed. It is almost impossible to tell which viruses a tuber has just by looking at it. A tuber that seems healthy can have multiple viruses replicated within its cells. The diagnostic symptoms for the viruses that do the most damage worldwide—potato leaf roll virus (PLRV), potato virus x (PVX), and potato virus y (PVY)—do not show up visibly in storage. The evidence shows up on the aboveground foliage after the tubers are planted. Unless great care is taken in multiplying seed tubers, a project ends up multiplying viruses. To test for viruses, CIP pioneered simple ELISA (Enzyme-Linked Immunosorbent Assay) detection kits; to eliminate viruses from seed stock, it promoted meristem multiplication. Nonetheless, it is a never-ending battle. Vectors for the twenty-eight known viruses that attack the potato include aphids, mites, nematodes, and fungi, in addition to the tuber seed itself. CIP cannot send a potato from one country to another without a phytosanitary certificate; it is just too dangerous.[2]

For high yields, farmers need young, virus-free tubers, tubers that are free of bacterial wilt and free from pests like weevils and tuber moths. In short, for the best results, farmers should start with fresh seed every season. That is a tall order. Recall that farmers need between two and three metric tons of potatoes to plant a hectare.[3] Storing such seed is a terrible headache.

Compared to traditional seed, minitubers reduce the storage space needed by roughly 75 percent. Minitubers are much younger than conventional tuber seed, just one season away from life as tiny seeds. By contrast, growers typically multiply out certified seed several times before they sell it to farmers. It may be virus free, but it is still much older than a minituber. TPS-generated minitubers are more diverse genetically than clonally multiplied seed; hence a field of minitubers is more disease resistant. Admittedly, the minituber is a great improvement on the old system. Even so, it does not eliminate the old system's deficiencies entirely. If farmers keep recycling minituber-generated potatoes year after year, they will eventually end up where they started—with large tubers of questionable sanitary status.

"This whole system is absurd and unnecessary," Upadhya said. "TPS is virtually virus free, it does not transmit bacterial wilt, and it is pest free.[4] If farmers grew potatoes exclusively from TPS, they could sell the entire crop every year; they would not have to hold over tubers for the next crop. To plant a hectare, all they need is a handful of TPS; no more cold stores, no more hauling potatoes to their fields."[5] It is truly a revolutionary concept. So I could see the revolution in action, Upadhya took me from his office to his demonstration plots.

TPS, Upadhya explained, is not sown directly into an open field as with corn or wheat. Instead, like the seed of its cousins the tomato and tobacco, it is planted densely in a seedbed and then transplanted into fields. The TPS seedlings are planted on the edge of a raised furrow, just above the irrigation channel. As the plant grows, dirt is piled up around it. Upadhya's transplanted seedlings could be harvested in ninety days; yields are from forty to forty-five tons. That leaves room to spare; in Peru, commercial farmers expect yields of between twenty and twenty-five tons.

In the 1980s, attempts to produce potatoes directly from TPS did not pan out. That was one reason an intermediate step, the minituber, had been added in the first place. But since then, the quality of TPS had improved in some remarkable ways. Noël Pallais, for example, had started screening seed for its size and density. He does this by putting TPS in a chamber and blowing air through it. The seed that scatters the most is the lightest, which means it is smaller and not as dense. By eliminating the lightweight TPS, about 40 percent, the best is left behind—what Pallais calls Super Seed.[6]

Upadhya gets high yields from TPS at a research station, but what about ordinary farmers? Can they get a commercially viable crop from transplanted TPS seed-lings? Farmers in Thai Binh province, located in Vietnam's Red River delta, do just that. The potato is a winter crop that fits in between two rice crops. Thai Binh's farmers plant small TPS seedbeds in September, transplant the rooted seedlings in October, and harvest a mature commercial crop between ninety and one hundred days later, in January and February. Yields are at least twenty-five metric tons per hectare. TPS is popular because it eliminates a nine-month storage period; stocks of tuber seed inevitably deteriorate in the humidity of the summer monsoons. Vietnam imports its TPS from India.[7]

Since the beginning, training has been a key aspect of the project's success. Peter Schmiediche, CIP's envoy for Southeast Asia, sent the details back to headquarters in Lima. According to his report, between 1994 and 1996, 3,121 Vietnamese farmers attended TPS training courses held throughout Thai Binh province; seedling nurseries were set up in 351 households. In 1996, some 4,000 farmers were involved in the TPS pilot project. As farmers refined their techniques, the amount of TPS they used to generate a hectare of potatoes declined from almost three hundred grams of seed to less than one hundred grams. Transplant survival rates are between 89 and 97 percent.[8] During the 1996–97 season, some 17,000 farm families were involved. By then, Noël Pallais and Mahesh Upadhya had provided the project with more potent ammunition: Super Seed. Its superior performance made it popular almost immediately. Even without the benefit of Super Seed, however, TPS did well enough to justify the investment made. Farmers paid the equivalent of eight hundred dollars for a kilo of TPS. For the 1995–96 season, yields from TPS-derived material averaged nearly twenty-four tons across thirteen sites in Thai Binh Province, compared to eleven tons for the local favorite. Even so, it was the Super Seed that convinced skeptics. Before its introduction, Vietnam still toyed with the minituber and storage. Now it's towing the TPS line that Upadhya advocates, producing a commercial crop directly from TPS transplants. Schmiediche's report to Lima concluded that with Super Seed "farmers could avoid storage almost entirely" and that potato production in the Red River delta "would be completely revolutionized." In 1997, Vietnam's Ministry of Agriculture allocated the equivalent of seventy-five thousand dollars to begin its own TPS production. The facility will be set up in the southern highlands, near Dalat.[9]

Vietnam was not the only place farmers had started down the TPS path. In the mountain community of Chacas, some three hundred kilometers north of Lima, farmers are producing their own TPS hybrid called Chacasina. The parents cross-pollinate naturally, as the female is self-sterile and the male parent is a prolific pollen producer. The community has invested in seedbeds, a nursery, and warehouses. In 1997, Chacas had a stockpile of over two hundred kilos of TPS, much more than it could possibly use in one season. Based on this success, CIP designed and illustrated a nontechnical guide for use by Andean campesinos; it shows how to sow, transplant, tend, sort, and store a potato crop produced directly from TPS.[10]

"They have millions of seeds," Pallais told me, "and TPS can be stored almost indefinitely." It is a good investment. In Peru, a kilo of TPS is worth about one thousand dollars. It provides security against droughts and El Niño rains. Chacas can start over again with TPS, if necessary. The community still stores some tuber seed; after all, keeping potatoes in Peru is much easier than in monsoonal Asia. And Chacas farmers still plant minitubers. What matters is that the community not only plants TPS directly, but it can produce its own seed. Overall, Chacas farmers now produce six times as many potatoes at a third the cost. Yields average over thirty metric tons per hectare, on par with harvests in the United States and Western Europe.[11]

"Before I leave this earth," Mahesh Upadhya remarked, "I want people to see the potato as a plant that reproduces sexually, not clonally." Perhaps the TPS revolution will not go as far as Upadhya dreams. What drives farmers to TPS technology is the cost and quality of tuber seed relative to yields. If tuber seed gets better and cheaper, then TPS suffers. This is less of a threat to minitubers, which outyield both TPS transplants and tuber seed.[12] Chacas may be an exception; perhaps the minituber will win out in the end. Even if it does, it will still promote the potato's sexual reproduction. That in itself is remarkable. A plant selected for its tubers is now being selected for its seed, which puts the potato on a different evolutionary trajectory. With potatoes, diversity is an advantage. The clonally produced crop is simply more vulnerable.[13] Old-fashioned sexual reproduction may still be the best game in town.

PART VIII

Sweet Potatoes

CHAPTER 26

Starch

"How do you like your starch?" I was in Soroti, a district capital in northeast Uganda, a day's drive from Kampala. Choices included *ugali*, an Ugandan version of cornmeal mush; fried potatoes; rice; or *eboo*. I tried the *eboo* and got two large, white-fleshed sweet potatoes covered with a peanut sauce. Compared to the orange-fleshed, watery types popular in the United States, the Ugandan varieties are larger and rounder, not so sweet, and flaky, due to a much higher dry-matter content. Uganda is Africa's largest sweet-potato producer; per-capita consumption is over ninety kilos a year, as high as it gets anywhere in the world.[1]

A New World crop, the sweet potato, *Ipomoea batatas,* originated in lowland South America on the edge of the Amazon basin. Archaeological remains show that ancient gatherers and farmers have exploited the crop for at least ten thousand years.[2] Sweet potato was a favorite in the Caribbean with the Arawaks. On his first voyage, Columbus brought sweet potato specimens back to Spain from Hispaniola.[3] In their chronicles, the Spanish conquistadors mention two sweet potato types: starchy cultivars with large roots, which natives called *aje*; and sweeter, more delicate cultivars with thin skins they called *batata*.[4] *Batata,* in its various linguistic forms, won the nomenclature battle. A sweet potato in both Spain and Portugal is still called a *batata*. From the Iberian Peninsula, Portuguese ships introduced sweet potatoes to trading stations in Africa, India, and the Far East. In Portuguese-speaking Guinea Bissau in West Africa, the sweet potato is still *patat;* farther south in Angola, once a Portuguese colony, it is *batata;* so too on Portuguese-speaking Timor in the Indonesian archipelago. An alternative route to Asia was on the Spanish galleons that crossed the Pacific between Acapulco in Mexico and Manila in the Philippines. To this day, most Filipinos call the sweet potato by its Mexican name, *camote*.[5]

How the sweet potato reached the South Pacific is still a mystery. From New Guinea to Easter Island, the crop was already entrenched at the time of first European contact. Perhaps the seafaring Polynesians brought the starchy root back to the Pacific Islands from the Americas in pre-Columbian times. Molecular analysis shows significant genetic difference between cultivars from South America and those from Papua New Guinea, which implies "many years of independent evolution in an isolated environment." On the other hand, within-region variance for a South American sample greatly exceeded that for a Papau New Guinea sample. Such variation reinforces the claim that the sweet potato originated in the Americas.[6]

Don't even think about calling a sweet potato a yam! Yams and sweet potatoes are from entirely different families; they do not even look alike. Yams are white-fleshed and often round, with tough, thick, black skins. A single large yam can weigh a couple kilos. Most of the yams produced in the Americas originated in Africa or tropical Asia.[7] In African markets, one would never confuse a pile of yams with a pile of sweet potatoes. The confusion in the United States comes from calling two sweet potato varieties by different names. According to CIP's sweet potato breeder, Ted Carey, what is called a "yam" in the United States is actually the elongated, moist, Puerto Rican–type sweet potato, which has a dark-orange flesh and smooth skin. By "sweet potato," Americans mean the larger, drier, chunkier, New Jersey types, with a rough, light-orange skin and flesh color. Less discerning shoppers use the terms interchangeably.[8]

A sweet potato likes it best with the mercury above twenty-four degrees centigrade (seventy-five degrees Fahrenheit); it gets chilly below ten degrees centigrade (fifty degrees Fahrenheit), a temperature that makes a potato feel healthy.[9] In Uganda, the sweet potato thrives in the hot, humid lowland plains to the north and east, a climate hostile to the spud.

CIP is widely known for its potato research. Less recognized is its general work on Andean roots and tubers, of which the sweet potato is a worthy exemplar. Worldwide, the potato's popularity far surpasses that of its sweeter cousin. The potato ranks fourth on the list of the world's most important food crops—after rice, wheat, and corn. The sweet potato is eighth. In terms of total tonnage, dry matter, caloric energy, and protein, that still puts the sweet potato ahead of both the tomato and the banana.[10] In Uganda, the sweet potato outclasses the spud. So, while I did not ignore Uganda's potatoes, the main item on my agenda was the sweet potato.

Switching crops can get confusing. Both are "potatoes," both come from under the ground, and both can be boiled. However, with respect to propagation, harvesting, and use, the differences are fundamental.

The potato is an annual crop that dies off each year. Farmers hold over tubers for the next crop. The sweet potato, by contrast, is a perennial. Its large roots store carbohydrates. Left alone, a stand of sweet potato maintains itself for years. In

Uganda, the foliage dies back during the dry season, but the unharvested roots survive underground. When the rains return, the old roots eventually resprout and send up new shoots. To start the next crop cycle, farmers take cuttings or "slips" from these vines and plant them in mounded hills. The vines take root. In four to five months, farmers can start harvesting their sweet potatoes. With potatoes, farmers often dig everything up all at once. With the sweet potato, families harvest piecemeal as needed. Roots can stay in the ground for two to three months without spoiling. Sweet potato vines grow so fast and cover the ground so densely that weeding is minimal. Uganda's sweet potato farmers do not use chemicals of any kind; except for the sweet potato weevil, pest damage is usually slight. By contrast, the country's potato farmers spray against late blight and fret over bacterial wilt. With sweet potato, a family can use both the roots and the "greens," or vines. The tender, nutrient-rich foliage can be eaten fresh; tougher leaves can be cooked like spinach; vines also make a good forage. The potato's merits are many, but edible foliage is not one of them.[11]

The sweet potato is versatile. In the United States, mashed sweet potato—with brown sugar, nutmeg, and bourbon added to taste—is a required dish at Thanksgiving, not to mention sweet potato pie. The mashed, fresh root can be added to bread dough or pureed into a nutritious baby food. In Indonesia, spices are added to the puree, and the pungent product is marketed as ketchup. Argentina's national desert, *dulce de batata,* is a sweet potato puree boiled down with an equal weight of sugar and a bit of vanilla. Given the root's water-soluble pectin, it is easy to make sweet potato jam. The potato, of course, can be mashed. But it rarely masquerades as ketchup, much less as a jam.[12]

Compared to a spud, a sweet potato is more likely to be dried or to end up as starch. For drying, fresh roots are sliced up and left in the sun. Once dehydrated, the pieces can be ground into flour, used as a base in porridge and stews, or processed into starch. Depending on a variety's dry-matter content, one hundred kilos of sweet potato roots will yield from twelve to thirty-seven kilos of flour. Cookies, donuts, and bread can be made with varying proportions of wheat and sweet potato flour. In beer, sweet potato flour can substitute for up to 50 percent of the malt.[13]

Starch is the single most important sweet potato product. The crop yields at least 30 percent more starch per kilo than rice, corn, or wheat. A processing unit starts with either dried chips or fresh roots. With chips, the dried pieces must be rehydrated, mashed, and mixed with water. Fresh roots, by contrast, are crushed into a pulpy mass, which is then mixed with water in a tub. Once the roots are mashed or ground up, the steps that follow are identical. The pulp is strained out and the starch gradually settles to the tub's bottom. Once the starch begins to firm up, the liquid is drained, and the starch is put in sacks to drip dry. Starch goes into food processing as a thickener, paste, or gel. In China's Sichuan Province, which has a population of some 113 million, transparent noodles made from sweet potato starch are a staple. Modified sweet potato starch is also used by the paper, textile, and chemical industries. To make alcohol, dried chips are first boiled in water, and then an enzyme is added to convert the starch to sugar; the next step is to ferment the sugar to alcohol and distill. As a biomass substitute for fossil fuel, sweet potato is better than corn. Half a hectare of roots yields some one thousand liters of ethanol alcohol; for corn, the figure is three hundred liters.[14]

The potato can be converted into starch too, but that is not its forte. The dry matter in a typical potato is about 23 percent.[15] In a sweet potato, the average is 30 percent; of this, 80 to 90 percent is carbohydrates, mainly starch and sugars.[16] In the United States, weight watchers have declared war on carbohydrates. But in many Third World countries, Uganda included, getting enough starch is the first step to adequate nutrition. On a fresh-weight basis, a sweet potato is 1.5 percent protein. That may be less than the grains, and it's less than in a potato, but it is superior to bananas and manioc, both staples in East Africa. Sweet potatoes have more fiber and more fat than potatoes. They have substantial amounts of vitamin C and moderate quantities of vitamin B1 and B2 (riboflavin).[17] The orange-fleshed types are rich in beta-carotene; eating them prevents vitamin A deficiency, and hence blindness.[18] No potato can make such a claim.

In Europe, lots of potatoes go to the pig sty. In Asia, it is the sweet potato that feeds the swine. But the roots must first be cooked or at least blanched. In China, the world's largest sweet potato producer, 30 percent of the crop goes to animal feed, mostly pigs. For a high-quality silage, sweet potato vines can be mixed in at a one to three ratio. Even so, for adequate weight gain, extra protein such as soybean meal should be added to the mix.[19]

CIP's main office for sub-Saharan Africa is in Nairobi, Kenya; it also has a liaison office in neighboring Uganda. Both countries work together in agriculture, which facilitates cooperation between the two CIP offices. The day after I arrived in Nairobi, I met CIP's sweet potato breeder, Ted Carey. A tall, lanky man in his late thirties, Carey was famous for his vegetable garden and his sweet potato trials. According to Carey, the distinguishing feature of Africa's sweet potato is its high dry-matter content. Farmers in East Africa conserve the crop by drying it; over time, they selected out the varieties that processed the best, a factor directly related to a root's dry matter. The dry-matter range in local varieties is between 32 and 36 percent. In the United States, where consumers prefer watery types, the dry-matter content is much less, only 18 to 24 percent.[20]

Worldwide, African sweet potatoes can help jack up the crop's dry-matter content, thus enhancing starch, flour, and alcohol production. There are other advantages. "African farmers selected sweet potato varieties under heavy disease and pest pressure," Carey emphasized. "Under low-input conditions, their cultivars do better than improved germ plasm from CIP headquarters." Tropical Africa has sweet potato landraces with high dry-matter content and virus resistance.[21] By crossing African types with varieties from CIP's collection, Carey and his partners expect to generate families with the best traits of both. A top priority is an orange-fleshed cultivar.

Vitamin A deficiency, which can cause permanent blindness, is chronic in tropical Africa, especially among children.[22] Unfortunately, white-fleshed varieties of the sweet potato have virtually no beta-carotene. In both Uganda and Kenya, it's the white-fleshed types most people like, both for fresh consumption and for processing.[23] The vitamin A–rich, orange-fleshed kinds tend to be mushy; dry-matter content rarely surpasses 24 percent. Given the lack of vitamin A in rural diets, getting an acceptable orange-fleshed type to farmers is a priority. The quickest path to selection has come from multilocational National Performance Trials (NPT), in which Uganda and Kenya exchange material collected from each other's farmers. Such trials helped Carey and his counterparts identify Kenyan cultivar SPK 004, which is both rich in vitamin A (3.44 milligrams per one hundred grams of sweet potato) and has an acceptable dry-matter content of 31.6 percent.[24] Thanks to CIP-backed regional cooperation, SPK 004 vines have been quickly distributed for on-farm trials.

By now, we were walking through Carey's sweet potato trials. Like the potato, the sweet potato plant sets true seed. Flowering occurs under short-day conditions, which is perfect for an equatorial country like Uganda. In the United States, the sweet potato is a long-day summer crop; consequently, the plant does not usually flower or set seed. A sweet potato plant is rarely self-fertile. This makes emasculation of the female parent unnecessary. The sweet potato produces a marble-sized seed ball that contains only three to four seeds. Over the course of a season, two parents generate plenty of flowers. To get enough seed for breeding, however, requires hand pollination. The sweet potato is a member of the morning glory family, and its flowers last only a couple of hours.[25] The crop's genetic diversity comes from both sexual reproduction and from somatic mutation. As Carey explained, with sweet potatoes, repeated cuttings from the same vines net some degree of mutation. To maintain trueness to type, breeders check both vines and roots.[26] Carey's field of sweet potatoes was a wild thing; to the untrained eye, the trials looked like a collection of separate species. Potato plants have different shapes, colors, and sizes below ground, but aboveground they are roughly the same. Sweet potato vines are unruly anarchists by comparison. Vines can be compact, sprawled out, or virtually upright; foliage can be a deep reddish-purple, blue-green, even a yellow-green. Leaves can be heart shaped, spade-like, round, or variegated; some even have spiky upright foliage that looks like elephant grass.[27]

When selecting a sweet potato, what characteristics did Carey and his team look for? A good yield is a necessary but not a sufficient criterion. Ugandan farmers need roots that recoup quickly after the dry season, supplying new planting material; they like sweet potatoes that put out a dense canopy, because the shade deters the weeds; and they like early-maturing roots that process and store well. By the end of the dry season, families are short of food. The earlier the next sweet potato crop can be eaten, the better.

CHAPTER 27

Food Security

"Without the sweet potato," said Robert Odeu, "there would be a terrible hunger." Odeu is from a Dokolo village in Soroti district in northeast Uganda. A decade of civil war killed off the villagers' cattle, a virulent mosaic virus devastated the manioc crop, and the parasitic striga weed attacked the corn and sorghum. Dokolo is not alone. Some 85 percent of Uganda's twenty million people live in villages; almost everyone depends on the sweet potato for food. When it is in season, families eat it every day at almost every meal. Responsible for feeding their families, women are typically in charge of the sweet potato plots.[1]

It was December, 1997. I was in Uganda with CIP scientists who worked on joint projects with Uganda's National Agricultural Research Organization (NARO). Before we left for Soroti, we had stopped at NARO's Namulonge Research Station, which did research on root and tuber crops. We had expected to meet with the station's sweet potato staff, but they had left for Serere the day before. Serere was the site of a three-day training session on sweet potato utilization; we were headed there the next day. The time spent at Namulonge, however, was not wasted. We met with G. W. Otim-Nape, who heads up NARO's manioc program. A distinguished scientist and agricultural activist, he has led the battle to save the crop.[2]

A starchy root from the New World, manioc has even more dry matter, fiber, and carbohydrates than a sweet potato. The crop is propagated vegetatively from the plant's woody stalks, which farmers cut down at harvest time and store over the dry season. Then, at planting time, they mound up the soil and insert the stalks, which resprout when the rains come. Manioc can be of either the sweet or bitter type. Sweet manioc can be cooked and eaten right after harvesting. The bitter types, however, contain cyanide and must be processed into starch or flour first. In Uganda, farmers plant sweet manioc almost exclusively; it used to be more popular than the sweet potato.[3] Then in 1988, a new mosaic virus attacked the plant. Efforts to contain the disease failed. Between 1987–89 and 1995–97, Uganda's manioc harvest fell by a third, from an average of 3.3 million metric tons a year to 2.2 million tons.[4] Carried by white flies, the disease quickly spread through the country's manioc belt. According to Otim-Nape, when the virus hits it infects every plant. Stalks collected from infected fields carry the disease, even those that look healthy. When farmers replant the next year, the stalks are already infected. "In the first year," Otim-Nape said, "yields are greatly reduced. The second year, farmers get almost nothing. The third year, they abandon the crop."[5] To fill the gap, farmers turned to the sweet potato. In northern Uganda, production rose by 43 percent; in eastern districts, including Soroti, it increased by 26 percent.[6]

NARO's manioc research used to be done at the Serere station, in Soroti. But the station had to be abandoned during the insurgency. Later, a NARO technician slipped back into Serere to see if any trials had survived both the war and the virus. He found a couple plants that looked resistant, took cuttings, hid them in his pockets, and brought them back to Namulonge. These plants turned out to be resistant to the mosaic virus. Working with the International Institute for Tropical Agriculture (IITA) in Nigeria, which has manioc as one of its research crops, NARO and its partners were able to breed, test, and distribute resistant varieties in just six years. Otim-Nape estimated that by 2001 Uganda's manioc production would be back on track. But this would not be the case in neighboring countries. The virus has spread west into the Congo and east into Kenya.[7] In Uganda, manioc is a very problematic plant. In the 1980s, it was attacked by green spider mites; then came a bacterial wilt, followed by mealybugs, and now a mosaic virus. The sweet potato, by contrast, is more robust. It is also early. Farmers can get a sweet potato crop in just four months. With manioc, it takes almost a year. For all these reasons, sweet potato is now East Africa's food-security crop.

And do not forget the striga. I had a chance to see the damage it did at Alupe Research Station in Kenya.[8] Alupe is down on the lowland plains, near Lake Victoria. A short, grassy-looking weed, striga penetrates the roots of its host plant and draws out the nutrients. Cereal crops such as corn and sorghum, both African staples, are especially susceptible. The result is a shriveled plant with very little yield, if any. Striga seed can stay in the ground for twenty years and still germinate around a host plant. That is because host plants release a chemical that triggers the parasitic weed. Sweet potato, however, is not a host, so striga does not bother it. Even when planted as an intercrop—and African farmers love to intercrop—sweet potato deters striga. Farmers plant corn first so it gets a head start; then comes the sweet potato. Its thick, viney foliage inhibits the low-lying striga, which cannot get enough sunlight to proliferate.

Soroti district is now at peace, but it will take years for farmers to rebuild their cattle herds. Add to this the pest-prone manioc plant, plus striga problems, and one can see why Soroti farmers have embraced the sweet potato. Consider its merits. The sweet potato is dependable. In just three months, as soon as the plant forms edible roots, a hungry family can start eating. And an established patch keeps producing, despite drought, for months. The sweet potato is flexible. Farmers can stagger production across the region's two rainy seasons, from April into June/July and from August into November/December. They get two crops, which they can leave in the ground to harvest piecemeal as needed, or they can harvest everything at once. While the supply lasts, families boil or steam the roots, eating them with pungent peanut sauces.[9]

Food security, however, is as much about storage as it is about production. When the dry season sets in, weevils start to proliferate. To keep the sweet potato crop from being destroyed, villagers harvest whatever is left in their fields. This surplus can tide them over during the long December-to-April dry season—if it can be kept from spoiling. Farm families minimize weevil damage by drying and storing what is left.

"Not much can be done about the weevil, except to avoid it," says Nicole Smit, who runs CIP's liaison office in Kampala.[10] She has lived in Uganda now for almost a decade, with her husband and two daughters. Her lab is at NARO's Namulonge Station. I had met her my first day in Uganda, at Entebbe airport. In Uganda, sweet potatoes are attacked by two weevil species, *Cylas brunneus* and *Cylus puncticollis*. So far, varietal resistance has proved elusive. Research on biological controls, including pheromone traps, has just started. According to Smit, a farmer's best defense is still prevention. "Weevils," she said, "are not good diggers; they have to find an exposed root, or get to a root through a crack in the ground." During the rainy season, weevils proliferate slowly and have more natural enemies, including a fungus disease, and the wet soil makes it hard for them to get at the roots. Once the dry season sets in, the balance shifts to the weevil's advantage. To reduce weevil damage, families should harvest the largest roots first and hill up the soil around the remaining roots to protect them. In western Uganda, where the dry season is not so bad, farmers complain about weevil damage but have learned to live with it. When they harvest, they just cut out the damaged parts of the plants. Soroti district, however, gets very dry; farmers must harvest, store, and process the crop or lose everything.[11]

I visited Dokolo village with CIP scientist Vital Hagenimana. A Rwandan, Hagenimana had studied agronomy in Canada, where he specialized in food processing and storage technology. We met Robert Odeu. On behalf of his village, Odeu worked on a joint post-harvest project sponsored by NARO and CIP. A short, thin man in shorts, sandals, and a colorful short-sleeved shirt, Odeu was quiet and dignified. What he wore did not matter; how he treated others did.

Odeu showed us the new storage pits redesigned by NARO and constructed from local materials. A typical circular pit is about a meter deep and a meter and a half wide. Once the pit is cleaned out, farmers compact the walls and cover the bottom and sides with dried grass. Roots are then packed into the pit in a pattern that permits sufficient air circulation. A straw roof goes over the top, braced from below by a bamboo frame. Thus protected, fresh roots can be kept two to three months.[12]

We walked by Odeu's manioc field; it was a disaster. "I knew the stalks were diseased when I planted them," he said, "but it was the only option I had." We headed for a rocky escarpment, which villagers use to dry crops and thresh grain.[13] The rocks were smooth and worn. Drying is the traditional way to preserve sweet potato. According to Hagenimana, men and women share all aspects of sweet potato production, from planting to harvesting. Tasks in drying, however, are gender based. For coarse chunks, or *inginyo,* women crush the roots with heavy wooden mallets and leave the pieces out to sun dry. Large neat piles were spread out on the rocks. To dry, a pile needs three hot sunny days. If it rains, the chunks start to ferment, which reduces the quality. For chips, or *amukeke,* men slice up the roots into round, flat pieces and spread them out to dry. Whether crushed or sliced, both keep for between four and five months. Dried sweet potato is boiled in sauces along with beans and vegetables. For the starchy staple *atapa,* women grind up dried chunks or chips into a coarse flour, which is rehydrated in water, boiled, mashed, and then eaten directly as a thick porridge. Milled sorghum or millet can be mixed in, along with some tamarind fruit, lemon, or mango.[14]

"We prefer the sliced *amukeke* chips," said Odeu, as we headed back to the village compound. "The quality is higher and they store better; unfortunately, processing chips is much more tedious." To speed up chip production, NARO has designed a durable iron slicer with an adjustable, hand-cranked blade. A small work group can slice about 180 kilos of sweet potato roots an hour, or about a metric ton a day. Three Soroti villages, including Odeu's, are trying slicers out.

Like most of Dokolo's waddle-and-daub circular compounds, Robert Odeu's had a bamboo frame and a thick straw roof. It was shaded by an enormous mango tree. Odeu also had two large, urn-shaped granaries—wide at the base and narrow at the top. Made from a mixture of mud, straw, stones, and gravel, they were large enough for two adults to fit inside. When dry, the exterior has a hard, almost ceramic-like surface that is resistant to termites. To keep out the rain and dampness, the urns are placed upright on platforms covered by a straw roof. Odeu said the urns took four days to make but will last at least five years. The straw roof, however, must be replaced from time to time. Millet can be stored for up to a year, *amukeke* chips for five months, and sorghum for three months. Off to the side sat Odeu's ten-year old daughter, crushing peanut shells with a heavy wooden roller.

She heaped them all on a round winnower made of reeds and bamboo; with a steady, graceful twist of her wrists, she separated the nuts and shells.

Odeu said he was a Christian, which meant he rarely sat around to "exchange ideas." In Dokolo, such exchange requires prodigious quantities of waraji, an alcoholic beverage that can be made from dried sweet potato, even from weevil-infested chunks and chips. Just add water and yeast and let the mixture ferment. A waraji can also be made from sorghum, millet, dried manioc, or even bananas. While sharing a conversation, participants drink from the same waraji pot, but with their own bamboo straws. One reclines or squats; no chairs are allowed. Most important of all, one must bring one's own waraji. No conversation ends until all the waraji is gone. Young men can start drinking when they have their own waraji supply, which is usually after they get a wife. Among women, only the elderly have time to drink. Young wives are too busy raising children, cooking for their husbands, or working in the fields.

Today was the last day of school in Soroti district. Report cards had just been handed out. Kids crowded around to show us how they did. But our visit was cut short. The sky darkened; we heard the roll of distant thunder. The rainy season was still in full force. That was good news. A severe drought had delayed the sweet potato crop by two months. All over Uganda, villagers prayed the rains would last through January, long enough for the roots to mature.[15] To judge from the storm upon us, the chances were good. Once the rain hit, it did so with tremendous force. Odeu urged us to leave immediately, report cards or not. A heavy downpour quickly turns the narrow trails to mud; there are no roads to most villages.

In Soroti, farmers can sell surplus sweet potatoes to truckers, who haul the loads to the Kampala market.[16] Truckers use custom-made, eighty-kilo plastic mesh bags with distinctive logos. Agents distribute the bags and get a commission on each bag filled. Truckers pick up the bags at collection points along the main road. They recognize which bags are theirs from the logo. The truckers pay the agents, the agents pay the farmers, and the farmers do all the work.

The Soroti-Mbale road is paved, which makes it the district's superhighway. It was filled with bicycles, people, and an occasional motor vehicle. Bicycles were the old-fashioned, low-maintenance, one-speed type. Made in Kampala, they seemed virtually indestructible. Four to five passengers could be balanced on a single bike. Most were pushed with their loads carefully tied on top; there was no room for a rider. Loads included bundles of wood, bunches of bananas, five-gallon plastic water jugs, one hundred-kilo bags of cooking charcoal, eighty-kilo bags of sweet potatoes or pineapples or cabbages, plus baskets of fish, not to mention mattresses, rugs, and bamboo mats. The road was a crowded pedestrian thoroughfare, busy with folks hauling farm implements to their fields, driving their cattle, or carrying water jugs and charcoal on their heads. And if the timing was right, a parade of children, mostly barefoot, trekked to and from school in colorful uniforms.

The next morning, I met Dai Peters. A Chinese American, Peters did graduate work in sociology at North Carolina State University. She left sociology, "which never concludes anything," and switched to resource management. At CIP, she had worked on a pilot project in Indonesia to produce flour from dried sweet potato.[17] She had been in Soroti for six weeks, assessing prospects for small-scale flour production. Peters has incredible energy and a great sense of humor. Her research in Soroti showed that local demand for a high-quality dried chip was expanding.[18] According to Peters, millers grind the chips into flour, which is sold in the towns, mostly to make atapa-based porridges. The flour can also be used in baking. Many types of cookies, as well as deep-fried *mandazi* dough, can be made almost exclusively with sweet potato flour. For cakes and bread, up to 50 percent of the flour used can come from dried sweet potato, but it has to be very white and very clean. Fresh, mashed, sweet potato can also be mixed directly into the dough. In local tests, consumers said such additions improved the color, texture, taste, and freshness of deep-fried buns, chapatis, and mandazis. The finished product is also less greasy, as the sweet potato mash reduces oil absorption.[19]

With the cooperation of NARO's post-harvest program, CIP sponsored a three-day processing workshop at Soroti's District Farm Institute in Serere. I spent an afternoon at the workshop with Dai Peters and Vital Hagenimana. More than fifty village women attended, plus members of community organizations and even some local bakers. They spent a day testing recipes and experimenting with different amounts of sweet potato flour; they tried out the slicing machine, debated the best oven construction, and examined the storage pits. In Soroti, prospects for sweet potato substitutes in baked goods are especially promising. There is an established demand, processing technology is available, and the high price of imported wheat makes substituting sweet potato flour a virtual necessity. I talked with James, a baker from Atira, who said he made about two hundred loaves of bread a month. His main expense is wheat flour, which costs 833 Ugandan shillings (1,000 shillings = one U.S. dollar) a kilo; he said he would pay up to 500 Ugandan shillings for sweet potato flour. The going rate for an eighty-kilo bag of fresh sweet potato root is about 4,000 shillings. Given a dry-matter conversion rate of 30 percent, that's enough to produce twenty-four kilos of sweet potato flour (80 kilos x 30 percent). At 500 shillings per kilo of flour, gross receipts per bag would total 12,000 shillings (500 shillings x 24 kilos). That leaves processors a profit margin of 8,000 shillings (12,000 shillings in receipts -4,000 shillings for roots) on each eighty-kilo bag.

We stopped for afternoon tea. I had "wet," or African tea, which is made by boiling leaves, water, and milk together in the same pot. We also had hard-boiled eggs, plus thick slices of fried sweet potato and large sweet potato cookies. Tea done, we left for Awoja village to inspect swine production.

Sweet potato, fresh or dried, is a staple for almost every village in Soroti. It also makes a good feed for hogs. According to Dai Peters, "keeping hogs is how villagers convert their surplus sweet potato into a source of capital; a hog is a little savings account with a short-term maturity."[20] For digestibility, fresh roots should be boiled, and the vines, which provide a good source of protein, have to be chopped up. Using roots as feed, however, does not endanger food security. As Peters explained, villagers start raising hogs in July, when the first sweet potato harvest begins. Roots are abundant into December. Hogs are sold off at Christmastime, just as the dry season gets underway. To judge from the scrawny pigs in Awoja, however, the project has a long way to go.

We talked with Simon, the village's representative. He said he bought two piglets five months ago. He estimated that they ate about two kilos of sweet potatoes a day. Apparently, it had not done them much good. Despite eating their way through three hundred kilos of cooked roots each, the pigs weighed only fifteen kilos apiece. Peters got out her pocket calculator. Simon said he paid 7,500 shillings for each piglet and expects to sell each one for 15,000 shillings. At 4,000 shillings for an eighty-kilo bag, Peters calculated that Simon was probably losing money. But then, Simon's arithmetic is different. He already has the sweet potato, so from his point of view, the feed is free. He is just turning it into a different, more secure, value-added form.

Peters had spent two months in Vietnam prior to her sojourn in Uganda. She claimed that pigs there weighed in at sixty to seventy kilos in just four months. As in Uganda, sweet potato was the main ration. Vietnamese farmers, however, also use the vines, and they add a protein supplement like blood meal. They worm their pigs, too, and keep them that way, confined in rustic sties. In Awoja, by contrast, pigs root around in the fields, damaging crops and contracting tapeworms. For hog production to be worthwhile, the Soroti project will have to improve the feed, promote confinement, and monitor animal health. The benefit for farmers is that pigs gain more weight faster. Manure can be recycled from the sty to fields.[21]

Food security in Soroti means storage, processing, and utilization. But front-end production cannot be ignored either. Villagers typically leave some sweet potato roots in the ground during the dry season. Months later, when the rains start, the roots of this hardy, perennial plant resprout. After a few weeks, the vines are strong enough for farmers to take cuttings, which they use to reestablish the crop.[22] This year, however, the rains stopped early and started late. The result was a long drought that killed off most of the sweet potato plants and jeopardized the food supply. Simon said he had six varieties in the ground last April, "but the rains stopped and they all died." Fortunately, he had taken the precaution of keeping some vines going near some marshes, where he could water them. So when the drought finally ended in October, he could start taking cuttings.

The drought was even worse in Kenya. "We managed to get multiplication going again, but it took much too long," noted Philip Ndolo, who heads up the sweet potato program at the Kenya Agricultural Research Institute (KARI).[23] To prevent another production crisis, Ndolo and his team now work with women's groups in fifteen communities near Alupe, in Kisumu district. This year, each member will plant healthy cuttings in a small 1.5 meter–square nursery bed prepared in advance. They will water the bed as needed through the dry season. Then, when the rainy season starts, they can take cuttings from the vines straight away, they will not have to wait for the old roots to resprout.

There is another advantage. According to CIP entomologist Nicole Smit, weevils often infest the roots and woody vines farmers leave unharvested. Months later, when they take cuttings from these older plants, they end up transferring weevils to their new fields. "But with cuttings from a nursery," Smit said, "farmers begin the next season with fresh, clean planting material less likely to be infested."[24]

Mrs. Mwanzi, of Okame village near Alupe, had several parcels of sweet potatoes laid out in neat squares and rectangles; the cuttings came from her own nursery.[25] She and her ten-year-old daughter had dug up the ground, hilled the soil, and planted the cuttings. She had a plot of SPK 004, an orange-fleshed variety whose cuttings came from the Alupe substation. "I like a lot of varieties," she said, "that are ready at different times." For that reason, she also staggers her planting, establishing new plots every couple weeks. Mrs. Mwanzi demonstrated her technique. She cut four viny shoots from a plant with her machete, each perhaps thirty centimeters (about a foot) long. She put two of them together, mounded up a hill of dirt with her hands, and then pushed the stems into the hill. She took a couple more cuttings and pushed them in from the side. "It is best to plant a lot of stems," she said, "in case some of them die." She transplanted her nursery cuttings as soon as the rains permitted, because as she said, "I like to get the first new roots early, to kill the hunger."

Mrs. Mwanzi planted SPK 004 because she had heard it was good for her children's health. She is not alone. What had blocked acceptance of orange-fleshed types was insufficient starchiness. Once this was solved, households embraced the newcomers. Just one hundred grams a day is more than enough to prevent vitamin-A deficiency. In 1999, CIP began a five-year effort to bring the benefits to East Africa's rural families. Called VITA, the project helps farmers to multiply and distribute the new cultivars. It encourages microenterprises to add orange-fleshed types to traditional foods and pastries. And it has designed a marketing campaign geared to mothers and schoolchildren. In rural areas, promoting orange-fleshed sweet potatoes is a more practical antidote to childhood blindness than distributing vitamin-A supplements.[26]

CHAPTER 28

Farmer Field Schools

"The problem with technology transfer is that it's top down," said Elske van de Fliert. "Farmers end up passive appliers of packages' developed by researchers." Van de Fliert ought to know. For three years, she studied farmer field schools in four villages of Central Java in Indonesia.[1] The "schools" were part of a national campaign to cut down on pesticide use in the country's rice paddies. A field school is participatory. Farmers learn how to work out practical solutions to basic on-farm problems. The school is not a building. "It can be a rice paddy, a sweet potato field, or a cabbage patch," van de Fliert emphasized, "the actual place where farmers work and grow their crops."

Two-thirds of Indonesia's population still lives in rural areas; agriculture employs half the adult workforce. Rice is the crop that feeds the country. Per-capita consumption of milled rice is 138 kilos, compared to about 10 kilos in the United States.[2] The worst rice pest is the plant hopper. In the past, its predators kept the population in check. Insecticides, however, changed the balance of power in the paddies. Plant hoppers reproduce much faster than their spider enemies. Dousing both with the same lethal concoction favors the hopper population, which recuperates and builds up resistance much faster. The spider population, by contrast, takes more time to bounce back. The result was a costly, escalating battle in Indonesia's rice paddies.

In 1985, a new plant hopper biotype emerged, which the old chemical arsenal could not contain. This time, however, Indonesia's Ministry of Agriculture called a cease-fire. It opted for a pest-management approach that gave beneficial insects like spiders a chance to recuperate. A crash program trained thousands of farmers in the cardinal principles of pest management: grow a healthy crop, protect friendly insects, monitor fields regularly. This cadre of farmer experts then returned to their villages to spread the faith. They taught their fellow farmers which insects were harmful and which were their allies. They demonstrated that insecticides were unnecessary until the number of hoppers per rice plant exceeded a strategic threshold. Between 1986 and 1990, the Indonesian government banned fifty-seven chemicals and phased out pesticide subsidies. In pilot projects, insecticide use fell by 60 percent. When farmers resort to insecticides now, they use chemicals specifically for plant hoppers, chemicals that leave the spiders unscathed. Despite the drop in pesticides, rice production increased from an average of thirty-four million metric tons for 1981–83 to over forty-three million metric tons for 1988–90. As a result, integrated pest management (IPM) became national policy, and not just for rice.[3]

Indonesia's IPM program is now in its second phase. The goal is to set up farmer field schools in fifteen thousand villages across the country's rice bowl, extending pest management to soybeans and horticultural crops such as cabbage, tomatoes, and hot peppers.[4] Farmers are expected to be active learners, to become experts on their own fields. The schools teach by doing; they focus on field problems and encourage farmers to make their own decisions.

When I met Elske van de Fliert in 1995, she was helping CIP set up field schools tailored to the sweet potato. CIP had a regional office in Bogor, West Java. Van de Fliert was from the Netherlands, where she had studied at Wageningen Agricultural University. Project partners included a nongovernmental organization known as the Yayasan Mitra Tani Foundation (Partner of the Farmer Foundation); Duta Wacana Christian University, where van de Fliert taught; and Indonesia's Research Institute for Legumes and Tuber Crops. A combination agronomist and anthropologist, van de Fliert spoke both Malay, the archipelago's lingua franca, and Javanese. Easygoing, yet determined and tough minded, she took a firm yet compassionate view of the country she had adopted. Van de Fliert and her Javanese husband, Yogi, lived in central Java near Yogyakarta with their two children. Yogi had learned Dutch studying film production in the Netherlands; he was making a documentary on farmer field schools.

The project had set up field schools in four villages that were trying out new sweet potato varieties. It was November, time for the harvest and a project evaluation. Van de Fliert invited me to go along. We left Yogyakarta at 5 A.M. the next morning and headed for Bendunganjati village in the Mojokerto district. It was a long day's drive away, across Mount Lawu volcano, which demarcates central from east Java.

Indonesia is the world's third largest sweet potato producer, after China and Uganda.[5] Sweet potato is a cash crop, not a food-security crop. With irrigation, farmers in east Java can manage three crops a year. They fit a sweet potato rotation between rounds of rice. A respite from monoculture is good for rice productivity, good for soil structure, and good for the soil's fertility.[6] The rotation with rice is likewise beneficial for the sweet potato, as flooded paddies help check the weevil population. Farmers plant their sweet potato cuttings on neatly spaced, mounded ridges, which they build up inside diked fields. They let the cuttings wilt for a couple days before planting them. A ridge of newly planted slips looks terrible. But once irrigated, or after a rain, the cuttings perk up. For irrigation, farmers let the water fill up between the ridges, but they do not submerge the plants, as with rice.

A crop of sweet potato is much less work than rice. Sweet potato vines spread quickly and densely, which cuts back on the weeding. At harvesttime, no threshing or milling is necessary. In fact, farmers often avoid harvesting sweet potatoes themselves. Instead, they sell to truckers while the crop is still in the ground. Once the price is set, the trucker pays for digging and bagging the roots. In east Java, a yield of twenty metric tons per hectare is common. By comparison, yields are about twelve metric tons in Vietnam.[7] If farmers expect to plant sweet potatoes in another patch nearby, they simply take cuttings from their own vines. If they are going from rice to sweet potatoes and have no cuttings of their own, they get some from another farmer. There are sweet potatoes in the ground somewhere all the time.

Between Solo and Sragen we passed through some of central Java's finest rice lands.[8] Farmers had diked their paddies in small, irregularly shaped configurations. They do not waste space. On top of the bunds, or dikes, they plant manioc, or king grass, which deters erosion and makes a good animal fodder. Most farmers follow rice with rice. November is the time to get the dry-season rice out and plant the wet-season rice crop. Every phase of production was underway. Farmers were cutting down the grain by hand, threshing it in portable five-horsepower machines, and then bagging it right in the field. Truckers would later pick up the rough rice and haul it to the mills. Most families store enough rice to meet their own needs. They hull it themselves and then spread it out to dry in the sun on bamboo mats, plastic tarps, or any unused surface, including a paved road. Once the rice is cut and bagged, farmers let their goats graze on the rice stubble that is left in the fields. Then another cycle gets underway. Farmers flood their paddies, puddle the soil, and then transplant the seedlings. Today, I saw no mechanical tiller; in central Java, the water buffalo still does the plowing. Farmers do not cultivate identical paddies from year to year. The size and configurations change constantly, especially given rotations to other crops. Farmers can make more money on sweet potatoes than on rice.[9] Even so, to be a farmer in Java means to plant rice at least once a year. Farmers like to be self-sufficient, and rice is dependable. The market will take all the rice they can produce; they always make a profit.

We crossed over the southern slopes of cone-shaped Mount Lawu volcano. Some 3,265 meters high (almost 11,000 feet), it towers over Sragen and its rice paddies like a jealous god.

We spiraled upwards. As the temperature cooled, tropical crops gave way to carrots and cabbages, to yard-long beans, shallots, and onions. We passed through a pine forest and then started our descent. The terraces spread down the slopes in wide steps, miracles of hydrologic mindfulness. Irrigation water channeled from mountain streams filled the top paddy and then emptied down from one paddy to the next in a gentle, steady stream. The terraced fields created a tapestry of crops and colors. We are all artists, if we can only find our medium. The farmers of Mount Lawu shape the land with their terraces and paint its surface with their crops.

Whether we stuck to the main roads or veered off to the villages, battalions of schoolchildren and bicycles filled the streets at recess. During my six weeks in Java, I lost track of the many color combinations schools had chosen for uniforms, but brightness was the rule: brown with orange, green with yellow, grey and blue, red and orange. School buildings were freshly painted and had well-tended grounds; there was no shabbiness or broken glass. "I never see a dirty uniform," I said to Wiyanto, "not even a wrinkled one." Wiyanto had just joined the staff of the Yayasan Mitra Tani Foundation; he had ten years experience teaching mathematics in junior high school. "No matter how poor a family is," Wiyanto said, "they would never send a child to school in a uniform that is not washed and ironed." Completing six years of elementary school is now almost universal in Java; the same is true for the big islands like Kalimantan, Sumatra, and Sulawesi. Wiyanto said the big change came in the early 1980s. Today, parents are eager for their children, both boys and girls, to read and write. "The difference comes with junior high," he said. "Most Javanese boys complete it, but lots of the girls drop out; their families don't think the extra education matters much for a housewife." Traditions die hard. Even so, only one in five girls wore a full-length dress or covered her head with the *hijab*. These are orthodox Arab customs, not Javanese ones. Underneath, it is easy to spot the jeans and sneakers.[10]

We had arrived in the village of Bendunganjati, Mojokerto district, where we stopped at the local Mitra Tani office. Besides Wiyanto, I met Rini Asmunati, coordinator of the field-school project. Asmunati spends her life on the road, shuttling between one project village and the next.[11] Her friendliness and good humor make her a favorite with the villagers, men and women alike. We headed for Sriyono's house, where Asmunati and I were to spend the night. The closest hotels were miles away. Project workers like Rini Asmunati and Elske van de Fliert always stayed in the villages.

Sriyono is the village's *peneliti lapangan*, or field trainer. It's Sriyono who works directly with Bendunganjati's field school. As Asmunati explained, he had attended a workshop a year ago with a small group of farmers from participating villages. The workshop taught the principles behind the schools, the activities involved, and the role trainers are expected to play. Experience shows that extension workers often make poor trainers; they find it hard to see farmers as equals and resist letting them make their own decisions. These days, extension personnel often lack farming experience of their own. Consequently, a farmer-to-farmer exchange is much more effective.[12] Farmer trainers like Sriyono speak the same language and share the same way of life as their fellow farmers. Like other field trainers, Sriyono is paid a salary. The farmers who attend the field schools, however, do so based on the project's intrinsic relevance, not because they get paid. Few farmers understand insect life cycles or which symptoms go with which diseases. Once armed with such knowledge, they can cut back on pesticides with confidence.

During the growing season, field-school participants meet together once a week, in the morning. They divide into small groups of five and then go out to the sweet potato plots with a farmer trainer. Special activities are set up for each meeting. For example, groups might inspect samples of sweet potato plants taken from different locations in a field. On large sheets of paper, participants write up what they observe with colorful magic markers; farmers with poor writing skills draw pictures. Because the dense vines make a good habitat, a sweet potato plot has an abundance of insects, both friends and foes. Group tasks include finding unfamiliar bugs and bringing them back to show the whole class later, and collecting larvae. Participants put the larvae in jars with leaves; subsequently, they observe which insects hatch and which leaves each pest eats. Groups bring back caterpillars and keep them until they become butterflies. They track the deadly impact that a systemic insecticide like carbofuran has on friendly insects. They observe the difference between systemic and contact insecticides. "The point is," said van de Fliert, "farmers discover things for themselves. They learn practical things they can show to other farmers. They learn to look closely, to see more, to solve problems, to be patient, collaborative, and precise."[13]

Sriyono brought over a notebook with a pink cover. Participants use it to track each plot of sweet potatoes: the size of each field, the date planted, the date harvested, and the yield. Farmers indicate which crop preceded the sweet potatoes— usually rice. They keep track of how much of the crop goes to home consumption, how much is sold, and at what price. For each week in the sweet potato's growing season, participants note how much labor went into the crop, the inputs used, and any expenditures. At the end of the season, farmers compare what they did. The green book, which is for farmer trainers like Sriyono, is more detailed. When he takes a group into a field, he keeps track of what they see. His book has a list of pests, disease symptoms, and natural enemies; with an x, he marks the ones the group observes. He also tracks an insect's frequency, noting whether the incidence is slight (+), some (++), or many (+++).[14]

While van de Fliert and Asmunati explained how the field schools worked, Sriyoni's wife served sweet, steaming-hot tea. It was poured into tall glasses and then covered with a lid. The lid is replaced between sips, which keeps the tea inside warm and the insects out. Next came thick slices of sweet potato, which had been dipped in flour and fried. The white-fleshed, starchy varieties popular in the village have a delicate taste. There were bowls of peanuts and salak. An oval-shaped, fig-sized fruit with a tough skin, salak is peeled back before eating, much like a lichee nut.

At dusk, the mosque sounded the day's last call to prayer. Villagers stopped, took up their prayer rugs, and hurried to the mosque; the men entered on one side, the women on the other. There was no minaret towering over the village. Like the people it served, the mosque was simple and unpretentious. Children ran up and down its stairs; they romped, played, pushed, and shoved in its courtyard. In Bendunganjati, Allah did not seem to mind the pranks. Nor was the village's dress code a strict one. Village women never covered their faces. Outside the mosque, few bothered with the *hijab;* they wore colorful Javanese caps instead.

Sriyono's house had a large, comfortable porch; we sat on the stoop and removed our shoes before going inside to eat. In the villages there is but one basic meal eaten three times a day. There is no special breakfast food. Most meals consist of white rice, a peanut sauce with hot chilies, cooked greens, crackers made from manioc flour, plus tempeh. Made from soybean starch, tempeh is fermented, cut into squares, and deep fried. The food was set out with a huge pot of rice. I never had a meal in Java that omitted the rice. Eating was informal. We took a bowl, a tablespoon, and a napkin, helped ourselves, and then sat down to eat wherever we wanted. Meals are followed by fresh spring water, tea, and bananas. Variations are possible. If there is meat, it is usually chicken. Muslims do not eat pork. Sauteed combinations of shredded cabbage and carrots can go on the rice. Common too are mung-bean sprouts, deep-fried hard-boiled eggs, and fried manioc balls, reminiscent of Brazil's *pão de queijo,* likewise made from manioc starch.

For cooking, a cement, tiled stove had been built along the back wall of the house. The stove had oval-shaped burners hollowed out on top for coals; grates went on top to keep pots and pans from touching the hot coals. In the corner sat large pots of cooking oil.

Two small bedrooms were built off the hallway. Sriyono insisted that as guests of honor, Asmunati and I take the rooms. Each room had a narrow cot with a mattress on top of a board. There was a small table and a chair. I slept immediately. My alarm was set for dawn, but I hardly needed it. Between the roosters, the bellowing of donkeys, and the cows, I had no trouble waking up. For safe measure there was the mosque and its loudspeaker. An enclosed washing area was beyond the kitchen, a survivor of the original house. For washing up, a small room with an open, tiled holding tank was built into the wall; a long-handled ladle hung next to the tank. A pipe brought fresh water to the tank from a nearby stream. In Sriyono's house, it's customary to bathe twice or sometimes three times a day. Take a ladle of cool water from the tank, pour it over your body, and soap up. Then wash off. Do not step into the tank! Instead, use the ladle to rinse off, scooping out water from the tank. That's refreshing at the end of a hot, dusty day, but it's a rude awakening at 4 A.M.

Bendunganjati's field schools were experimenting with new sweet potato varieties. The vines either came from the project or were obtained in exchange with other villages. We set out to harvest and assess the trials. Around Bendunganjati village, hillsides are terraced, and intercropping is popular: three meters of corn, then five to six meters of sweet potato. The corn went in first, so it could keep ahead of the vines. We discussed sweet potatoes. The rainy season was late, the soil had cracked, weevil damage was probably bad. Sweet potato prices were discouragingly low. Farmers were leaving the crop in the ground as long as possible, hoping prices would go up.

There was plenty of help harvesting. Villagers first cut back the sweet potato vines with machetes. The vines were rolled back like sod and then tied into bundles. For "twine," villagers pared back the central stem of a banana leaf. Then we weighed the vines. How much foliage a variety generates is no small consideration, as farmers use the vines for fodder. According to Sriyono, the crop had been in the ground for 150 days, or about five months. The sweet potatoes were planted in mounded ridges. To harvest them, he stuck a machete in at the base of the mound and then pushed up, which loosens the roots. The roots were big, which everyone liked, but they had considerable weevil damage, which was a disappointment. We took out infested or diseased roots and put them in piles. Later, we weighed them to get an estimate of the total yield.

The roots went into heavy plastic-mesh sacks. Sriyono used a balance, which the villagers had lugged into the field, to weigh the roots. It had a ring at the top and a hook on the bottom. Through the ring went a pole, held by two men, one on each end. The hook was inserted through the top of the mesh bag of roots. The men with the pole then stood up gradually, while others moved the balance to register the weight. The results showed which varieties had the highest yields and the least weevil damage. Seeing is believing.

It was the end of the dry season. By now, many farmers had used up their allotment of irrigation water. The soil had cracked, making weevil damage likely. Java's sweet potato weevil is the common species, *Cylas formicarius*. Its habits are much better known than those of Uganda's weevils, which are a different species.[15] The sweet potato reacts to weevils by releasing an enzyme. The substance does little damage to weevils, but it is toxic to both animals and humans. Infected roots should be collected and discarded. Much of the time, such roots are left out in the fields, which promotes weevil proliferation. If a plot is rotated to paddy rice, this is not a problem, as the weevils are killed off by flooding. But if sweet potato follows sweet potato in an infested field, the second harvest will be much worse than the first.

Field schools promote rotations and field sanitation. Farmers learn to check vine cuttings for weevils prior to planting. To track weevil incidence, the schools introduced pheromone traps. A field school, however, is supposed to listen to farmers, not impose its own objectives. Weevil damage in Java is usually manageable. In most years, farmers do not consider it excessive. Of much greater concern is the price the sweet potato crop fetches and the demand for what they produce.[16]

When farmers sell off a standing crop for one lump sum, they avoid the cost of harvesting, transportation, and storage. The tradeoff, however, is a much reduced profit margin. If a harvest exceeds a trucker's expectations, farmers get no additional payment. If the yield falls short, or weevil damage is great, truckers often pay less than the agreed-upon price. To drive a harder bargain, farmers wanted an accurate way to estimate a field's yield. With this in mind, field schools taught farmers to sample their sweet potato plots and extrapolate to yields. In addition, farmers visited the big-city markets to find out firsthand how the sweet potato trade is set up.[17] Elske van de Fliert and Wiyanto took me to the sweet potato depot in Mojokerto, about an hour's drive from Bendunganjati.

Open-air markets in Java's main towns are crowded with hundreds of booths, kiosks, and stalls. Each part of a market has its own specialties. The dry-goods section, for example, had stalls for bamboo baskets, blinds, and mats; there was a booth for ornate paper fans, one for brooms, and one for brushes. Some stalls stocked only soaps and detergents. At the sewing-machines kiosk, one could get clothes mended or altered. Some stalls sold recycled products, such as kerosene lanterns made from old headlights and oil jars. There was a row of barber shops. The baked-goods section included not only bread and pastries but a bewildering collection of noodles, krispies, and crackers, made from rice and manioc starch. The narrow alleys lined with produce had the biggest crowds, the most clutter, the best smells, and the most vivid colors. I started with the hot peppers separated into piles of green, orange-yellow, and red. There were baskets of gingerroot and turmeric; strings of onions, garlic, and shallots; piles of cabbages, carrots, tomatoes, and eggplants; plus an assortment of greens, including amaranth. I passed by the bean sprouts, the green beans, and the yard-long beans, and then by the peanuts, the avocados, and the gigantic pumpkin squashes. I went on to the yams, the taro root, and the manioc, then to the sweet potatoes and the spuds. Finally, I headed for the fruits: stacks of coconuts, jackfruits, watermelons, and mangoes; piles of papayas, pineapples, and salak, of lichee nuts and passion fruits.

Several blocks from the main market in a large, open field was the town's sweet potato exchange. Truckers hauled the roots in from the surrounding villages and then sold them here to traders. As in most of Java's outdoor markets, women handled the business. They bought the loads and then looked through them, taking out diseased and damaged roots. They resold this sorted, higher-quality product to Jakarta truckers, who paid a premium price. What happened to the weevil-infested roots? An enterprising woman had worked out a way to recycle the waste and make a profit. At minimal cost, she bought infested roots from the traders. Most of the time, weevils do not destroy the entire root. So she cut out the good part. From one hundred kilos of bad roots, she salvaged about twenty kilos. This she sold back at a profit to the traders. As for the weevil-infested waste, she dried it and sold it to an insecticide company. Evidently, it ended up in mosquito-repellant coils.[18]

Until the field schools spurred an interest in marketing, farmers had no idea how high the markup was at each link in the trading chain. Renting trucks and marketing the crop directly had never occurred to them. Nor had processing. The demand for fresh roots is likely to fall as the country becomes more urbanized. So far, the biggest single outlet, which takes about 15 percent of the sweet potato crop, is the island's ketchup industry, centered in Jakarta.[19] Consumers do not realize the sweet potato is a main ingredient; it's a trade secret. Ketchup brands that use tomatoes exclusively, like Hunts and Del Monte, are much more expensive. But what of Indofood's "Saus Tomat"? It looks red enough, but rumor has it that the product is mostly sweet potato. Likewise with the chili sauces. The labels boldly display peppers and garlic, but never tomatoes. Such sauces are never deep red, but a yellow-orange instead, surely the telltale sign of the sweet potato. If labeling laws ever catch on in Jakarta's supermarkets, the demand for the sweet potato is likely to plummet. After studying the market, the project concluded that sweet potato flour offered the best processing prospects.

Wheat flour is imported and expensive, approximately $350 per metric ton. By comparison, a ton of fresh sweet potato flour can be produced for about $200; fresh root sells for about $50 a ton. How good is the flour? The faster the processing and drying, the whiter the product. Roots are first washed, thoroughly peeled, and shredded. Once the excess water is pressed out, the pulp is spread out on straw mats to dry. The last step is milling and then sifting the flour. Village women tried out the flour in donuts, noodles, cakes, breads, and various snacks. To get the right color, taste, and texture, they had to adjust the recipes and the percentages of different types of flour. To compete with wheat flour, the sweet potato product has to be absolutely white and the savings must be at least 25 percent. The higher the dry matter in a variety, the higher the flour production per kilo of fresh roots.[20]

The field-school project had started with a focus on pest management, but in the end, it had to consider both how the crop was marketed and how it was processed. A field school's agenda necessarily reflects the crops in question, for each crop has its unique problems. With rice, cutting back on insecticides was an obvious objective. With sweet potatoes, by contrast, farmers spent little or nothing on chemicals.[21] So the rationale behind pest management had to be different. In Java, rice is always in demand, farmers can sell all they want, and they always make money on it. With sweet potatoes, demand is the crop's key problem. If CIP and its partners ignore this simple truth, all the research on yields, pest resistance, and drought tolerance will be just a waste of time. Improved varieties and better production technology will only help farmers produce more of a crop they cannot sell. With sweet potatoes, end use is the issue. Working out the fine points of flour production is much more important than adding a couple metric tons to the crop's yield potential.

"A field school's objective," Elske van de Fliert always reminded me, "is to help farmers work out their own solutions, not tell them what to do." That approach made sense to pest-management activist Merle Shepherd, who had preached against dousing Java's rice with insecticides for years. From Clemson University, he spearheaded a project on rotation crops that typically follow rice.[22] He and his Clemson team worked out management protocols that reduced or eliminated insecticide use on soybeans, onions, chili peppers, and cabbage. I met Shepherd at Indonesia's National Agrarian University in Bogor, the project's headquarters. The "p" in pest management, Shepherd was fond of saying, stands for people. "The key," he said, "is not host plant resistance, it is village-level organization."

With respect to the sweet potato, the field schools have taken on a broader crop-management focus. By the end of 1998, more than eighty farmers, extensionists, and development workers had become field trainers; some 161 farmers had participated in sweet potato field schools. Over the next five years, with backing from CIP and Indonesia's Directorate of Food Crops Production, the project will have trained 12,000 farmers in thirteen provinces.[23]

CHAPTER 29

Feeding China

"**G**ood food does not come just from grain," said CIP postharvest specialist, Chris Wheatley. "To meet the food demands of the future, alternative crops must be exploited." I met Chris at CIP headquarters for East and Southeast Asia in Bogor, Indonesia.[1] He had just returned from a field trip to China. A tall, gangly Englishmen in his forties, Wheatley spent as much time as he could in the field. His work on crop utilization, sweet potato starch, flour, and feed is location specific. To Wheatley, sitting around in Bogor was mostly time wasted.

The World Watch Institute worries that development has eroded China's capacity to feed itself. It predicts massive grain imports.[2] In Wheatley's views, such calculations ignore the role that root and tuber crops play in feeding the country. Corn is not the only crop that can meet the demand for animal feed or for starch. Estimates of China's food prospects typically ignore the sweet potato. Yet, it can substitute for corn, especially in pig feed, and the root is used in starch production.[3] As China has urbanized, the demand for fresh sweet potato has dropped.[4] Promoting new utilization schemes can help take up the slack. As for the potato, it has benefited from China's new affluence. Urban dwellers like the potato's color, its taste, and its sophistication. As a french fry, it is a favorite at new fast-food outlets. At home, the potato mixes well with gingerroot and soy sauce, and it is easy to prepare.[5]

No country exploits roots and tubers more than China. For 1997–99, China's sweet potato harvest topped 102 million metric tons a year, which is 84 percent of the world's harvest.[6] Over 40 percent of this goes to animal feed, mostly to hogs, and China is the world's largest pork producer.[7] Between 1977 and 1994, China's pork production increased fivefold, from 5.8 million metric tons to 30 million metric tons; per-capita pork consumption climbed from six to twenty-four kilos. It takes four kilos of grain to produce one kilo of pork, and pork accounts for three-fourths of China's meat production.[8] So if China fed its pigs mostly on sweet potatoes, the impact on projected corn needs would be considerable.

As to the potato, production has mounted rapidly. Between 1989–91 and 1997–99, China's potato harvest went up by almost a third, from an average of 31 million metric tons to 48 million tons. It is now the world's largest producer. Much of the increment is due to added acreage, which increased from an average of 2.8 million to 3.2 million hectares.[9] Overall, for the past thirty-five years, the growth rate in China's potato sector has averaged an impressive 3.9 percent a year, far exceeding any grain crop.[10] China may end up a big grain importer, but it should be self-sufficient in its roots and tubers.

Between 2000 and 2020, demographers expect the world to add 1.5 billion people, half of them in Asia. China's population will increase by some 160 million, to total over 1.4 billion.[11] Feeding more people is only part of China's challenge. In the north, in China's wheat belt, the water table has dropped, making irrigation water scarce. Urban sprawl, new industries, and highways are destroying prime farmland.[12] Justifiably, World Watch keeps tabs on China's food production. One way China can keep its grain imports manageable is to foster root and tuber crops. By and large, that comes down to sweet potatoes and potatoes, for China grows only 3.6 million metric tons of manioc, 1.4 million tons of taro roots, and much less by way of yams.[13] With this in mind, CIP and China have stepped up collaboration.

CIP's work with China goes back two decades. In 1978, the Chinese Academy of Agricultural Sciences sent a scientific delegation to CIP; in turn, a CIP delegation visited major potato-growing regions in China. Regular visits by senior scientists followed. In 1985, the Chinese Academy and CIP signed an accord to set up a regional CIP office in Beijing. Subsequently, with the addition of the sweet potato to CIP's mandate, cooperation intensified via germ plasm exchange, joint research projects, workshops, and training.[14]

Sichuan and Shandong provinces each produce about 17 million metric tons of sweet potatoes a year; together, they account for almost 40 percent of China's production.[15] In Sichuan, postharvest priorities include sweet potato starch, flour production, and animal feed. Since 1978, township enterprises have invested in food processing. So even though dietary preferences have shifted, sweet potato output has held steady, bolstered by demand from the starch industry. In 1980, for example, about 60 percent of Sichuan's sweet potato crop was consumed fresh. By the early 1990s, the figure was just 5 percent. In the meantime, the percentage that went to starch processing rose from 5 to 20 percent, and the percentage used in fodder jumped from 30 to 70 percent. For China overall, between 1985–87 and 1992–94, sweet potato starch production more than doubled, from an average of 410,000 metric tons a year to 943,000 tons.[16]

According to Chris Wheatley, most of Sichuan's starch is currently sold as wet cakes for noodle production. Starch that is off-white and filled with ash impurities makes an unattractive product. As a replacement, urban consumers turn to more expensive, wheat-based pastas. That means an added demand for wheat products that China is hard-pressed to meet. Sichuan farmers also use sweet potato as feed, particularly for swine. Whether as starch or feed, sweet potato can substitute for more expensive grains, but only if the final product is good enough.[17]

For the past decade, CIP and the Sichuan Academy of Agricultural Sciences have worked on improving starch quality. According to their 1995 report, Sichuan had some 7,000 starch-production units, many of them household- or village-level enterprises. Such units produced an estimated 140,000 metric tons of starch; the conversion rate of roots to starch was 15 percent.[18] Manual processing has given way to small-scale mechanized equipment, such as root washers, grinders, raspers, and drum separators. CIP and its Sichuan collaborators identified the steps in extraction—such as mesh size during separation—that had the greatest impact on quality. A white, pure starch fetches a higher price and improves, in turn, the quality of products down the line.[19] Transparent noodles are a case in point. (See fig. 14.)

With backing from CIP, the Sichuan Academy developed a motor-driven screw extruder into which the hot sweet potato dough is pressed. "The result," said Wheatley, "is a uniform, high-quality product with better market acceptance, a product with just the right viscosity, opaqueness, and thickness." In Santai County, the demand for sweet potato–processing equipment has more than doubled. In 1996, for example, the sale of locally made root washers, starch separators, and extruders totaled over $180,000. In the meantime, the proportion of the 91,000 metric tons of sweet potatoes that Santai County processed increased from 36 percent to 76 percent, or to some 69,000 metric tons. The residues were recycled into animal feed. Santai County raised 110,000 hogs, a 70 percent increase over 1989.[20]

Besides its use as starch, sweet potato makes a high-quality flour. Food scientists at the Sichuan Academy of Sciences developed an "instant" sweet potato flour that can be mixed with soybeans and sesame in a popular porridge. Sliced roots are steamed for five minutes, which fixes the starch and prevents decoloration. The slices are then dried and milled into a fine, white flour. In taste tests, products made with sweet potato flour got high marks from consumers.[21]

Sichuan province is China's largest hog producer. In 1994, it raised over sixty million swine, almost 20 percent of China's total. Some 90 percent of this was household production of fifteen hogs or less per year. More sweet potato goes to the feed lot than for any other use. The rationale behind CIP's collaboration in Sichuan is to improve the quality of sweet potato feed, thus boosting the weight gain per kilo of feed and hence adding to on-farm profits. In Suining County, a typical case, hog production accounts for almost half of all farm income. Suining farmers usually

Fig. 14. Sweet potato noodles drying, Sichuan Province, China.
Courtesy CIP and C. Wheatley, from CIP, *International Potato Center Program Report 1991*, p. 132.

intercrop one row of corn between two rows of sweet potatoes. Hog manure provides a natural source of fertilizer. The sweet potatoes feed the hogs, and the hogs feed the soil with organic nutrients: between 180 and 220 buckets of manure per mu (one mu = 140 square meters). Hogs eat precooked, fresh sweet potato; they eat sweet potato vines and feed on starch-production waste. Feeding exclusively on sweet potato, however, is inefficient. Rapid weight gain requires a protein supplement, such as soybean dregs, oil seeds, or grain bran. Such additives make a big difference. On-farm experiments in Suining County showed that it took farmers a year to raise a hog exclusively on sweet potatoes; the daily weight gain averaged just 264 grams. By contrast, supplements reduced the cycle to just eight months, much closer to the U.S. average of six months. The daily weight gain averaged between 381 and 517 grams, for an overall savings of between 86 and 150 kilos of sweet potatoes per hog. Analysis showed that the value added by weight gain more than offset the cost of a protein supplement.[22]

China's yields of rice and wheat are at or above U.S. levels. With the sweet potato, however, there is still room to maneuver. Yields are seventeen metric tons per hectare, compared to twenty metric tons in South Korea and twenty-four metric tons in Japan.[23] A key factor is the quality of the vine cuttings farmers plant. With CIP's support, Shandong province began to multiply vines from test-tube, tissue-culture cuttings. They used ELISA (Enzyme-Linked Immunosorbent Assay) tests for virus detection. The project dates back to 1987, when CIP conducted a small course on viruses. An ELISA kit is a polystyrene plate with multiple small holes. To check for a virus, field researchers take a cutting from a sweet potato vine, macerate it into a paste, and then put a tiny fragment into one of the holes. Next, they add a couple drops of antiserum. If the serum finds virus-protein molecules with which it can bind, the color of the liquid in the hole changes. Using different antisera, researchers can test for several viruses simultaneously.[24] This durable technology has greatly simplified virus-free multiplication of both sweet potatoes and potatoes. A tissue-culture workshop followed the next year.[25]

To generate virus-free vine cuttings, Shandong agronomists start multiplication from meristems, essentially the same technology used with potatoes. In Shandong, sweet potato can be planted twice a year, so the multiplications add up quickly. Starting with just five hundred plantlets the first spring, the project produced enough roots to cover 6.7 hectares in vines by the second spring. In turn, cuttings from these vines generated 100 hectares of roots. When harvested, the yield was thirty metric tons per hectare, or three thousand metric tons overall. By the third spring, the project had enough stock to plant over 13,000 hectares. Thanks to this effort, the province's farmers began to plant virus-free cuttings in 1994. The impact on yields was dramatic; the average yield advantage across nine sites and five varieties was over 40 percent. The added output requires no additional fertilizer, water, or change in farming methods. Almost any farmer who uses virus-free cuttings gets higher yields. In 1998, some 3.1 million Shandong farmers planted half a million hectares of virus-free sweet potatoes, which accounted for almost 85 percent of the province's total. The result was an additional four million metric tons, valued at $167 million a year.[26] In Shandong, sweet potato starch, especially for noodles, is a growth industry. Currently, half of Shandong's output goes to local starch making.[27]

Sweet potato starch has multiple transformations: into noodles, vermicelli, and sheet jelly; into refined starch for the food industry; into starch derivatives, such as amylophosphate, amylum acetate, and soluble starch, for industrial use; and into maltose and sugar residues for beverages and brewing.[28] That being the case, selecting for high dry-matter content, which directly adds to the amount of starch extracted, is a top priority for CIP and Chinese breeders. According to Dapeng Zhang, CIP breeder and project leader, the objective is to "produce more usable material in every sweet potato." The recovery rate for starch averages about 15 percent by weight of the unpeeled roots. Therefore, varieties with enough dry matter to boost starch recovery by just 5 percent could add a third to the total extracted— that is, from 15 percent by weight to 20 percent by weight.[29]

China's farmers rarely plant just sweet potatoes. Instead, they grow them in rotations with grains, oil seeds, and legumes such as soybeans and peanuts. In some zones, farmers intercrop, with sweet potatoes on the ridges and corn or peanuts planted in the furrows between. Rotations help improve soil structure, control pests, and foster soil fertility. Intercropping maximizes the use of space, distributes labor, and increases a field's overall productivity.[30]

China is the largest global user of CIP germ plasm.[31] For the potato, the exchange began in 1978 when CIP-24 entered China as an in-vitro plantlet.[32] The Wumeng Agricultural Research Institute in Inner Mongolia did the propagation and field trials. This hardy, drought-tolerant cultivar produced larger tubers and higher yields than local varieties. In 1993, it was produced on over 150,000 hectares in four provinces, mostly in the north, under rain-fed conditions.[33] For its part, China does multilocational trials of CIP cultivars. Selection criteria include resistance to late blight, plus early maturity and drought tolerance.[34]

Almost 40 percent of China's potatoes are grown in the north as a main crop during the summer. The next year, farmers repeat potatoes or rotate to wheat, oats, or buckwheat. Farther south, the potato is sown as a fall or winter crop after wheat or paddy rice.[35] To fit the potatoes in, farmers need early-maturing types. In 1991, CIP's Philippine-based UPWARD (Users' Perspective with Agricultural Research and Development) network began a potato-production project for the rice-based cropping systems of Zhejiang province in south China. As in India's Ganges valley, potatoes follow the monsoonal season and are harvested in the winter. Early varieties coupled with new production technology have cut back the time it takes to grow potatoes; the project has also moved seed multiplication to the highlands, where disease pressure is less, and it has organized more than 350 training sessions for eighteen thousand farmers. A rotation to potatoes after rice improves subsequent rice yields. By combining the two crops, food production per land area has more than doubled. About half the potato harvest is cycled into hog feed, which both adds to the protein intake of farm families and generates manure for fertilizer.[36]

Across the Formosa Strait, Taiwanese farmers have planted potatoes in a rice-based system for years. In 1994, I visited farms near Taichung with horticulturalist Jocelyn Tsao, who worked for the Taiwan Agricultural and Research Institute. According to Tsao, seed quality deteriorates rapidly in Taiwan's hot, humid climate. If farmers do not restock with new seed, yields fall after just a couple seasons. We met with local farmers, including Mr. Hsieh. An experimenter who had his own greenhouse, Hsieh did on-farm potato trails for Tsao; he was also a leader in the local Farmers Cooperative. Hsieh said he planted two potato crops and a rice crop each year. The potatoes go in back-to-back, the first crop in September and the second in December. The rice follows in late April, timed to the rains. Hsieh claimed he had little trouble with bacterial wilt, because he got his second potato crop out by the end of March, "before the increase in soil temperatures begins to favor the disease." Hsieh also plants a green manure. The day before he harvests his rice, he broadcasts legume seeds over the rice fields. Once the rice is harvested, the rice straw left in the field protects the tender legume shoots. Two weeks prior to planting the September potato crop, he incorporates the green manure and the rice straw. This improves the soil's organic content and its capacity to absorb water. Mr. Hsieh said he made more on his potato crop than on his rice. He renews his potato seed stocks regularly.[37]

In China's southern provinces, by contrast, the poor quality of seed potatoes keeps yields low. To break the impasse, CIP has worked on TPS production with its project partner, the Wumeng Agricultural Institute in Inner Mongolia. The market for the project's TPS is in southern provinces, such as Guizhou, Yunnan, and Sichuan. Farmers sow TPS in nursery beds and then transplant the seedlings.[38] In the south, intercropping potatoes with other crops is commonplace. In Sichuan province, the country's largest potato producer, 80 percent of the potato crop is intercropped with corn—the potatoes in the furrows, the corn on the ridges.[39]

How important will China's roots and tubers be in 2020? Let's start with potatoes. Between 1983 and 1996, China's consumption grew by 4.6 percent a year. Farmers struggled to keep pace with the demand, adding a million hectares to potato cultivation. In the years ahead, such rapid growth is likely to drop. Nonetheless, even at a reduced rate, China's potato harvest is expected to increase 80 percent by

2020, reaching some 88 million metric tons. Most of the added output will come from higher yields, which are projected to rise from today's 15 metric tons per hectare to about 24 metric tons. That still leaves ample room for improvement. Yields in the United States, Germany, the Netherlands, and France are closer to 40 metric tons. If China closed the gap in yields, it could add another 30 to 60 million metric tons of potatoes to its 2020 total.[40]

On a fresh-weight basis, the sweet potato is China's most important crop after rice.[41] During the 1990s, sweet potato production leveled off as fresh consumption dropped. But assuming industrial demand picks up as predicted, China's sweet potato harvest should go by up a third, reaching some 136 million metric tons in 2020. The gains will come exclusively from yields, as the total area devoted to the crop is likely to fall.[42] By 2020, China's combined potato–sweet potato harvest is expected to reach some 224 million metric tons. That is the dry-matter equivalent of 60 million tons of grain.[43]

China can feed itself, but only with an empirical approach targeted to the cropping systems and constraints of its rural counties.[44] Where does the potato fit in, and what is the market for the sweet potato? What are the best ways to multiply and distribute seed, to increase yields, to protect the crop? Given China's diversity and the penchant its farmers have for intercropping and rotations, standardized solutions are not the answer. The "best set of practices" is necessarily "site and crop specific."[45]

PART IX
Conclusions

CHAPTER 30

Potato Lessons

"**S**eed does not lie; it can only give back the quality it has." That seems like a small truth for modern times. A great truth, however, can spring forth from a tiny seed. The potato has lessons for today worthy of its ancient lineage, of its ancestoral ties to culture, art, and time.

The potato teaches us to start with a strategy grounded in the basics. From seed production and pest management to genetic improvement and storage, the key is to take the diversity imposed by place, by farming traditions, and by climate seriously. It does not matter how big the problem is. Projects applied this "think small" rule to community-based seed production, to insect pests, to TPS technology, and to the late-blight wars; the rule was applied to tracking demand for Andean crops, to processing Uganda's sweet potato, even to raising pigs in China. In each case, local knowledge and hands-on experience shaped the solutions. Without such a practical approach, CIP and its partners risk defining unrealistic or irrelevant objectives.

In Bolivia, PROINPA sent out research teams to ask farmers about rotations, pest problems, and production practices, about yields, costs, profits, and marketing; they sampled potato plots by climatic tier and by community. Based on this information, PROINPA rank-ordered priorities, drawing up a research agenda and a plan of action for each. For high altitudes, breeding concentrated on frost tolerance; for the temperate valleys it was late-blight resistance; for the yungas it was resistance to bacterial wilt. For weevils, the project developed simple but effective control strategies. In the upper valleys, PROINPA worked on rotations to reduce nematode buildup; for seed production, it fostered protected beds. With community support, the project cleaned up old Andean landraces, returning virus-free seed to the villages the samples came from. Wherever PROINPA worked, it emphasized on-farm trials and farmer participation.

Ecuador's FORTIPAPA was likewise strong on the basics. Riobamba's research team had mapped out Chimborazo Province district by district. They knew the size and distribution of farms in each district. They knew rainfall and frost patterns valley by valley; they knew where the best soils were. The project collected data on farming practices: on how long farmers recycled their seed, on the varieties each village favored, and on each district's pest and disease problems. They knew how much of the potato crop farmers usually sold, how much farm families ate themselves, and how much they held over for seed. Once FORTIPAPA had a clear picture of a district's potato problems, it could work out a realistic plan of action. The project's redoubt was seed multiplication, in this case from stem sprouts. The technology allowed farmers to greatly expand how much seed each tuber generated.

To protect Andean crops, FORTIPAPA realized that collecting germ plasm was not enough. The project had to preserve Andean knowledge about how the crops were used, their distinctive traits, and their cooking characteristics. This rich heritage provided the script for FORTIPAPA's television series, "Cook with Class," which spurred urban demand for the old crops. In the meantime, market studies revealed what urban consumers knew about such crops. The result was accurate estimates of household demand for oca, ulluco, arracacha, and mashua.

In Tunisia, the tuber-moth project switched to less toxic chemicals. With the threat of pesticide poisoning thus minimized, the project turned to long-term solutions. The first step was to monitor the tuber moth's life cycle in Tunisia's potato fields. Armed with this site-specific knowledge, the CIP-INRAT project devised practical, inexpensive controls. By avoiding the tuber moth at its peak, by keeping fields moist, by hilling, and by rotating to other crops, many farmers eliminated insecticides almost entirely. This was not the case in Egypt's delta, where winters are warmer and frosts rare. The tuber moth peaks before the potato crop can be harvested; there is no escape. To get insecticides out of the potato's life cycle, the project pioneered organic substitutes. For GV, Egyptian researchers used a virus strain common to the delta's native tuber-moth population; for Bt, they isolated a strain of the bacterium in delta soils and then multiplied it. Farmers took the lead, testing both GV-based and Bt-based products. They used them in the field and on the potatoes they stored in their *nawallas*. Delta farmers verified for themselves that the substitutes worked, even though they did not kill on contact like insecticides.

In Uganda there was no formal seed sector. So CIP and NARO turned to a much simpler, farmer-based approach. The flush-out system uses up seed stocks each year, which prevents disease buildup. To contain bacterial wilt, the project began with on-farm field tests. Recommendations followed: farmers should start with clean seed and plant whole tubers; they should hill up around the potato plants, uproot wilted plants, and rotate crops.

To tame the late-blight dragon, CIP has turned to biotechnology and global research networks. But it has not forgotten that the real battle is in each farmer's field. With CIP's backing, farmers fight the disease armed not just with resistant varieties, but with knowledge about seed health, the microbes involved, and the potency of the fungicides used.

CIP worked on true potato seed for years. Success came with the give-and-take between research centralized at CIP and the down-to-earth approach taken by India's national program. Indian farmers and research teams selected good TPS parents; they worked out mass pollination, devised wooden-frame peg boards for uniform spacing, and added a step, the minituber. In West Bengal, a hard-working extension service collected on-farm data and tracked the minituber experiment. Farmers are now learning to produce their own minitubers from TPS directly. In Vietnam, farmers wanted to skip the minituber step, eliminating monsoonal storage entirely. They planted TPS in seedbeds and then transplanted the seedlings directly to their potato fields. The result is a potato crop produced from sexual seed rather than from last year's tubers. In Peru, Chacas farmers learned to cross TPS parents and produce their own true seed. Now, they use all the seed technologies simultaneously: they transplant TPS, produce minitubers, and store tuber seed.

Before CIP and NARO worked on Uganda's sweet potato, they documented which varieties farmers liked and why. They learned how farmers got vines to plant; the production techniques they used; and how households prepared, preserved, and consumed sweet potato products. The project helped village women keep nursery beds healthy through the dry season; it worked on storage and on the best way to process a high-quality, dried chunk or chip; it experimented with how to incorporate sweet potato into popular snacks and baked goods; and it looked at weight gain in hogs fed a sweet potato diet. In the meantime, breeders had found an orange-fleshed, vitamin A–rich variety that farmers liked. CIP now coordinates East Africa's VITA project.

The examples above teach us that solutions require participation. As CIP's Javier Franco put it, "Doing things the way other people want them done is not an answer, and it is not a very good definition of development." Successful projects consulted with farmers and incorporated their priorities; otherwise, they risked solving problems of little practical interest. In Indonesia, farmer input changed the focus of the field schools from weevil control to marketing and utilization. This does not mean farmers are smarter than researchers. Many farmers do not understand insect cycles, much less late-blight sporolation. Even so, projects must address problems that farmers want solved. When a project does this, participation is less problematic. In Bolivia's high valleys, the demand for protected beds exceeded PROINPA's ability to supervise their construction and use. Ecuador's Andean communities were multiplying seed at a prodigious rate. Midnapore's farmers planted all the minitubers the project could supply. Uganda's seed growers wanted more seed for multiplication than the station could supply. In Soroti, every compound seemed to have a sweet potato nursery.

A project strong on participation usually had a training component. At Uganda's District Farm Institute in Serere, village women attended a three-day workshop on sweet potato processing. No lesson seems more fundamental than the lesson that people learn by doing. To control the potato weevil or the tuber moth, to manage bacterial wilt or late blight, farmers had to change old habits or try things they had never done before. It seems simple enough to put traps in a field, use a biological powder, track rotations, or look through a microscope. Such innovations, however, can mean more work, require new skills, or imply new knowledge. How many weevil traps are needed per hectare? How much GV powder is needed to cover a ton of tubers? When is a TPS seedling ready to be transplanted? How likely is a late-blight outbreak? For such practical questions, hands-on training is the solution. In Indonesia, the "school" farmers went to was the very field where they gained their livelihood—a rice paddy, a sweet potato field, or a cabbage patch. The teachers were other farmers; villagers learned from each other. As a result, training added to a community's confidence and its capabilities.

Think small, build participation, bolster basic education. These are fine principles for village life, but how relevant are they to modern problems? From sustainable agriculture to family planning and energy efficiency, such principles are fundamental. Feeding China is a good example.

The world has a great stake in China's future, and China has an enormous stake in its farmers. Agriculture is still the core of the country's economy and the mainstay of gainful employment. What would happen if yields fell, if too much prime farmland were lost to industry, or if the income gap between China's villages and its cities kept growing? The country already has some 120 million migrants stalking the cities looking for work.[1] What would happen if China's farmers headed for the cities as quickly as Brazil's did a generation ago? In just fifteen years, between 1965 and 1980, Brazil's urban population increased from half to two-thirds of its population, a jump of 17 percent.[2] What if a comparable shift took place in China? China's urban population would climb from 376 million in 2000 (30 percent of its current 1.254 billion people) to 647 million in 2015 (47 percent of a projected 1.377 billion people), for a net urban increase of 271 million people.[3] These are enormous numbers. The current population of the United States is just 274 million.[4] Who would feed these millions of new city dwellers? How would they be employed?

American agribusiness will never feed China. Only Chinese farmers can. Rice is China's most important crop, so it makes a good case for analysis. For 1997–99, China's farmers produced some two hundred million metric tons of rice a year, or about 34 percent of the world's total. If China's rice harvest dropped by 10 percent, could U.S. rice exports make up a twenty million metric ton shortfall? Not by a long shot. For 1997–99, rice production in the United States averaged around nine million metric tons, of which about three million tons were exported. It would take almost 90 percent of all the world's rice exports to cover a 10-percent drop in China's rice harvest.[5]

Globalization and U.S.-style development are expected to save China, creating employment and prosperity. A wealthy China could afford to import all the food it needed. That seems like blind faith to me. Anyway, the arithmetic does not work out. If burgeoning cities and depleted resources required China to match Japan's per-capita grain imports, China would need some three hundred million metric tons a year. That far surpasses the current total of global grain exports. A more modest one hundred million metric tons still comes to more than half of the world's current exports. Such demand would bring sharp price hikes. Even if enough grain were available, could China afford to buy it?[6]

Big development cannot deliver on all its promises. An automobile industry is a key pillar in a modern economy. Can it play such a role in China? In 1996, China had about 1.8 million passenger cars, one for every 670 people; the United States, by contrast, had 123 million automobiles, or about one car for every two people.[7] Given the gap, General Motors, Mercedes-Benz, Chrysler, and Ford all want to build cars in China. But have they considered the consequences? What if per-capita automobile ownership in China equaled that of the United States? China would have over 600 million passenger cars, which exceeds the number of automobiles currently on the world's highways. Global demand for gasoline would double, as would CO_2 emissions.[8] And where would the cars go? China's population and its best farmland are already squeezed into about 40 percent of its territory.[9] Long before China produced 600 million cars, it would run out of food, not to mention gasoline.

Even without the automobile, "development" has already exacted its toll on the country's environment. Five of the cities with the worst air pollution in the world are in China. The level of "total suspended particulates" of soot and dust in the winter peaks at between four hundred and eight hundred micrograms per cubic meter, many times the World Health Organization's guideline of sixty to ninety micrograms. In southern China's Guangdong province, a hub of the country's industrial boom, 60 to 90 percent of the precipitation is acid rain. The damage done to China's forests and crops is put at $2.8 billion a year.[10] By some estimates, pollution and environmental degradation subtract from China's GDP faster than big development can add to it.[11] Factories and urban sprawl have taken their toll on the country's farmland.

Given its limited resources and huge population, China's capacity to feed itself is far more important to its future than an automobile industry. When it comes to development, the country's farmers are better allies than Wall Street brokers. To shore up its agriculture, China does not need the International Monetary Fund or free trade. It needs to think small, seek local solutions, build participation, and educate. This implies a strategy that builds up the country's agriculture systematically, county by county and province by province. Given how large China's population is and how small the margin for error, it is the only sustainable path to the future. How Suining County did this with the sweet potato will have to be multiplied many thousands of times to fit different crops, climates, and circumstances.

Suining agronomists filed two reports, one on diversifying the market for sweet potato starch, the other on the sweet potato's role in hog production.[12] The same eight-member research team did both reports. Consider product diversification. The team started with "desk research," using extension data to estimate Sichuan's total sweet potato output and the percentage that was processed. They identified which counties had the highest sweet potato production. Then came market research. They visited small-scale enterprises that specialized in starch production, in noodles, and in flour. They talked with owners. They studied the quality of the products produced and the technology used. When they knew the facts, they calculated what the potential demand for starch by local industries was. They had bakers try out sweet potato flour. In the meantime, an engineering team had greatly improved the quality of flour production. They steamed root slices for five minutes prior to drying and milling. They worked on an "instant flour" for household use in desserts and snacks or for mixing in with soybeans and sesame for porridge. Taste panels were set up in selected villages and towns. The result was an action plan that improved technology, attracted local investment, and utilized a local resource efficiently.

The same was true for hog production. Using on-farm interviews, the team described the household production system. This included how much of a family's income came from animal husbandry and how much from crop farming. The team identified the main crops produced, the most popular intercropping patterns, and the predominant rotations. The team looked at all aspects of hog production, from feed to weight gain and weaning rates. They looked at each task by the age and gender of the household members responsible. What resulted was a plan to improve the quality of sweet potato–based fodder. This included using fresh vines, blanched roots, recycled starch residues, and a protein supplement.

These are small steps. When it comes to industry, we all believe in multiplier effects. Why not in agriculture? In Sichuan's Santi County, the demand for locally made root washers, starch separators, and extruders more than doubled. Hog production increased by 70 percent. Rapid weight gain more than offset the cost of protein supplements like oil seeds and rice bran. Small-scale agriculture has technological spinoffs too. In-vitro tissue culture, ELISA virus detection, TPS production, and molecular breeding are all biotechnologies.

An impressive macroanalysis will not feed China. Soil quality, the irrigation system in place, and farm characteristics are all location specific. China's food problem is not a single problem. Each region and each province faces a unique set of limitations and challenges. The constraints in the north, in an arid, wheat-growing interior province like Gansu, are radically different from the situation in the south, in a richer, wetter, rice-growing province like coastal Fujian.[13] Farmers in Gansu apply less that 50 kilos of nitrogen; in Fujian, the average exceeds 250 kilos.[14] To assess China's food prospects, begin with its provinces, and within each province, analyze by county.

Just such an analysis was done for Fujian province. County-level campaigns sought to build up the province's grain base. To do this, low-yielding fields had to be salvaged. How? By building terraces, by erosion control, and by improved drainage. Then the soil's structure had to be improved by adding organic matter, changing crop rotations, and applying lime. In addition, the province had to reduce storage and transit losses; it wanted to improve seed production, notably for rice; it needed better fertilizer and more efficient water use; it wanted more "ecological agriculture" that combined aquaculture with rice production.[15] There is nothing remarkable here. To better feed itself, Fujian has a plan that assumes that changes made locally will add up to large changes overall, that a modest gain made on each rice paddy and sweet potato plot will add up impressively.[16]

To meet its food needs, China will have to take its agriculture as seriously as it takes its industrial projects. Agriculture must be built up plot by plot. This means local research on the crops, irrigation systems, and cultural practices that characterize farming systems all over China. It means small-scale food processing, from sweet potato noodles to french fries and soybean curd. It means product development and better marketing. It means user-friendly computer software to track district-level inputs, crop mixes, and production. It means scientific research, from hydraulic engineering and genetics to forestry and agronomy. It means science in the service of agriculture and ordinary farmers. And it means less waste. If China thinks small, finds wisdom in its problems, collaborates with its farmers, and learns by doing, it will find its way. A future based on watered stocks and currency speculators, on leveraged buy outs and get-rich schemes, may be no future at all.[17]

CHAPTER 31

The Patented Potato

"**H**ave you heard about the patented potato?" an irate friend of mine asked me. Since I was writing about potatoes, research, and Third World farmers, surely I knew about this sinister version of the spud. My links to an international organization like CIP seemed suspicious.

The patented potato in question came from Monsanto, not CIP. Marketed in the United States under the "New Leaf" label, the Monsanto spud is genetically altered to carry a Bt strain (*Bacillus thuringiensis tenebrionis*) toxic to the Colorado potato beetle.[1] The Bt toxin ends up in every cell of the plant, including the tubers. When the Colorado beetle feeds on a New Leaf plant, it soon bites the dust. When consumers buy and eat a New Leaf potato, they also get the Bt gene that is encoded therein. There is no escape.

Bt is a bio-insecticide. In Egypt, delta farmers used a different strain, *Bacillus thuringiensis kurstaki,* against the tuber moth.[2] Bt-based products are considered kosher in organic gardening.[3] Even Rachel Carson recommended Bt as a replacement for DDT.[4] Spraying a potato with Bt, however, is quite different from eating one with Bt inside. Is such a potato safe? Apparently.[5] When humans eat a potato with Bt incorporated into it, the toxin's DNA protein is broken down by stomach acids before it can be digested. That is why it is considered safe, even in large quantities.[6]

Does a Bt potato make you nervous? If so, consider the chemical regimen to which the spud is subjected in the United States. A special *New York Times* potato report went through each step. In the spring, farmers start with a soil fumigant. If nematodes are a problem, they "douse the soil with a chemical toxic enough to kill every trace of microbial life." Then, at planting time, a systemic insecticide is sprayed on the ground. Young seedlings absorb the chemical, which protects them against insects foolish enough to feed on their leaves. Then come the herbicides, followed by a barrage of fungicides against A-2 type late blight. To protect against aphids, which transmit virus diseases, farmers might use an organophosphate like Monitor. If so, they are best advised to avoid a freshly sprayed field, as Monitor exposure can cause neurological damage. Some Idaho farmers refuse to eat the very potatoes they produce. For household consumption, they grow potatoes with fewer chemicals on a separate plot.[7] A potato raised in Uganda is more fit to eat than many an Idaho baker.

Monsanto's New Leaf allows farmers to cut back on insecticides. The rub is that from start to finish a New Leaf plant puts the Colorado potato beetle under intense pressure. How long will it take for the beetle to develop resistance? Not long.[8] Monsanto advises farmers to include corridors of nonresistant plants as a refuge. In this way, beetles with resistance still find mates without, thus delaying the onslaught of a beetle population more robust than ever. According to the *New York Times,* farmers out west rarely bother with an insect refuge. Much less do they collect beetle samples and test them according to "resistance monitoring protocol."[9]

The potato is not the only crop engineered to carry a Bt toxin. Bollgard Bt cotton kills off both the bollworm and the budworm; Yieldgard Bt maize is lethal to the notorious corn borer; the Bt in tobacco kills the hornworm. Like the tuber moth, all these pests are lepidopteran members of the moth-butterfly family. For protection, the altered cotton, corn, and tobacco plants all incorporate the same Bt toxin, kurstaki. In 1997, Bollgard cotton and Yieldgard maize were planted on over five million hectares in the United States. That same year, a scientific team reported that 1.5 out of 1,000 diamondback moth larvae had a gene that carried resistance to four different Bt toxins; the rate was a thousand times higher than expected. In the meantime, farmers in Texas, Louisiana, and Mississippi have all reported major failures of Bollgard cotton.[10]

This is bad news for organic farmers. It is also bad news for Third World farmers. As a spray, Bt is insect specific rather than systemic; it rarely harms predators or other insects.[11] When incorporated into a plant, however, Bt is more concentrated and more potent. Bt corn, for example, may endanger the monarch butterfly.[12] Used sparingly, traditional Bt-based products could hold up for years. Unfortunately, large-scale deployment of transgenic Bt crops is sabotaging this valuable ally. The worst case would be generic cross-resistance to Bt toxins of every stripe.[13] If that happens, will agrochemical companies compensate Idaho's organic gardeners or Egypt's potato farmers?

If Bt loses its clout, can agrochemical companies find new toxins to incorporate into their spuds? Probably. But if New Leaf is any example, it won't come cheap. Much of what Idaho farmers save on chemicals they pay out in technology fees—

in the case of New Leaf, an extra forty to sixty dollars per hectare. Worse still, when harvested, the crop does not entirely belong to them. Monsanto has patented New Leaf potatoes. Any farmer who keeps part of his crop for next year's tuber seed is violating federal law.[14]

What is good for Monsanto may be good for its stockholders but not necessarily for agriculture. Biotechnology can serve the public good, at least in theory. In practice, the lion's share of corporate transgenic investment has gone into engineering crops for herbicide resistance—or better said, for resistance to the specific herbicides a company happens to sell. For Monsanto, this means Round Up; for Ciba-Geigy it means their Basta herbicide; for the France-based firm Rhône-Poulenc, it means Buctril. Their patented versions of crops like cotton, soybeans, and corn can only be doused with their specific, trademark herbicides. The weeds die, but the tender crop seedlings survive. The catch is that Monsanto's new soybeans, or Rhone-Poulenc's cotton, only go with Round Up or Buctril respectively. The incorporated resistance is herbicide specific. If Round Up is used on Rhône-Poulenc cotton, it will kill it—likewise if Buctril is spayed on Monsanto's transgenic soybean.[15]

Companies claim that with transgenic crops herbicide use is more efficient. That may be true in a narrow sense. Overall, reliance on such crops is expected to boost herbicide use, making a bad problem worse.[16] In 1999, half the U.S. soybean crop was of the transgenic, "Round Up Ready" type.[17] Relying on a single herbicide builds up resistance in weed species faster, kills off more soil microbes, reduces soil fertility, and can even reduce the germination rate of subsequent crops.[18] Transgenic plants could cross-pollinate with intrusive wild grassy relatives, thus adding hybrid vigor and herbicide resistance to weeds. The result could be ecological disruption as new super weeds, or herbicide-resistant volunteers from the last crop, crowd out the competition.[19] Even more powerful herbicides would then be required. As herbicide use mounts, the residues will show up in soybean products, cotton-seed oil, and animal feed, to the detriment of human health. Moreover, the shift in plant metabolism and genetic structure required to confer herbicide resistance itself poses risks.[20]

Whatever promise genetic engineering holds, its corporate genesis has contributed little to either public health or sustainable agriculture. New agricultural technologies "should not make pest and weed problems worse, erode crop diversity, mine the soil, or add to pollution."[21] CIP's work comes much closer to this standard than a corporation with one eye on the stock market. CIP's mandate is to "develop, adapt, and expand research on potato and sweet potato production in developing countries."[22] Sponsorship of sustainable technology is part of its job description. Moreover, CIP goes about its business in a way quite different from that of Monsanto or Dupont.

CIP does not collect royalties on its technology or patent its varieties. It does not buy up seed companies to market its potatoes. Its mission is to share technology and research, not monopolize them. When CIP was founded in 1971, in-vitro potato propagation was considered an advanced technology. Today, thanks to CIP-sponsored research networks, rapid multiplication is used by national programs from Bolivia's Toralapa research labs to Kenya's quarantine station at Maguga.[23] In Ecuador, projects used tuber sprouts for multiplication, in Uganda, the Kalengyere station took stem cuttings from mother plants.[24] Even farmers can work with tissue culture and rapid multiplication. In 1979–80, the Dalat research station in south Vietnam received seventeen improved cultivars in test tubes from CIP. Once farmers saw how technicians multiplied seedlings at the station, they copied the techniques themselves. They set up makeshift laboratories to generate mother plants from nodal cuttings. From the mother plants they took stem cuttings, which they rooted in pots made from banana leaves. In just a couple seasons, they were turning out some two hundred thousand rooted seedlings a year. In 1983, twelve such farmer-run operations sold nearly three million potted seedlings to local cooperatives for direct planting.[25] CIP has also disseminated techniques for low-cost virus detection. When I first went to CIP in 1988, ELISA kits and serology were elite technologies. Today such kits are found in labs throughout the Third World.[26] In China's Shandong province, for example, ELISA technology helped the Xushou Research Station multiply out virus-free sweet potato stock. For bacterial wilt, CIP pathologist Sylvie Priou recently worked out a cheap, reliable way to screen seed tubers for infection prior to planting. Using a $200 NCM (nitrocellulose membrane) ELISA kit, seed projects can now sample and test 2 metric tons of seed, enough to generate a thousand hectares of bacterial wilt-free potatoes. The new kit can detect ten bacteria per milliliter of extract, which makes the test a million times more sensitive than prior technology was. The kit comes with a video and instruction manual.[27]

CIP's first postharvest team included a contingent of anthropologists. They brought a "farmer-back-to-farmer" approach to CIP. To this day, CIP projects are strong on farmer-based diagnosis, participation, and evaluation. CIP has promoted its farmer-centered methodology from Bolivia and Ecuador to Tunisia, India, and China. This may seem like a self-evident approach now. But back then, a problem's social context seemed much less important to scientists than its technical side. Storing potatoes in diffused light is an example; it was one of CIP's first on-farm success stories. The inspiration came from the way farmers built the inner courtyards of their homesteads. They raised the roof up above the walls so that light was admitted indirectly. CIP scientists used a similar design for diffused-light potato storage. Constructed from local materials, the rustic stores caught on quickly. The key was farmer input and on-farm testing. Without this, what CIP designed might have been superior technically, but it would have cost too much in time, labor, and materials.[28]

Yields of both potatoes and sweet potatoes are much better if farmers start with clean planting material. CIP's promotion of disease-free propagation reflects its focus on seed systems and their deficiencies. That was the rationale behind Bolivia's rustic beds, Tunisia's seed program, Uganda's flush-out system, and Shandong's vine multiplication. It is likewise the key to true potato seed. CIP worked on breeding TPS for years and solved key technical problems, such as how to select good parents and how to break the seed's dormancy. Even so, before the new technology caught on, Indian researchers had to reshape it. TPS meets the sustainability test. A plot of TPS plants is more genetically diverse and more disease resistant than a plot of clones. TPS does not transmit late blight, bacterial wilt, or viruses.

From the beginning, CIP stressed cooperation in research. Its original charter was based on agreements between the government of Peru and North Carolina State University. Today, CIP has an impressive list of academic collaborators. In the United States, this list includes Cornell, Louisiana State University, the University of Georgia, North Carolina State University, and the University of Wisconsin; in Europe, it includes Wageningen University in the Netherlands, the University of Birmingham in the United Kingdom, and Tübingen University in Germany. CIP currently lists over 165 research partners in forty-five countries; in addition to public and private universities, the list includes national programs, development agencies, international organizations, and NGOs.[29]

CIP was the first center to make regional research networks the hallmark of its strategy—before the fax machine, E-mail, and the Internet made such collaboration fashionable.[30] CIP still takes a decentralized approach. In 1999, it had seventy scientists posted to regional or liaison offices in fourteen countries.[31]

To insure access to its technology, CIP sponsors in-country training for national scientists, extension workers, and teachers. In 1998, some 1,280 people were trained in specialties that ranged from ELISA-kit virus detection and rapid multiplication to disease and pest management, farmer field schools, postharvest utilization, and sustainable crop management. In addition, forty national scientists received training at CIP headquarters in Lima. As backup, CIP's communications department puts out a steady stream of do-it-yourself manuals, training bulletins, IPM brochures, and scientific publications.[32]

For 1997–99, CIP's budget averaged approximately $22 million a year. The largest share, some $13 million annually, came from European countries; at $4.5 million, the Swiss Agency for Development and Cooperation was the single largest donor. Contributions from the United States Agency for International Development averaged $2.1 million; for Japan, contributions were $1.6 million; and for developing countries collectively, the total came to $523,000 a year. Canada averaged $657,000, Austrailia $155,000. The rest came from private foundations, international organizations, and the World Bank.[33] How much is $22 million? Not much.

A small liberal arts college in the United States has a bigger budget than that. One college more or less makes little difference. CIP, however, has made a great difference to the potato and its prospects, to small farmers, and to the world's food supply.

To gauge its impact, critique its programs, and evaluate its management, including how it makes decisions and spends its money, CIP is reviewed by an independent panel. Since 1972, four panels have assessed CIP, most recently in 1995. Members are selected by the Technical Advisory Committee of the World Bank. CIP's management is scrutinized regularly and more rigorously than American universities, not to mention public agencies, NGOs, and international organizations. Systematic peer review assures donors that the funds contributed are accounted for. To keep on track, CIP is expected to set its priorities in advance by program, project, and projected budget. In 1997, for example, it had its 1998–2000 plan ready for distribution.[34]

When CIP became an international center, its first task was to collect and classify potato varieties.[35] Using Lima as a base, it amassed thousands of rare Andean cultivars, including wild species. This genetic inheritance does not belong to CIP. It is held in trust for all nations under the auspices of the Food and Agricultural Organization (FAO) of the United Nations. Over the past twenty-five years, this collection has helped the potato take on late blight, stand up to viruses, withstand pest attacks, and survive frosts. Whether late blight or nematodes, CIP breeders do not work alone. They have allies and partners in scores of countries. Research networks take promising new cultivars and evaluate them. The testing sites today span latitudes and regions. The result is a free flow of germ plasm under CIP auspices that benefits local farmers. In China's province of Inner Mongolia, farmers plant CIP-24; in Uganda, farmers like Cruza-148, a variety from Mexico. For late-blight resistance, Peruvian farmers are replacing Canchán with Chata Roja, a more resistant, farmer-selected variety. Bolivia's PROINPA now has Gendarme and Revolución, varieties that withstand frost. CIP has bred a potato with tiny sticky hairs on its leaves and stems. The trait, which could help deter insects, comes from a wild potato species, *Solanum berthaultii*. Common potatoes have shallow root systems, which makes them finicky about water. CIP's work on a deep-rooted potato may help the spud put up with longer dry spells.[36]

CIP is a go-between for germ plasm exchange. In the fight against late blight, for example, PROINPA worked with tuberosum-andigena hybrids that came from national programs in Mexico, Colombia, and India. How much CIP germ plasm a variety has is mostly irrelevant. What matters is that farmers get a good crop. CIP-24 is a case in point. It began in the 1950s as a cross between a German potato and a wild species. This material was in turn crossbred with Argentine cultivated potatoes. From this came a new variety released in Argentina as Achirana. The new spud did not catch on with Argentine farmers. Nonetheless, a CIP scientist saw

Achirana, liked it, and had a sample sent back to Lima in 1976. Later, after the variety was tested for disease resistance, CIP sent a sample to China. After multiplication, testing, and evaluation, it was released to farmers in Inner Mongolia. By 1993, C-24 was in twelve developing countries from Madagascar to Bhutan.[37] So who should get credit for CIP-24? Should it be the Germans who made the first cross, the Argentines for releasing Achirana, CIP for circulating it, or the Chinese for multiplying the seed?

To make good use of the potato's genetic inheritance, CIP has doubled its biotechnology budget.[38] Much of the work concentrates on ways to directly transfer genes from one potato species to another. To make this pay off, scientists must first know where the genes are and what they control. CIP relies on molecular techniques to map the genes. An example is the Andean cultivated species *S. phureja*, known for its quantitative, r-type late-blight resistance. CIP hopes to genetically localize the genes involved and then transfer them to popular varieties. Hence, biotechnology need not narrow the potato's genetic base.[39] How different is this from inserting a Bt gene into a potato cultivar? Not very different technically, but there is a great difference biologically. The Bt gene comes from an organism totally alien to the potato. By contrast, in the case of r-type resistance, the genes come from *S. phureja*, another potato. The r-gene transfer is "natural," as conventional hybridization could also introduce the gene, but not without years of patient selection and backcrossing.[40]

When I visited Egypt's delta, farmers did not have a Bt potato. They might have one soon—and from CIP, not Monsanto. In collaboration with Belgium's Plant Genetic Systems, CIP has engineered a potato with the Bt strain *Bacillus thuringiensis berliner,* which is toxic to the tuber moth. Field trials showed zero larvae per plant, which is very high resistance. Nontarget insects escaped unscathed.[41] So far, Sangema and Cruza-148, popular varieties in Uganda, have been transformed and tested.

Because a Bt potato comes from CIP does not exonerate it of the dangers discussed above. At least CIP scientists used a different Bt strain. Even so, a Bt spray is much more versatile and less problematic than a Bt spud.[42] Farmers can use a spray on any variety they happen to plant. But if they want a Bt potato, the selection is necessarily restricted. And what if the only option were New Leaf? This helps explain why CIP is in the Bt potato business. A private company could insert a Bt toxin into a Bolivian cultivar and then patent the result. Bolivian farmers could end up paying royalties to use the very genetic resources they helped preserve. So to be on the safe side, CIP has to keep up with the competition. Perhaps it will be forced to patent its technology. If it did, its policies would be shaped by the Convention on Biological Diversity, which obligates CIP to make its "materials, products, innovations, and technologies" freely available to its partners in developing countries.[43]

To hawk their herbicides, agrochemical companies have genetically altered some of the world's most important crops. Critics claim they did so without adequate concern for human health or biosafety. Nonetheless, this does not make biotechnology and transgenic plants the enemy. Both raw potatoes and bananas, for example, have been genetically altered to carry the hepatitis-B vaccine. Just ten hectares of transgenic bananas would suffice to vaccinate all the children in Mexico.[44] Crops with higher vitamin content and essential amino acids could improve nutrition. If the nitrogen-fixing traits of legumes could be crossed into grains, farmers worldwide could cut back on chemical fertilizers. The world needs plants that do well in saline soils, that hold up despite drought, and that resist frosts. Such engineered characteristics could bolster yields and reduce crop failures. Built into the plant, insect and disease resistance could cut pesticide use to the benefit of agricultural workers, consumer health, and the environment. As with Bt, however, the risks, costs, and benefits must be carefully weighed.[45]

For a balanced assessment, who would you trust? A for-profit agrochemical company tied to its stockholders or an organization like CIP, responsible to international donors, NGOs, national programs, and Third World farmers? CIP is not perfect. It makes mistakes. Trying to breed a tropical potato did not pay off. Maybe its Bt potato is not a good idea. But CIP is far more accountable than an agrochemical company. And it has a better record. CIP has done basic research in pest management for years. It has promoted locally made bio-insecticides. The objective of its late-blight research is to cut down or eliminate fungicides, not to mix a more potent concoction. In Bolivia, CIP and its partners reduced nematode levels with crop rotations and intercropping, not by killing off the soil's microbes. And do not forget the traps and baits, the community-focused way projects worked on the Andean potato weevil, or how a farmer field school teaches its principles. From pest management to late blight and molecular genetics, CIP has been the ally of science in the service of the world's small farmers.[46]

CHAPTER 32

The Macro Level

“**B**abies aren't made at the macro level,” chided anthropologist T. Scarlett Epstein. Given the penchant for global summits on population growth, one would almost think international agencies controlled fertility. Epstein recommended a client-led approach to family planning and fewer conferences. “Babies,” she reminded the policy makers, “are made in the minds and bodies of ordinary people living their everyday lives.” Before global agencies designed a new family-planning blitz, Epstein suggested they do a little homework, what she called “culturally adapted” market research. If not enough couples practice contraception, then clinics needed to know why. To find this out, they need to recruit and train local research teams.[1]

For family planning, the cultural context in which people make decisions is of paramount importance. Northwest Europe's population doubled between 1750 and 1850. It did so because more people got married younger and fewer of their children died. After 1850, total fertility rates began to drop as women had fewer children, but the exact pattern was different in each country. Today, much of Asia is in the midst of a similar fertility transformation. Understanding the dynamics is not a matter of a simple experiment in food supply and reproduction. In human societies, marriage customs, a household's division of labor, literacy rates, and economic opportunity are all part of the equation. To be successful, a clinic has to understand what drives demand; it has to keep in touch with its clients. The same is true in agriculture.

Most CIP projects incorporate a version of Epstein's market research. The “farmer-back-to-farmer” slogan is based on respect for the client's point of view. It took a while to learn this lesson. Between 1977 and 1980, CIP conducted farm-level surveys and on-farm experiments in Peru's Mantaro valley. The interdisciplinary team included a contingent of anthropologists and rural sociologists. They arrived at conclusions not unlike Epstein's. The report criticized CIP's top-down approach. At CIP, "development" meant that farmers adopted the "technological packages" scientists had designed for them in advance. Improved seed, for example, was the orthodox cornerstone of CIP's scientifically approved yield-improvement package. In the Mantaro valley, however, on-farm research showed that the local seed was not so bad. Replacing it with expensive seed from outside was simply not a top priority. What farmers did need was a better way to store seed. From this came CIP's collaborative work on diffused-light storage and a more hands-on, farmer-friendly approach to testing its technology.[2]

The Mantaro team found little evidence that technology could be transferred directly to farmers "without local refinement or adaptive research." Experience showed that farmers quickly identified which aspects of a new technology suited their needs. They were better at such practical adaptation than either researchers or extension workers. This was not the kind of conclusion academic experts back in the United States wanted to hear. They rejected CIP data because it came from "on-farm experiments." In its place, they proposed a research design much too complicated for a farmer's field. The mania for "scientific rigor," conventionally defined, came at the expense of relevance.[3] Norman Borlaug, who won the Nobel Prize for his research on wheat, once remarked that every discipline needs a reality check. Good crop research, he argued, "needed contact with the land, with crops, with farmers." Without such interplay, research gravitates to fancy high-tech centers where it becomes "affluent, oversophisticated, and complacent."[4] At CIP today, biotechnology is only part of the story; on-farm trials, training, and farmer field schools all show that CIP still remembers where potatoes come from.

A country's success in family planning does not depend on how wealthy it is, or how fast a country's GDP is growing. In India, Kerala is one of the country's poorest states. Yet, it has a total fertility rate comparable to Canada's. What made the difference was a state government dedicated to community health, rural education, land reform, and food for all. With potatoes, success does not depend on how sophisticated a project's research design is. Solutions have to work on farms, not just in laboratories or at experimental stations. Whether FORTIPAPA's weevil traps or India's minitubers, what matters is the way a project combines local research with on-farm training to improve productivity.

A global economy is touted as the fast track to the future, but like it or not, that future will be shaped by what happens to the world's two billion villagers.[5] Global analysts are prone to forget where babies really come from. They like their numbers big and their statistics general; they find specifics confusing. GDP is a case in point. The premise is that GDP growth benefits the majority. When the GDP is up, it is a good sign for development; a drop spells trouble. That is a half-truth at best, even in the United States.

In the 1990s, the American economy regained its dominance; its growth rate exceeded that of any industrialized nation. How much did this stellar performance contribute to the lives of ordinary folk? Much less than the figures on GDP suggest. Between 1990 and 1998, the GDP of the United States increased by almost 50 percent, from $5.7 trillion to $8.5 trillion. By contrast, the country's median household income was up by just half the GDP rate. In the meantime, the richest 20 percent of American families gained at everyone else's expense. They made 44.2 percent of the country's aggregate income in 1990 and 47.2 in 1997. The super-rich did even better. The top 5 percent earned 17.1 percent of family income in 1991 and 20.7 percent in 1997.[6] Corrected for inflation, the United States had sixty-eight thousand millionaires in 1995, five times more than in 1979; the average inflation-adjusted income of the richest 5 percent of U.S. families was $123,000 in 1975 and $189,000 in 1995.[7] Today, America's income distribution looks more like Mexico's than it does like Germany's or the United Kingdom's.[8]

A rise in GDP means someone is getting more, but more of what? And who gets it? In 1970, middle-class families (the third and forth quintiles) earned 41.4 percent of aggregate income; in 1997, the figure had dropped to 38.7 percent. The percentage of children living in families with incomes below the poverty line increased from 15 percent in 1970 to more than 19 percent in 1997, despite the fact that, even adjusted for inflation, the GDP had doubled.[9] For the majority, the American standard of living deteriorated. In 1970, a new car cost 38 percent of a young couple's income; today it takes half their income; an average house today costs four times a young couple's income, compared to just twice as much in 1970. In the meantime, a greater percentage of Americans are working than ever before, including young mothers.[10] Compared to 1970, the United States has more broken families, higher divorce rates, more drug trafficking, and more people in jail.[11]

A statistic like GDP masks such elementary facts. The activities that generate GDP are not all equal. Economist Herman Daly calls all the GDP hype "boundless bull."[12] GDP is simply money changing hands; it says "virtually nothing about whether life on the street is getting better or worse."[13] The activities that generate GDP are not all equal, any more than goods sold at a discount are: along with the great deals we get lots of junk. The same is true with GDP. Even so, when it comes to the final tally, the *Wall Street Journal* treats it all the same—good, bad, or indifferent. That cannot be right. Should making computers, killing people in car accidents, building churches, and buying cigarettes all be lumped together in the same happy GDP family? Of course not. One has to count the costs in depletion, pollution, and disruption.[14] Not to make such distinctions is to hide from the facts of life, to forget where babies come from.

The Irish potato famine was not a simple matter of running out of food. To understand what happened requires some social accounting. The renter regime, free-trade ideology, and the politics of taxation were as much to blame as the potato ever was. Similarly, without honest accounting, what an increment in GDP adds up to is hard to figure out. Salaries, after all, are not paid at the macro level. Much of what gets included in GDP is actually the cost of "borrowing resources from the future and fixing the blunders of the past." From GDP we ought to subtract resource depletion and our lost leisure time, and we ought to adjust it for income equality. Then we should add in the value of work around the house, even though no money changes hands.[15] In short, how the statistic is constructed is as important as the statistic itself. It is the component parts that matter most.

The twenty-first century opened in the midst of a knowledge revolution and a new global order. The old realties seemed overturned; the world was a different place. But I am not so sure. The old polarities are still at work, cast in terms the Brandt report emphasized two decades ago: rich countries and poor ones, the north against the south.[16] What matters in such a world is not just stock prices and growth indexes, but results that make a difference to ordinary people: a plot of land, schools, job training, health clinics, and a safe environment. When such criteria are used, countries that are lackluster in GDP terms look much better. With respect to infant-mortality reductions, Costa Rica, Jamaica, and Sri Lanka rank among the world's ten most successful countries, despite slow growth in GDP and per-capita income.[17]

Vested interests stand behind the GDP line, the media included. The only time the locals make the news is when they screw up. Everyday life—the place where people raise their children, earn their money, and pursue their happiness—does not matter. Reporters do not know what it is about or how to cover it. They prefer a storyline like GDP, with its "appearance of empirical certitude combined with expert authority."[18] Anyway, it's cheaper to just wait for the next election, the next killing, or the next drug bust. And it is much less work.

What the networks define as news is as problematic as what gets included in the GDP.[19] It is also dishonest. The "McLaughlin Group," "Crossfire," and the "Capital Gang" are shows. They go on the road and stage performances with the same material, questions, and jokes used over again somewhere else. TV journalists make no decisions, are not held accountable for their predictions, and have no public responsibilities. They are actors who read scripts.[20] Lots of folks could seem smart, too, if someone else wrote the lines. In the 1990s, the media conglomerates have gone global. They are great advocates of open media markets.[21] The news they hawk is far removed from the neighborhoods and local communities that are force-fed their story lines. Much like the composition of GDP, what ends up in the news distorts reality. Japan is where terrorists spray nerve gas in the subways. Brazil is where gangs kill homeless children and sell their body parts. America is where kids kill each other at school. This is the diet we are fed regularly. Small wonder stress therapists warn their clients against the news.

The media's version shortchanges the micro world, where the problems must be solved. Without respect for local culture, for the primacy of place, or for community action, solutions will be inadequate. To the imbalance that global thinking creates, we need a counterweight. In this respect, CIP's balancing act is instructive. Molecular breeding is done at CIP. Getting healthy seed to farmers, by contrast, is a local effort, as are protected beds for Bolivia's spuds or household nurseries for Uganda's sweet potatoes. From pest management to compaigns against late blight, farmers must adapt the technology to suit their own needs. Both scientific rigor and relevance must be considered. Through CIP's efforts, biotechnology ends up at the service of ordinary farmers. In the end, high-tech research and local projects must work in tandem. Science does not have all the answers. Nor does village life, narrowly focused as it is on household economy. Participatory research adds weight to a project's recommendations. But it is extra work and it must be done right.[22]

Loaning money to poor people is likewise lots of work, but that does not make it any less worthwhile. For over thirty years, the Grameen Bank in Bangladesh has loaned small sums to its clients. In 1976, the bank loaned $27 apiece to forty-two people; in 1997, its total assets in property and loans came to $429 million.[23] Its target group is households with less than a quarter-hectare of cultivated land, or the equivalent. Members are organized into groups of five; six such groups make up a "banking center." This center is not a building; it is wherever members decide to hold their weekly meeting. The bank goes to its members, not the reverse. Bank staff attend the meetings and make loans, which are recovered in weekly installments.[24] In 1998, the Grameen Bank had sixty-six thousand centers in over thirty-eight thousand villages. Of its 2.3 million members, some 95 percent are women. Over the years, the cumulative value of its loans has topped $2.4 billion; the repayment rate is 99 percent. Loans go to weaving, pottery, matmaking, and farming—especially growing rice, bananas, and potatoes. The bank has helped peddlers and shopkeepers buy supplies; it has helped families get rickshaws, bicycles, and pushcarts.[25]

Low cost housing is the Grameen Bank's greatest success to date. The practical $300 house it designed won an international award. When the bank's founder, Muhammad Yunas, applied for a loan to fund the project, Bangladesh's Central Bank turned him down because "housing did not qualify as income-generating," and besides, the poor were a bad credit risk. Yunas insisted they would pay back. "This is the only chance they have in life," he argued. "They cannot afford to fail." He got his loan.[26]

Actually, it is the rich who turn out to be the deadbeats. What caused the Savings and Loan (S&L) debacle in the United States was a profligate loan policy. The banks doled out billions to insiders and fat cats whose schemes had little by way of collateral behind them. The S&Ls violated just about every banking principle in the book. When the real-estate boom went bust, the U.S. banking system almost collapsed. Congress had to shore it up with a $150 billion subsidy from U.S. taxpayers. The total cost of the bail out exceeded half a trillion dollars. Despite the massive fraud, the justice department did little by way of investigation; hardly anyone went to jail. But hey, that's the macro level. No sense getting worked up over a little white collar crime.[27]

Why don't we hear about CIP, the Grameen Bank, or new ways to measure GDP on the nightly news? Why did the networks so quickly forget about the S&L scandal or the bankrupt hedge fund at Long Term Capital Management?[28] Because they have a predefined point of view, much like GDP. They simply do not know where a Grameen Bank fits. As to the hedge fund collapse, it hits too close to the corporate home base; it's much easier to distract folks with Y2K bugs and Confederate flags.[29] Anyway, the macro level will never solve our problems, not even reborn as global orthodoxy. Global trade will not make us all rich, the worldwide web will not make us all smart, and e-mail will not make us all friends. They all ignore where babies come from.

The benefits of America's leap forward in GDP did not trickle down all the way. In 1997, the bottom 40 percent of wage earners made just 14 percent of America's aggregate income, down from 17.6 percent in 1970.[30] Growth in GDP does not by itself make for better schools, stable families, or decent jobs. Growth alone will not solve America's problems—or China's or Indonesia's. China's rapid growth may so deplete its resources that its future prospects will be greatly diminished.

The promise of global capitalism is that employment will grow fast enough to clear the market of the goods produced; demand will match supply. Consequently, we will all be better off. So far, that promise has not been kept. In 1971, the Fortune 500 had global sales of $721 billion; today the figure is $5.2 trillion. Such firms account for a third of all manufactured exports, three-quarters of commodity trade, and four-fifths of the trade in technology and management services. Even so, the total work force employed by the Fortune 500 has increased but little since the 1970s. Moreover, they have shifted jobs from high-paying industrial countries to low-paying Third World countries.[31]

In 1979, the average labor input needed to build a Ford was 40 hours; in 1993, it was just 25.4 hours. At Chrysler, it fell from 41 to 32.5 hours; at GM the average dropped from 41 to 32.5 hours. A Ford plant in 1979 made 920 Granadas a day with 4,270 workers; in 1990, it made 1,200 Escorts with 1,800 workers. Compared to 1979, Ford today builds a world car. The component parts are made globally; its work force is international. Between 1980 and 1994, North America lost thirty-two car and truck assembly plants; employment in the automobile industry dropped by 180,000 jobs. Of course, American companies had to get lean and mean to face the competition. Motorola is another example. It moved its computer-chip manufacturing to Malaysia in the early 1970s. Even though its work force has grown to 142,000, only half those jobs are in the United States. In Malaysia, the basic pay at a Motorola plant is $100 a month.[32]

This is not a recipe for a new world order; it is a recipe for disaster. With downsized corporations seeking the lowest paid workers, who is supposed to buy what gets produced? Overcapacity in the global automobile industry is put at 25 percent. The industry could produce seventy-nine million vehicles a year but can only sell fifty-eight million. Automobiles are not alone. Tires, computers, consumer electronics, pharmaceuticals, chemicals, the aircraft industry, and textiles face the same dilemma. We could end up with idle workers, unused production, inadequate demand, and falling wages.[33]

This is not to dismiss growth or to ignore the aspirations people have for development. It is simply to assert that the big payoff has been oversold. The alternative, "bringing the bottom up," has been undersold. That means adding to wages on the low end, it means social equity, and it means rebuilding the infrastructure that balanced prosperity requires.[34] To do this, we need to think about development the way CIP thinks about the potato. There is no single solution that applies across potato plots from Bolivia's altiplano to India's Ganges valley and Egypt's delta. No production technology, pest-management strategy, or seed system fits all situations. For the potato, the starting point is the challenge posed by different climates, cultural practices, and seed-production logistics. Why can we not take this wisdom and apply it? Whether it is planting potatoes, organizing a health clinic, or setting up a school, the context is not an extraneous variable; it is the heart and core of the entire equation, the place where babies are born, where the world's future begins.

It is time to let globalists solve their own problems. In the meantime, like the Grameen Bank, like CIP's potatoes, let's start with the world as it is. Whether education or energy, job training or productivity in agriculture, let's start where people are, with solutions they can see.

How can this be done? Consider the wealth the global economy invests in itself. Global bond trading was worth $30 billion a day in the early 1980s and now is worth over $500 billion a day. In 1983, the world's five major central banks held $139 billion in foreign-exchange reserves against a daily trade of $39 billion. Today, the daily currency trade is estimated at $1.2 trillion. Between 1980 and 1991, the total value of international bank loans quadrupled to $3.6 trillion. Direct foreign investment by corporations is valued at almost $2 trillion. Every month, global traders swap some $24 trillion worth of financial assets, in loans, bonds, stocks, and currencies.[35] By comparison, what it costs to support a research center like CIP or the Grameen Bank is spare change. The world has overinvested in its automobiles and petro-chemicals and has underinvested in its people. It is time for a change.

Investment firms have created the portfolios that fuel the world economy. Can we not create social funds to build schools and public-health clinics, to invest in agriculture and village life, to rebuild urban transportation, and to create new energy systems? The Grameen Bank pays 5 to 6 percent interest on the bonds that supply its funds. TIAA-CREF, one of America's largest retirement funds, tells clients to expect a lifetime return that averages about 4 percent. Properly secured, retirement contributions to a global social fund could be as safe as U.S. Savings Bonds. Small investors are not like the hucksters at the hedge funds; they need security, not risk. Could we not create funds from which countries, cities, and communities could borrow to build a better habitat for humanity—funds denominated in major currencies, funds guaranteed by the world's great banks?

This "we" is all of us. The global middle class can be the backbone of such a fund. Retirement contributions or bonds sold at work could be invested directly. Multinational corporations, from Microsoft and Time-Warner to Sony and Siemens, could match worker contributions or invest a fraction of their profits. The daily trade in stocks, bonds, currencies, and commerce should be taxed. Given the huge volumes involved, the tiniest levy would add up fast. Computers can track the traffic and make sure finance capital pays its fair share. The fund would be managed by the groups, companies, and nations that contribute to it and by those who borrow from it.

When the year 2000 threatened the world's computer system, the wealthy rallied to fix it. They did not want their electrical grids, communications systems, or bank accounts put at risk. The global bill for fixing Y2K problems, both real and imagined, is estimated at $600 billion.[36] Building a future on abandoned children, desperate people, and a depleted environment threatens our future every bit as much. Multinational corporations expect the GATT (General Agreement on Tariffs and Trade), the IMF (International Monetary Fund), and the World Trade Organization (WTO) to protect their interests. But what about the rest of us? What if the prosperity a global trading system promises turns out to be as elusive, or as poorly distributed, as the alleged benefits of GDP growth? A necessary counterweight must be added that brings the bottom up, that allows families, communities, and organizations to invest in the future.

In the 1830s, Charles Dickens saw the onslaught of the industrial age. To him, the great specters stalking humanity were ignorance and want.[37] They still are. To secure the future, the world needs CIP and its research, it needs its Grameen banks, and it needs to get resources to people and projects that make a difference. Let us think small, learn by doing, build participation, and educate our children. If we do, we might find out where babies come from. We might yet give birth to a world this century's children will want to call home.

APPENDIX

Potato Websites

1. Consorcio para el Desarrollo Sostenible de la Eco-región Andina (CONDESAN)
www.condesan.org

2. Consultative Group on International Agricultural Research
www.cgiar.org

3. Food and Agricultural Organization of the United Nations (FAO)
www.fao.org

4. Global Initiative on Late Blight (GILB)
www.cipotato.org/gilb.htm

5. International Food Policy Research Institute (IFPRI)
www.ifpri.cgiar.org

6. International Potato Center (CIP)
www.cipotato.org

7. International Rice Research Institute
www.cgiar.org/irri

8. Michigan State University: Potato Web Portal
www.potato.msu.edu

9. National Potato Council
www.npcspud.com

10. North Carolina Sweet Potato Commission
www.nscweetpotatoes.com

11. Ohio University Late Blight Fact Sheet
ohioline.ag.ohio-state.edu/hyg-fact/3000/index.html

12. Oregon State University
www.bcc.orst.edu/lateblight

13. Oregon State University: Malheur Experimental Station
www.cropinfo.net

14. Potato Association of America
www.ume.maine.edu/PAA

15. Purdue University: Center for New Crops and Plant Products
www.hort.purdue.edu/newcrop/default.html

16. University of Idaho: Late Blight
www.uidaho.edu/ag/plantdisease/lbhome.htm

17. United States Department of Agriculture: Chemical Usage 1999 Field Crops
Summary
usda.mannlib.cornell.edu/reports/nassr/other/puc-bb/aghco500.txt

NOTES

Introduction

1. Plutarch, *The Lives of the Noble Grecians and Romans.*
2. Fernand Braudel, *Civilization and Capitalism: 15th–18th Centuries. The Limits of the Possible,* vol. 1, *The Structures of Everyday Life.*
3. William Bryant Logan, *Dirt: The Ecstatic Skin of the Earth.*
4. Lewis Thomas, *The Lives of a Cell: Notes of a Biology Watcher.*
5. Timothy Foote, *The World of Bruegel 1525–1569.*

Chapter 1. Art and Agriculture

1. Paul G. Bahn, *The Cambridge Illustrated History of Prehistoric Art,* pp. xv, 83–141. Also see Robert Hughes, "Behold the Stone Age," and Michael D. Lemonik, "Ancient Odysseys."
2. Mario Ruspoli, *The Cave of Lascaux,* p. 17; Jean-Marie Chauvet, Éliette Brunel Deschamps, and Christine Hillaire, *Dawn of Art: The Chauvet Cave,* pp. 121–24. The hand stencils at Cosquer Cave are dated to twenty-seven thousand years ago; see Jean Clottes and Jean Courtin, *The Cave Beneath the Sea: Paleolithic Images at Cosquer,* pp. 166–67.
3. Bahn, *Prehistoric Art,* pp. xiii–xvi, 70–81; Richard Rudgley, *The Lost Civilizations of the Stone Age,* pp. 176–83. For an overview of how old ice-age art is, see Paul G. Bahn and Jean Vertut, *Journey through the Ice Age,* pp. 58–76.
4. Robert J. Wenke, *Patterns in Prehistory,* pp. 156–75.
5. On this point, see Stephen Jay Gould, *The Panda's Thumb: More Reflections in Natural History,* pp. 56–57.
6. Rudgley, *Lost Civilizations,* pp. 157–60; Alison S. Brooks et al., "Dating and Context of Three Middle Stone Age Sites with Bone Points in the Upper Semliki Valley, Zaire."
7. Consider the cultural knowledge that stands behind various pathways to domestication; see David R. Harris, "Alternative Pathways Toward Agriculture."
8. Jack R. Harlan, *Crops and Man,* p. 17–22. Also see Les Groube, "The Taming of Rain Forests: A Model for Late Pleistocene Forest Exploitation in New Guinea." For such interventions in ancient Amazonia, see Anna C. Roosevelt et al., "Paleoindian Cave Dwellers in the Amazon: The Peopling of the Americas," pp. 381–82. Contemporary practices are documented in Susana Hecht and Alexander Cockburn, *The Fate of the Forest: Developers, Destroyers, and Defenders of the Amazon,* pp. 34–37; and Charles R. Clement, "A Center of Crop Genetic Diversity in Western Amazonia: A New Hypothesis of Indigenous Fruit-Crop Distribution."
9. On how early "agriculture" might be, see Daniel E. Vasey, *An Ecological History of Agriculture,* pp. 24–26.
10. Harlan, *Crops and Man,* pp. 20–21; Rudgley, *Lost Civilizations,* pp. 137–41. In general, see Richard Rudgley, *Essential Substances: A Cultural History of Intoxicants in Society.*

11. See Dolores R. Piperno and Deborah M. Pearsall, *The Origins of Agriculture in the Lowland Neotropics*, pp. 7–8. In general, see Thomas Killion, ed., *Gardens of Prehistory: The Archaeology of Settlement Agriculture in Greater Mesoamerica.*

12. On paleoscience, see Rudgley, *Lost Civilizations*, pp. 86–105.

13. Harlan, *Crops and Man*, pp. 46–48, 60, 225; David R. Harris, "An Evolutionary Continuum of Plant-People Interaction," pp. 16–23.

14. On how fragile a modern ear of corn is, see Walton C. Galinat, "Maize: Gift From America's First People," p. 48. On the domestication of corn, see Charles B. Heiser Jr., *Seed to Civilization: The Story of Food*, pp. 92–104; Wenke, *Patterns*, pp. 252–55; and Walton C. Galinat, "The Origin of Maize: Grain of Humanity."

15. David R. Harris was an early revisionist; see "The Origins of Agriculture in the Tropics," pp. 185–92. On the limitations of wild wheat as a food crop, see Wenke, *Patterns*, pp. 236–38. On ancient alcohol production, see Maguelonne Toussaint-Samat, *A History of Food*, pp. 34–35; and Rudgley, *Lost Civilizations*, pp. 137–38.

16. Toussaint-Samat, *History of Food*, pp. 9–10. In the Americas, the domestication of roots, tubers, and squashes predates corn. See Piperno and Pearsall, *Lowland Neotropics*, pp. 158–60, 248–49; and Bruce D. Smith, "The Initial Domestication of *Cucurbita pepo* in the Americas 10,000 Years Ago."

17. Harlan, *Crops and Man*, pp. 130–33; Harris, "Origins of Agriculture," pp. 181–88. On the origins, propagation, and processing of these and related crops, see Piperno and Pearsall, *Lowland Neotropics*, pp. 109–29; for a brief history and description of such crops, see Jonathan D. Sauer, *Historical Geography of Crop Plants: A Select Roster*, pp. 179–81. Carl O. Sauer likewise argued for the primacy of roots and tubers; see *Agricultural Origins and Dispersals*, pp. 19–28.

18. On the role roots and tubers played in ancient American agriculture, see Piperno and Pearsall, *Lowland Neotropics*, pp. 1–6, 109–28; and Michael Goulding, Nigel J. H. Smith, and Dennis J. Mahar, *Floods of Fortune: Economy and Ecology Along the Amazon*, p. 23, generally, pp. 19–25. For taro root and yautia (cocoyam), also see Heiser, *Seed to Civilization*, pp. 148–51; Anna C. Roosevelt, *Moundbuilders of the Amazon: Geophysical Archaeology on Marajó Island, Brazil*, pp. 25–26; and Richard Evans Schultes, "Amazonian Cultigens and Their Northward and Westward Migration in Pre-Columbian Times."

19. Robert E. Rhoades and Douglas E. Horton, "Past Civilizations, Present World Needs, and Future Potential: Root Crop Agriculture Across the Ages," pp. 8–11; Harlan, *Crops and Man*, pp. 203–204, 209–10. Also see Jonathan D. Sauer, *Historical Geography*, pp. 179–81.

20. Harlan, *Crops and Man*, pp. 182–86.

21. Toussaint-Samat, *History of Food*, pp. 126–28; Harlan, *Crops and Man*, pp. 166–68.

22. Richard Critchfield, *The Villagers: Changed Values, Altered Lives*, pp. 3–39.

Chapter 2. Cycles

1. Jane MacLaren Walsh and Yoko Sugiura, "The Demise of the Fifth Sun," p. 34. On the Aztec calendar, see Anthony F. Aveni, *Empires of Time: Calendars, Clocks, and Cultures*, pp. 253–77. On the historical lack of precision in the Western calendar, see David Ewing Duncan, *The Calendar.*

2. On Babylonia, see Jane B. Sellers, *The Death of Gods in Ancient Egypt*, pp. 186–91; on China, see Aveni, *Empires*, pp. 317–18. By combining the five elements with the twelve signs of the zodiac, the Chinese produce a sixty-year cycle; see Lama Anagarika Govinda, *The Inner Structure of the I Ching*, pp. 52, 58–59.

3. See the glossary in José Argüelles, *The Mayan Factor: Path Beyond Technology*, pp. 199–200.

4. See Edwin C. Krupp, *Echoes of the Ancient Skies: The Astronomy of Lost Civilizations*, pp. 1–22.

5. David Souden, *Stonehenge Revealed*, pp. 30–31, 118–27; John North, *Stonehenge: Neolithic Man and the Cosmos*, pp. 221–27, 573–75.

6. North, *Stonehenge*, pp. 519–49.

7. Alexander Marshack, *The Roots of Civilization*, pp. 81–110; Rudgley, *Lost Civilizations*, pp. 100–104.

8. Rudgley claims that the constellation Pleiades has been recognized for at least forty thousand years; see Rudgley, *Lost Civilizations*, p. 100. The most ancient myths may be linked to the stars; see Giorgio de Santillana and Hertha von Dechend, *Hamlet's Mill: An Essay on Myth and the Frame of Time*.

9. Anthony Aveni, *Stairways to the Stars: Skywatching in Three Great Ancient Cultures*, pp. 41–48, 165–69.

10. Krupp, *Echoes*, pp. 1–3.

11. Daniel Quinn, *The Story of B*, pp. 163–90.

12. Rudgley, *Lost Civilizations*, pp. 107–16, 136–41.

13. Gould, *Panda's Thumb*, p. 55.

14. Anglican archbishop James Ussher (1581–1656) dated creation to 4004 B.C.; see Wenke, *Patterns*, p. 11. Also see Stephen Jay Gould, *The Flamingo's Smile: Reflections in Natural History*, pp. 126–38.

15. de Santillana and von Dechend, *Hamlet's Mill*, pp. 58–75.

16. Sellers, *Death of Gods*, pp. 172–73, 192–209.

17. See Aveni, *Empires*, pp. 208–209; and Argüelles, *Mayan Factor*, pp. 45–47. For a discussion of much larger Mayan time cycles, see David Freidel, Linda Schele, and Joy Parker, *Maya Cosmos: Three Thousand Years on the Shaman's Path*, pp. 61–64.

18. "Astronews," *Astronomy* (September, 1999), p. 28.

19. See Gould, *Panda's Thumb*, pp. 125–28; and Wenke, *Patterns*, pp. 136–56, 157–75. Controlled fire is dated to at least 1.89 million years ago. See Glynn Ll. Isaac, introduction to *Koobi Fora Research Project: Plio-Pleistocene Archeology*, p. 5; and Randy V. Bellomo and William F. Kean, "Evidence of hominid-controlled fire at the FxJj 20 site complex, Karari Escarpment."

20. Wenke, *Patterns*, p. 169.

21. Hank Wesselman, *Spiritwalker: Messages from the Future*.

Chapter 3. Columbian Exchange

1. See Charles W. Petit, "Rediscovering America"; and Wenke, *Patterns*, pp. 196–224.

2. See Wenke, *Patterns*, pp. 212–13; and Rick Gore, "Ancient Americans," pp. 93–96.

3. Wenke, *Patterns*, p. 213. Also see Harlan, *Crops and Man*, pp. 218–19; and Bahn, *Prehistoric Art*, p. 161.

4. See Goulding, Smith, and Mahar, *Floods of Fortune*, pp. 20–21; and Roosevelt et al., "Paleoindian Cave Dwellers," pp. 381–82.

5. Bahn, *Prehistoric Art*, pp. 114–15.

6. See Wenke, *Patterns*, pp. 213–14; Harlan, *Crops and Man*, pp. 218–19; and Donald Ugent and Linda W. Peterson, "Archaeological Remains of Potato and Sweet Potato in Peru," pp. 1–9.

7. Harlan, *Crops and Man*, p. 218.

8. See Petit, "Rediscovering America," pp. 60, 62; and Goulding, Smith, and Mahar, *Floods of Fortune*, pp. 19–21.

9. Jared Diamond, *Guns, Germs, and Steel: The Fate of Human Societies*, pp. 295–321, 334–53. Also see Wenke, *Patterns*, pp. 202–204, 470–73; Harlan, *Crops and Man*, pp. 201, 218;

and Les Groube et al., "A 40,000 Year-Old Human Occupation Site At Huon Peninsula, Papau New Guinea."

10. Time-Life, eds., *The Human Dawn*, pp. 164–65

11. Compare the evidence in Harlan, *Crops and Man*, pp. 168, 221; also see pp. 180–82, 197, 201–204.

12. For a discussion of carbon-14 dating, see James Trefil, "Architects of Time," p. 50; and Harlan, *Crops and Man*, pp. 164–66.

13. Harlan, *Crops and Man*, pp. 167–69.

14. Ibid., pp. 159–213.

15. Noel D. Vietmeyer, "Forgotten Roots of the Incas," p. 95.

16. Donald Ugent, Tom Dillehay, and Carlos Ramírez, "Potato Remains from a Late Pleistocene Settlement in Southcentral Chile," p. 16.

17. Carlos M. Ochoa, "The Andes, Cradle of the Potato."

18. See J. G. Hawkes, introduction to *The History and Social Influence of the Potato*, p. x. Also see Harlan, *Crops and Man*, p. 221; Lawrence Kaplan and Lucille N. Kaplan, "Beans of the Americas," pp. 62–65; and Smith, "Initial Domestication."

19. See Harlan, *Crops and Man*, p. 225; Jean Andrews, *Peppers: The Domesticated Capsicums*, p. 10; and Galinat, "Origin of Maize," p. 4.

20. See Geoffrey W. Conrad and Arthur A. Demarest, *Religion and Empire: The Dynamics of Aztec and Inca Expansionism*, p. 169; and Wenke, *Patterns*, p. 533. After 1492, Spanish colonists introduced cows, sheep, pigs, and goats, plus the horse, the donkey, and the mule.

21. For a listing and description of New World beans, see Jonathan D. Sauer, *Historical Geography*, pp. 73–80. Also see Alfred W. Crosby Jr., *The Columbian Exchange: Biological and Cultural Consequences of 1492*, pp. 172–73; and Kaplan and Kaplan, "Beans of the Americas." To this day, indigenous farmers sow tapary beans in the Sonoran Desert; see Gary Paul Nabhan, *Gathering the Desert*, pp. 107–21.

22. For a list of food plants of American origin, see the appendix in Nelson Foster and Linda S. Cordell, eds., *Chilies to Chocolate: Food the Americas Gave the World*, pp. 163–67; also see Harlan, *Crops and Man*, pp. 222–32. For squashes, see Jonathan D. Sauer, *Historical Geography*, pp. 46–50.

23. Wenke, *Patterns*, pp. 253–55; Galinat, "Origin of Maize," p. 4. For an excellent discussion of early crop domestication in the Americas, see Piperno and Pearsall, *Lowland Neotropics*, pp. 109–66.

24. See the appendix in Foster and Cordell, *Chilies*, pp. 163–67. On chocolate, see Allen M. Young, *The Chocolate Tree: A Natural History*.

25. These less familiar crops, including the eight potato species, are discussed in detail in Noel D. Vietmeyer, study director and scientific editor, National Research Council, *Lost Crops of the Inca: Little-Known Plants of the Andes with Promise for Worldwide Cultivation*. On the number of potato species, see chapter 4, note 54.

26. The quote is from Sophie D. Coe and Michael D. Coe, *The True History of Chocolate*, p. 129; also see Harlan, *Crops and Man*, pp. 230, 234–35. In general, see Jonathan D. Sauer, "Changing Perception and Exploitation of New World Plants in Europe, 1492–1800," pp. 813–26; and Charles H. Talbot, "America and the European Drug Trade," pp. 833–41. On the Aztecs, see Bernard R. Ortiz de Montellano, *Aztec Medicine, Health, and Nutrition*, pp. 181–92.

27. Food and Agricultural Organization of the United Nations (FAO), *FAO Quarterly Bulletin of Statistics* 1999, 3/4, pp. 18, 16, 25.

28. See Jean Andrews, *Peppers*, pp. 1–10; and Jean Andrews, "The Peripatetic Chili Pepper: Diffusion of the Domesticated Capsicums Since Columbus." The statistics on pepper production are compiled for 1997–99 from FAO, *Quarterly Bulletin*, 3/4, 1999, p. 89.

29. Toussaint-Samat, *History of Food,* pp. 168–69.

30. Jonathan D. Sauer, "Changing Perception," pp. 824–25.

31. William H. McNeill, "American Food Crops in the Old World," pp. 51–52.

32. Neill McMullen, *Seeds and Agricultural Progress,* pp. 154–55. On the importance of corn for the future, see Christopher R. Dowswell, R. L. Paliwal, and Ronald P. Cantrell, *Maize in the Third World.*

33. The quote is from Elizabeth Rozin, *The Primal Cheeseburger,* p. 107, in general, pp. 104–109. Also see Alan Davidson, "Europeans' Wary Encounter with Tomatoes, Potatoes, and Other New World Foods," pp. 4–10.

34. FAO, *Quarterly Bulletin,* 3/4, 1999, p. 80.

35. See Crosby, *Columbian Exchange,* p. 182; and Charles B. Heiser Jr., *Nightshades: The Paradoxical Plants,* p. 33.

36. Redcliffe Salaman, *The History and Social Influence of the Potato,* pp. 62–68.

37. J. G. Hawkes and J. Francisco-Ortega, "The Potato in Spain During the Late 16th Century."

38. Paul Grun, "The Evolution of Cultivated Potatoes," pp. 47–48.

39. Salaman, *History,* pp. 253–54.

40. Ibid., pp. 189–90, 231, 334–36.

41. Crosby, *Columbian Exchange,* pp. 183–84; Davidson, "Europeans' Wary Encounter," pp. 13–15.

42. McNeill, "American Food Crops," pp. 45–50.

43. M. Bergman, "The Potato Blight in the Netherlands and Its Social Consequences (1844–1847)," p. 392.

44. On Ireland's demographic losses, see Thomas P. O'Neill, "The Organization and Administration of Relief, 1845–1852," pp. 254–55.

45. For descriptions of the city and estimates of its size, see Walsh and Sugiura, "Demise of the Fifth Sun," pp. 36–37. Also see Michael D. Coe, *Mexico: From the Olmecs to the Aztecs,* pp. 163–67; and Wenke, *Patterns,* pp. 515–18.

46. For Mexico's population, see Coe, *Mexico,* p. 164; also see Leslie Bethell, "A Note on the Native American Population on the Eve of the European Invasions." For the Inca Empire, see Wenke, *Patterns,* p. 546.

47. For population estimates of Spain, France, and England, see John Lynch, *Spain Under the Hapsburgs,* vol. 1, *Empire and Absolutism 1516–1598,* p. 101.

48. Angel Rosenblat considers all the estimates; see "The Population of Hispaniola at the Time of Columbus," pp. 48–54; his own estimate is around one hundred thousand.

49. Sherburne F. Cook and Woodrow Borah, *Essays in Population History: Mexico and the Caribbean,* pp. vii–viii. A more conservative estimate is sixteen million; see Thomas M. Whitmore, *Disease and Death in Early Colonial Mexico: Simulating Amerindian Depopulation,* pp. 206–207.

50. David Noble Cook, *Demographic Collapse: Indian Peru, 1520 to 1620,* p. 114.

51. See Goulding, Smith, and Mahar, *Floods of Fortune,* p. 23; and William M. Denevan, "The Aboriginal Population of Amazonia," p. 29. The pre-Conquest population of Marajo Island is estimated at between one hundred thousand and two hundred thousand; see Roosevelt, *Moundbuilders,* pp. 38–39.

52. Crosby, *Columbian Exchange,* pp. 45–46.

53. See Alfred W. Crosby Jr., "Metamorphosis of the Americas," p. 85; and John W. Verano and Douglas H. Ubelaker, "Health and Disease in the Pre-Columbian World," p. 215.

54. Crosby, *Columbian Exchange,* pp. 43–44, 79–80.

55. The story is retold in Walsh and Sugiura, "Demise of the Fifth Sun," pp. 34–41; also see Crosby, *Columbian Exchange,* pp. 48–49.

56. Crosby, *Columbian Exchange,* pp. 50–56.

57. James Lang, *Conquest and Commerce: Spain and England in the Americas,* pp. 22–23.

58. Crosby, "Metamorphosis," pp. 86–87.

59. Herman J. Viola, "Seeds of Change," pp. 12–13.

60. See Crosby, *Columbian Exchange,* pp. 47, 53–54; and Crosby, "Metamorphosis," p. 73.

61. Barnabé Cobo, *A History of the Inca Empire,* pp. 27, 31; in general, see pp. 20–38, and Lang, *Conquest and Commerce,* pp. 22–24.

62. Such knowledge is impressively documented for the U.S. Southwest by naturalist Gary Paul Nabhan; see his books *The Desert Smells Like Rain* and *Gathering the Desert.*

63. Verano and Ubelaker, "Health and Disease," pp. 217–19.

64. See Crosby, *Columbian Exchange,* pp. 165–202; and McNeill, "American Crops," pp. 43–59.

65. Compiled from the United States Bureau of the Census, *Statistical Abstract of the United States (SAUS)* 1997, p. 266; and *SAUS* 1999, p. 278.

66. Consider the story of ebola virus reported by Richard Preston in *The Hot Zone* or the many examples provided by Laurie Garrett in *The Coming Plague: Newly Emerging Diseases in a World Out of Balance.* Also see Christopher Bright, "Invasive Species: Pathogens of Globalization."

Chapter 4. Potato Facts

1. On the natural superiority of the potato, see Earl J. Hamilton, "What the New World Gave the Economy of the Old," p. 858, generally, pp. 854–55, 857–59. On the labor needed to transplant a hectare of rice, see James Lang, *Feeding a Hungry Planet: Rice, Research, and Development in Asia and Latin America,* p. 4.

2. International Potato Center (CIP) and FAO, *Potatoes in the 1990s: Situation and Prospects of the World Potato Economy,* p. 29; also see FAO, *Quarterly Bulletin* 1999, 3/4, p. 38. The Spanish acronym CIP stands for Centro International de la Papa.

3. See CIP, "New Rice and Wheat Varieties Ignite Asian Potato Production," in *International Potato Center Annual Report 1995,* pp. 18–20. On the green revolution in wheat and rice, see Lang, *Feeding a Hungry Planet,* pp. x–xii, 9–13. For a summary of the factors favoring the potato in monsoonal Asia, see Gregory J. Scott, Mark W. Rosegrant, and Claudia Ringler, *Roots and Tubers for the 21st Century: Trends, Projections, and Policy Options,* pp. 3–4, 11–12, 19–20. The days to maturity in the Ganges valley are noted in S. K. Bardhan Roy et al., "Intensification of Potato Production in Rice-Based Cropping Systems: A Rapid Rural Appraisal in West Bengal," p. 208.

4. CIP-FAO, *Potatoes in the 1990s,* p. 29; and FAO, *Quarterly Bulletin* 1999, 3/4, p. 38.

5. Compiled from International Rice Research Institute (IRRI), *World Rice Statistics 1990,* p. 2; and FAO, *Quarterly Bulletin* 1999, 3/4, p. 18.

6. The comparison includes Belgium-Luxembourg, France, Germany, Italy, the Netherlands, Portugal, Spain, and the United Kingdom. See CIP-FAO, *Potatoes in the 1990s,* p. 29; and FAO, *Quarterly Bulletin* 1999, 3/4, p. 39.

7. The 1997–99 per-capita potato-consumption estimates asssume the same utilization percentages as for 1991–92. All figures are compiled from CIP-FAO, *Potatoes in the 1990s,* p. 31; and FAO, *Quarterly Bulletin* 1999, 3/4, p. 39. For 1999 population totals, see Population Reference Bureau (PRB), *World Population Data Sheet* 1999, pp. 8–9.

8. CIP-FAO, *Potatoes in the 1990s,* p. 31. These utilization trends continue; see Scott, Rosegrant, and Ringler, *Roots and Tubers,* pp. 7–13.

9. CIP-FAO, *Potatoes in the 1990s,* pp. 29, 31; FAO, *Quarterly Bulletin* 1999, 3/4, p. 38; for

per capita consumption and utilization, see National Potato Council (NPC), *Potato Statistical Yearbook 1999*, pp. 46, 68.

10. CIP-FAO, *Potatoes in the 1990s*, pp. 2, 13–14.

11. For the calorie estimate, see IRRI, *IRRI Rice Almanac, 1993–1995*, p. 7. The production figures are compiled from FAO, *Quarterly Bulletin* 1999, 3/4, pp. 16, 18, 25, 38, 79, 56, respectively. The population figure is a 1997–99 average.

12. IRRI, *Rice Almanac*, p. 7.

13. FAO, *Quarterly Bulletin* 1999, 3/4, pp. 18, 38.

14. Kamal Uddin Ahmad and Mumtaj Kamal, *Wealth From the Potato*, p. 24; A. Rastovski and A. van Es, eds., *Storage of Potatoes: Post-Harvest Behavior, Store Design, Storage Practice, and Handling*, pp. 34–35.

15. FAO, *Quarterly Bulletin* 1999, 3/4, pp. 18, 38.

16. See D. K. Salunkhe, S. S. Kadam, and S. J. Jadhav, *Potato: Production, Processing, and Products*, p. 2, table 1; and Ahmad and Kamal, *Wealth*, p. 24.

17. William G. Burton, *The Potato: A Survey of its History and of the Factors Influencing its Yield, Nutritive Value, Quality and Storage*, p. 146.

18. Per one hundred grams of edible potatoes, rice (milled), corn, and hard wheat, the percentages of fat are, respectively, 0.1, 0.5, 1.0, and 2.0; when cooked the figures are 0.1 for boiled potatoes, 0.3 for boiled white rice, 0.8 for cornmeal mush, and 1.6 percent for bread. See Jennifer A. Woolfe, *The Potato in the Human Diet*, pp. 24–27.

19. Salunkhe, Kadam, and Jadhav, *Production, Processing*, p. 14.

20. Burton, *Potato*, p. 146.

21. See Woolfe, *Human Diet*, pp. 25, 44–45; and Burton, *Potato*, pp. 166–67, 180. Estimates of dietary requirements for vitamin C vary. The forty milligrams per day used here is for the United Kingdom, Canada, and Australia; see Jim Mann and A. Stewart Truswell, eds., *Essentials of Human Nutrition*, pp. 223–24.

22. Woolfe, *Human Diet*, p. 27.

23. See H. Valentine Knaggs, *Potatoes as Food and Medicine*, p. 9; and Mann and Truswell, *Human Nutrition*, p. 223.

24. Ahmad and Kamal, *Wealth*, pp. 30–31.

25. Salaman, *History*, p. 135.

26. Burton, *Potato*, p. 166.

27. Ibid., p. 154; for comparisons, see Woolfe, *Human Diet*, pp. 26–29.

28. Ahmad and Kamal, *Wealth*, pp. 25–26

29. Rastovski and van Es, *Storage*, p. 44; in general, see Salunkhe, Kadam, and Jadhav, *Production, Processing*, pp. 12–22.

30. Knaggs, *Potatoes as Food*, pp. 10–12.

31. Burton, *Potato*, p. 146; Rastovski and van Es, *Storage*, p. 44; and Mann and Truswell, *Human Nutrition*, p. 69.

32. Knaggs, *Potatoes as Food*, p. 4; Burton, *Potato*, p. 172; and Hilaire Walden, *The Potato Cookbook*, pp. 9–10.

33. Burton, *Potato*, pp. 178–80.

34. Ibid., pp. 172–81; Woolfe, *Human Diet*, pp. 34–35, 45.

35. See Heiser, *Nightshades*, pp. 39–42. Also see the drawings in Salunkhe, Kadam, and Jadhav, *Production, Processing*, p. 13.

36. See Bill B. Dean, *Managing the Potato Production System*, pp. 59–65; and Douglas Horton, *Potatoes: Production, Marketing, and Programs for Developing Countries*, pp. 34–40.

37. Fern Marshall Bradley, *Rodale's Garden Answers*, p. 158, generally, pp. 158–60.

38. Ibid., p. 158.

39. How much tuber seed farmers use to plant a hectare depends on both the spacing between rows and the size of the seed planted. Two tons per hectare is a reasonable estimate for developed countries; see the spacing, seed size, and tonnage data in Bill B. Dean, *Production System,* pp. 47–48. In Peru, CIP's estimate is between two and three metric tons; see CIP, *Proyecto Chacasina: produzcamos papa con semilla sexual,* p. 15, picture 64. For the world's 1997–99 average potato yields, see FAO, *Quarterly Bulletin 1999,* 3/4, p. 38.

40. FAO, *Quarterly Bulletin 1999,* 3/4, pp. 38–39.

41. Based on an interview with Peter Schmiediche, Bogor, Indonesia, October 30, 1995, p. 32. All notes and interviews are on deposit at the Vanderbilt University Library, where they are organized by country. Pages for each country file are organized sequentially.

42. See Lang, *Feeding a Hungry Planet,* pp. 9–10, 141–42.

43. For a discussion of flavor principles, see Rozin, *Primal Cheeseburger,* pp. 85–109.

44. See Harlan, *Crops and Man,* pp. 19–20, 230; Heiser, *Nightshades,* pp. 173–75; and Jerome E. Brooks, *The Mighty Leaf,* pp. 50–59, 96.

45. Brooks, *Mighty Leaf,* pp. 11–27.

46. Ibid., pp. 70–71.

47. Heiser, *Nightshades,* pp. 166–72; Jerome E. Brooks, *Mighty Leaf,* pp. 191–278.

48. Heiser, *Nightshades,* pp. 129–57.

49. G. Lisińska and W. Leszczyński, *Potato Science and Technology,* pp. 32–34. Also see Burton, *Potato,* pp. 157–59; and Heiser, *Nightshades,* p. 43.

50. On the tree tomato and *naranjilla,* see Heiser, *Nightshades,* pp. 111–15, 115–23; also see Vietmeyer, *Lost Crops,* pp. 267–75, 307–16.

51. See Heiser, *Nightshades,* p. 2.

52. For drawings, see Roger Tory Peterson, *Field Guide to Wildflowers of Northeastern and North-Central North America,* pp. 72, 324.

53. J. G. Hawkes, *The Potato: Evolution, Biodiversity, and Genetic Resources,* pp. 6–9. For an appreciation of the potato's diversity, see Carlos M. Ochoa's well-illustrated *Las Papas de Sudamérica: Perú.*

54. I list eight cultivated species, but experts differ in how many they count. Hawkes lists *S. goniocalyx* (Limeña) as a subspecies of *S. stenotomum* (Pitiquiña); hence, he counts seven cultivated species. Ochoa lists *S. goniocalyx* as a separate species for a total of eight. Technically speaking, *S. hygrothermicum* can be added to make nine. Adapted to the lowland rainforest, *S. hygrothermicum* is very rare today, if not extinct. Hawkes lists it as a subspecies of *S. phureja;* Ochoa lists it separately. See Hawkes, *Evolution, Biodiversity,* pp. 179–80; and Carlos M. Ochoa, *Las Papas,* pp. 15, 25, 503. For a brief discussion of *S. hygrothermicum,* see Vietmeyer, *Lost Crops,* p. 100. On the genetic diversity of the potato in the Lake Titicaca basin, see Hawkes, *Evolution, Biodiversity,* pp. 3–5; Carlos M. Ochoa, *Las Papas,* p. 503; and Carlos M. Ochoa, "Andes."

55. Lang, *Feeding a Hungry Planet,* p. 5.

56. Hawkes, *Evolution, Biodiversity,* pp. 58–61.

57. Photoperiod sensitivity also varies by variety, especially in the tuberosum subspecies; see the discussion in Burton, *Potato,* pp. 53–56. Also see Horton, *Production, Marketing,* pp. 36–37.

58. See Zósimo Huamán, A. Glomirzaie, and W. Amoros, "The Potato," p. 21; and Zósimo Huamán and D. P. Zhang, "Sweetpotato," pp. 29–30.

59. On the potato's spread and its nomenclature, see Salaman, *History,* pp. 130–41; and Heiser, *Nightshades,* pp. 32–34. On Luso-Brazilian commercial ties with England, see James Lang, *Portuguese Brazil: The King's Plantation,* pp. 115–47.

Chapter 5. The American Potato

1. The 1870 per-capita consumption estimate is from Roger W. Gray, Vernon L. Sorenson, and Willard W. Cochrane, *An Economic Analysis of the Impact of Government Programs on the Potato Industry of the United States*, p. 10; the 1910 per-capita estimate, as well as the potato production statistics, are from United States Bureau of the Census, *Historical Statistics of the United States, Colonial Times to 1970*, pp. 330, 515. A hundredweight (cwt) = 100 pounds = 45.359 kilos.

2. Gray, Sorenson, and Cochrane, *Economic Analysis*, pp. 9–20.

3. Rozin, *Primal Cheeseburger*, pp. 145–52.

4. Joseph Guenthner, Biing-Hwang Lin, and Annette E. Levi, "The Influence of Microwave Ovens On the Demand for Fresh and Frozen Potatoes"; for per-capita consumption, see National Potato Council (NPC), *Potato Statistical Yearbook 1999*, p. 46.

5. Compiled from U.S. Bureau of the Census, *Historical Statistics*, p. 12.

6. Arthur Hawkins, "Highlights of a Half-Century in Potato Production." On Russet-Burbank's continued dominance, see NPC, *Potato Statistical Yearbook 2000–2001*, p. 54.

7. Daniel Dean, "Our Changing Potato Industry," pp. 110–11.

8. E. J. Kahn Jr., *The Staffs of Life*, pp. 132–33. Conversions are at the rate of 27.2 kilos to a bushel.

9. Gray, Sorenson, and Cochrane, *Economic Analysis*, p. 24.

10. Daniel Dean, "Changing Industry," p. 110.

11. Ibid., pp. 111–12.

12. Compiled from NPC, *Potato Statistical Yearbook 2000–2001*, pp. 50–51, 62–63; conversions are at the rate of twenty-two cwt per metric ton.

13. Kahn, *Staffs of Life*, pp. 134–35.

14. Versions of the story can be found in Jonathan D. Sauer, *Historical Geography*, pp. 152–54; Hawkes, *Evolution, Biodiversity*, p. 199; Larry Zuckerman, *The Potato: How the Humble Spud Rescued the Western World*, pp. 235–36; and Salaman, *History*, pp. 165–66.

15. The story is recounted in James Zalewski, "Russet Burbank: Is It Here to Stay?" pp. 20–22.

16. Ibid., p. 21; and Thomas S. Walker, "Patterns and Implications of Varietal Change in Potatoes," pp. 11, 17.

17. Thomas S. Walker, "Varietal Change," pp. 16–20; Zalewski, "Russet Burbank," pp. 21–22.

18. For recent figures on fresh, per-capita consumption see NPC, *Potato Yearbook 1999*, p. 46. Also see Thomas S. Walker, "Varietal Change," p. 4.

19. Guenthner, "Microwave," pp. 46–47.

20. NPC, *Potato Yearbook 1999*, p. 46. For the 1996–98 figures on metric tons of fries, fresh spuds, chips, and flakes consumed, see NPC, *Potato Yearbook 2000–2001*, p. 64; there are approximately 45 metric tons in 1000 cwt.

21. See NPC, *Potato Statistical Yearbook 1996*, p. 31; NPC, *Potato Yearbook 1999*, pp. 48–49; NPC, *Potato Yearbook 2000–2001*, p. 54; and Zalewski, "Russet Burbank," pp. 21–24.

22. Thomas S. Walker, "Varietal Change," p. 25, generally, pp. 16–22.

23. Zalewski, "Russet Burbank," pp. 22–24; Thomas S. Walker, "Varietal Change," pp. 18–19.

24. Walker, "Varietal Change," p. 26.

25. CIP, *Potatoes for the Developing World*, pp. 80–84.

26. Joe Sowokinos, "To Chip or Not to Chip"; also see Richard G. Novy et al., "Nor-Valley: A White-Skinned Chipping Cultivar With Cold-Sweetening Resistance," pp. 101–105.

27. Zalewski, "Russet Burbank," p. 8.

28. Based on an interview at Golden Flake's factory in Nashville, Tennessee, June 10, 1999.

29. See NPC, *Potato Yearbook 1996*, p. 31; NPC, *Potato Yearbook 1999*, p. 44; NPC, *Potato Yearbook 2000–2001*, p. 54; and Thomas S. Walker, "Varietal Change," pp. 15–21. Also see the list of U.S. favorites in Robert E. Rhoades, "The Incredible Potato," pp. 680–81.

30. Toussaint-Samat, *History of Food*, pp. 725–26.

31. Ibid., p. 725.

32. For an Amsterdam-quality fry, try B. Frites, 1657 Broadway at 47th Street in New York City. As raw material, they use Marrist Piper but have promised to switch over to Bintje in 2000. Their fries are doubled cooked in soybean oil in a Belgium-style frier.

33. See Snack Food Association (SFA), *50 Years: A Foundation for the Future*, pp. 11–18; and Kahn, *Staffs of Life*, pp. 136–37. The potato chip is but one episode in the story of processed food; see Harvey Levenstein, *Paradox of Plenty: A Social History of Eating in America*, pp. 101–18.

34. Based on a survey of the chips sold at The Health Nuts, Broadway at 99th Street in Manhattan, December 14, 1998.

Chapter 6. Andean Agriculture

1. John V. Murra, "Andean Societies Before 1532," pp. 61–62.

2. Lang, *Conquest and Commerce*, p. 7; Elías Mujica, "Terrace Culture and Pre-Hispanic Traditions."

3. Aveni, *Stairways*, pp. 46–48, 147–76; Krupp, *Echoes*, pp. 270–76. The description of the Temple of the Sun follows Garcilaso de la Vega, *The Royal Commentaries of the Inca Garcilaso de la Vega*, pp. 75–77.

4. John Howland Rowe, "Inca Culture at the Time of the Spanish Conquest," pp. 224–27; Jean-Pierre Protzen, "Inca Quarrying and Stonecutting," pp. 183–96.

5. John V. Murra, *The Economic Organization of the Inka State*, pp. 13–14, 66–80; de la Vega, *Commentaries*, pp. 122, 124–25.

6. For the estimate of the road system's size, see the careful calculations in John Hyslop, *The Inca Road System*, pp. 223–24. Also see Victor W. von Hagen, "America's Oldest Roads," pp. 17–21; and John Howland Rowe, "Inca Culture," pp. 229–33.

7. See de la Vega, *Commentaries;* Pedro de Cieza de León, *The Discovery and Conquest of Peru;* and Cobo, *History*.

8. Cieza de León, *Discovery and Conquest*, pp. 117–20.

9. See David Guillet, "Terracing and Irrigation in the Peruvian Highlands," p. 409; Vietmeyer, *Lost Crops;* pp. 3–7; and de la Vega, *Commentaries*, pp. 115–17.

10. John Howland Rowe, "Inca Culture," pp. 210–11, 233, 293–97; Murra, *Inka State*, pp. 5–20.

11. de la Vega, *Commentaries*, p. 140. Also see Murra, *Inka State*, p. 15.

12. For an excellent analysis of the Inca, see Conrad and Demarest, *Religion and Empire*, pp. 84–151.

13. See, for example, Augusto Cardich, "Native Agriculture in the Highlands of the Peruvian Andes," pp. 28–34; and Paul Kosok, *Life, Land, and Water in Ancient Peru*.

14. See Omar Sattaur, "The Lost Art of Waru Waru," and Alan L. Kolata and Charles R. Ortloff, "Tiwanaku Raised-Field Agriculture in the Lake Titicaca Basin," pp. 109–20.

15. C. R. Ortloff, "La ingeniería hidráulica chimú," pp. 93–94.

16. Clark L. Erickson, "Archaeological Methods for the Study of Ancient Landscapes of the Llanos de Mojos in the Bolivian Amazon," pp. 66–95.

17. On verticality, see Yoshiro Onuki, "The 'Yunga' Zone in the Prehistory of the Central Andes: Vertical and Horizontal Dimensions in Andean Ecological and Cultural Processes," pp. 339–56; and Murra, "Andean Societies," pp. 63–67.

18. The climatic divisions described here roughly follow Javier Pulgar Vidal, *Geografía del Perú: Las Ocho Regiones Naturales*, pp. 51–111, 209–10. For agriculture by zones, also see Murra, "Andean Societies," pp. 63–67.

19. Onuki, "'Yunga' Zone," pp. 339–53.

20. John Howland Rowe, "Inca Culture," pp. 291–92.

21. See Murra, *Inka State*, pp. 9–12; and John Howland Rowe, "Inca Culture," pp. 210–12. On *chicha*, see Hugh C. Cutler and Martín Cárdenas, "Chicha, a Native South American Beer." The quote is from Cobo, *History*, p. 28.

22. Vietmeyer, *Lost Crops*, pp. 15–17.

23. See Cardich, "Native Agriculture," pp. 22–25; de la Vega, *Commentaries*, p. 118; Vietmeyer, *Lost Crops*, pp. 139–61, 181–89.

24. See Murra, *Inka State*, pp. 3–12, 45–49; and Carl O. Sauer, "Geography and Plant and Animal Resources," pp. 338–39.

25. Murra, *Inka State*, p. 7; Vietmeyer, *Lost Crops*, pp. 99–100. These frost-resistant species are called rucki (Vietmeyer), luki (Murra), and sometimes luk'y.

26. On the steps in making *chuño*, see Mauricio Mamani, "El *chuño*: preparación, uso, almacenamiento," pp. 241–45.

27. Vietmeyer, *Lost Crops*, pp. 57–65, 129–37.

28. This is discussed in Carl Troll, "The Cordilleras of the Tropical Americas: Aspects of Climatic, Phytogeographical, and Agrarian Ecology." On the Nasca lines, see Anthony F. Aveni and Helaine Silverman, "Between the Lines: Reading the Nasca Markings as Rituals Writ Large."

29. John V. Murra, "El control vertical de un máximo de pisos ecológicos en la economía de las sociedades andinas"; also see Murra, "Andean Societies," pp. 63–66.

30. Norio Yamamoto, "The Ecological Complementarity of Agro-Pastoralism: Some Comments."

31. See the site chronologies in Piperno and Pearsall, *Lowland Neotropics*, pp. 204–205, 248–49, 274–75; for drawings and a discussion of early crops, see Piperno and Pearsall, *Lowland Neotropics*, pp. 109–66; for comparisons of crop origins in different zones, see Deborah M. Pearsall, "The Origins of Plant Cultivation in South America." On climatic change, see Ugent and Peterson, "Archaeological Remains," pp. 3–4. Roosevelt dates some of the levees on Marajo Island to eight thousand years ago; see *Moundbuilders*, p. 11.

32. See John V. Murra, "'El Archipiélago Vertical' Revisited," pp. 3–13.

33. See John Howland Rowe, "Inca Culture," p. 211; Cardich, "Native Agriculture," pp. 34–35; and de la Vega, *Commentaries*, p. 117.

34. Gary Paul Nabhan, "Native Crops of the Americas: Passing Novelties or Lasting Contributions to Diversity?" pp. 147–50.

35. Ibid. Also see Vietmeyer, *Lost Crops*, pp. 93–103. For a discussion of how many potato species there are, see chapter 4, note 54.

36. Karl S. Zimmerer, *Changing Fortunes: Biodiversity and Peasant Livelihood in the Peruvian Andes*, pp. 1–25.

Chapter 7. Lost Crops

1. Based on an interview with Carlos Arbizu, CIP headquarters, May 12, 1997, pp. 6–7. All notes and interviews are on deposit at the Vanderbilt University Library, where they are organized by country. Pages for each country file are numbered sequentially. Also see reporter Laurie Ochoa's interview with Arbizu in "Potato: Saving Peru's Dwindling Diversity."

2. See Noël Pallais, "Origin of the International Potato Center," p. 230. On CIP's early history, see Kenneth J. Brown, *Roots and Tubers Galore: The Story of CIP's Global Research Program and the People Who Shaped It.*

3. For more information about the CGIAR, check out the website at www.cgiar.org.

4. Arbizu, interview, CIP headquarters, May 12, 1997, pp. 3–5.

5. Vietmeyer, *Lost Crops*.

6. For a feature story on Andean roots and tubers, see CIP, "'Lost' Andean Roots and Tubers Travel Ancient Trade Routes," in *International Potato Center Annual Report 1994*, pp. 16–20.

7. Vietmeyer, *Lost Crops*, p. 3.

8. See Lang, *Conquest and Commerce*, pp. 48–57; and Geoffrey J. Walker, *Spanish Politics and Imperial Trade, 1700–1739*, pp. 5–11.

9. James Lang, *Inside Development in Latin America: A Report from the Dominican Republic, Colombia, and Brazil*, pp. 111–18.

10. Salaman, *History*, pp. 71, 36–37.

11. Geoffrey J. Walker, *Spanish Politics*, pp. 137–73.

12. Lang, *Commerce and Conquest*, pp. 74–76.

13. Chilies, tomatoes, and chocolate are the main ingredients in *mole poblano;* see Coe and Coe, *True History*, pp. 216–19.

14. Vietmeyer, *Lost Crops*, p. 7.

15. FAO, *FAO Production Yearbook* 1995, pp. 26–27.

16. This description of maca follows Vietmeyer, *Lost Crops*, pp. 57–65; and C. F. Quirós and R. Aliaga Cárdenas, "Maca," pp. 182–86, 189. Many of the crops discussed in this chapter are described in Heinz Brücher, *Useful Plants of Neotropical Origin and Their Wild Relatives*, pp. 4–53; and in Martín Cárdenas, *Manual de plantas económicas de Bolivia*.

17. See the tables in Woolfe, *Human Diet*, pp. 28–29; also see Quirós and Cárdenas, "Maca," p. 186.

18. CIP, *International Potato Center Annual Report 1998*, p. 21; Michael Hermann and J. Heller, "Andean Roots and Tubers at the Crossroads," p. 9.

19. Quirós and Cárdenas, "Maca," p. 183.

20. Based on Vietmeyer, *Lost Crops*, pp. 67–73, 83–91, 105–13. Also see the brief descriptions in Carlos Arbizu, Z. Huamán, and A. Golmirzaie, "Other Andean Roots and Tubers."

21. Vietmeyer, *Lost Crops*, pp. 105–13.

22. Ibid., pp. 83–90.

23. Ibid., pp. 67–73.

24. Patricio Espinosa and Charles C. Crissman, "Aspectos del consumo urbano de las raíces y tubérculos andinos y actitud del consumidor en Ecuador," p. 3.

25. Patricio Espinosa et al., *Raíces y tubérculos andinos. Cultivos marginados en el Ecuador*, pp. 33–35.

26. Based on an interview with Patricio Espinosa, CIP-Ecuador, April 29, 1997, pp. 15–16.

27. See Patricio Espinosa, ed., *Recetario de las raíces y tubérculos andinos*.

28. Espinosa et al., *Raíces y tubérculos*, pp. 6–11.

29. Espinosa and Crissman, "Aspectos del consumo," pp. 9–12; Espinosa et al., *Raíces y tubérculos*, pp. 21–35, 62–63, 75–76, 94–95, 108–10.

30. Espinosa, interview, CIP-Ecuador, April 29, 1997, pp. 14–17.

31. Espinosa and Crissman, "Aspectos del consumo," pp. 3, 7.

32. Michael Hermann, "Arracacha," pp. 78–89, 96, 108–109, 112, 119–22.

33. Vietmeyer, *Lost Crops*, pp. 75–81.

34. Arbizu, interview, CIP headquarters, May 12, 1997, p. 7.

35. See Martin Sørensen, Wolfgang J. Grüneberg, and Bo Örting, "Ahipa," pp. 16, 21–24, 29, 32, 42, 51–53, and 58; Martin Sørensen, *Yam Bean*, pp. 51–54; and Vietmeyer, *Lost Crops*, pp. 39–45.

36. See Alfredo Grau and Julio Rea, "Yacón," pp. 202, 209, 216, 218–20, 222, 230, 232–34; and Vietmeyer, *Lost Crops*, pp. 115–23. On yacón's use in Japan, see CIP, "Yacón in Hokkaido, Japan," p. 11.

37. For starch production, see Espinosa et al., *Raíces y tubérculos*, pp. 144, 153–58; also see Vietmeyer, *Lost Crops*, pp. 27–37.

38. The figures on Andean root and tuber consumption are from Hermann, "Andean Roots," pp. 8–9; also see Hermann, "Arracacha," p. 89. On achira in Vietnam, see Michael Hermann, N. K. Quynh, and D. Peters, "Reappraisal of Edible Canna as a High-Value Starch Crop in Vietnam." Also see CIP, "Achira in Vietnam: An Intriguing Example," in *International Potato Center Annual Report 1992*, p. 15.

Chapter 8. Wild Potatoes

1. J. G. Hawkes, "The Domestication of Roots and Tubers in the American Tropics," p. 484. Taro root *(Colocasia esculenta)* is not one of the crops domesticated in the New World. Adapted to wet soils, its mainstay today is the South Pacific; its original home is probably Southeast Asia. Taro root's equivalent in the Americas is yautia, any of the various species of *Xanthosoma*, which includes cocoyam or malanga *(Xanthosoma sagittifolium)*. All of these are from the aroid family, *Araceae*. Arrowroot *(Maranta arundinacea)* grew wild in the New World; evidence of its cultivation shows up in some of the earliest archaeological sites. At the time of the Conquest, it was still used as an antidote to poison-arrow wounds but was no longer cultivated as a food crop. On these points, see Heiser, *Seed to Civilization*, pp. 148–50; Hawkes, "Domestication," pp. 489–91; and Piperno and Pearsall, *Lowland Neotropics*, pp. 115–16, 126–28.

2. Vietmeyer, *Lost Crops*, pp. 124–21. On Andean grains, see Mario Tapia et al., *Quinua y kañiwa: cultivos andinos*.

3. For a note on potato species, see chapter 4, note 54.

4. Vietmeyer, *Lost Crops*, pp. 93–103.

5. Carlos M. Ochoa, *The Potatoes of South America: Bolivia*, p. xxix.

6. Zósimo Huamán, "Conservation of Potato Genetic Resources at CIP," pp. 2–4. Also see Zósimo Huamán, "Ex situ Conservation of Potato Genetic Resources at CIP."

7. Huamán, Glomirzaie, and Amoros, "Potato," pp. 22–24.

8. Ibid., p. 22–23.

9. Carlos M. Ochoa, *Bolivia*, pp. xvii–xxix.

10. Bill Hardy, "Carlos Ochoa—Potato Prize Winner," pp. 8–9; Ken Ringle, "Raider of the Lost Spud."

11. See Carlos M. Ochoa, *Peru*, and Carlos M. Ochoa, *Bolivia*.

12. Vietmeyer, *Lost Crops*, p. 1.

13. Based on a field interview with Don Carlos Huascar, Toralapa-PROINPA, March 1, 1996, pp. 15–17. All notes and interviews are on deposit at the Vanderbilt University Library, where they are organized by country. Pages for each country file are numbered sequentially.

14. Farmers like Don Huascar are described as "germ plasm conservationists" and typically plant twenty or more varieties; see Mario Tapia and A. Rosas, "Seed Fairs in the Andes: A Strategy of Local Conservation of Plant Genetic Resources."

Chapter 9. No Easy Answers

1. See Herbert S. Klein, *Bolivia: The Evolution of a Multi-Ethnic Society*, p. 228; and Jonathan Kelley and Herbert S. Klein, *Revolution and the Rebirth of Inequality: A Theory Applied to the National Revolution in Bolivia*, pp. 67–104.

2. For comparative figures on land reform, see Kelley and Klein, *Revolution*, p. 97.

3. For an account of this period, see Klein, *Bolivia*, pp. 227–68.

4. See Alma Guillermoprieto, *The Heart that Bleeds*, pp. 185–91; and Harry Sanabria, *The Coca Boom and Rural Social Change in Bolivia*, pp. 167–70.

5. Lang, *Feeding a Hungry Planet*, pp. 90–99.

6. The acronym PROINPA stands for Programa de Investigación de la Papa; IBTA stands for Instituto Bolivariano de Tecnología Agropecuaria. COTESU has since changed its name to the Swiss Agency for Development and Cooperation, or COSUDE (Agencia Suiza para el Desarrolo y la Cooperación).

7. Based on a workshop presentation by Julio Laredo, Sucre, Bolivia, April 4, 1990, p. 4. All interviews and notes are on deposit at the Vanderbilt University Library, where they are organized by country. Pages for each country file are numbered sequentially.

8. Hernán H. Zeballos, *Aspectos económicos de la producción de la papa en Bolivia*, pp. 27–30, 100–101.

9. The 1975–79 potato data is compiled from FAO, *Production Yearbook* 1977, p. 110; and FAO, *Production Yearbook* 1980, p. 113. For the 1985–89 data, see FAO, *Production Yearbook* 1987, p. 137; and FAO, *Production Yearbook* 1990, p. 90. The population data is from Thomas M. McDevitt, *World Population Profile: 1996*, p. A-8.

10. The CIP estimate is 2.5 metric tons; see CIP, *International Potato Center Annual Report: 1996*, p. 26.

11. For example, see CIP, *Potatoes for the Developing World*, pp. 22–23, 94–95.

12. John Tucker, "The Value of Seed Potato Certification to the Potato Industry," pp. 41–42. Also see Daniel Dean, "Changing Industry," pp. 109, 112; and B. Baribeau, "The Tuber-Unit Seed Plot in Quebec."

13. H. C. Moore, "Evidence that Certified Seed is Improved Seed." Also see William Stuart, "The Value of Good Seed Potatoes," pp. 84–85.

14. United States Department of Agriculture (USDA), *Agricultural Statistics* 1936, p. 160.

15. Tucker, "Potato Industry," p. 41. The conversion rate used is 27.2 bushels to a metric ton.

16. See, for example, R. L. Plaisted et al., "Andover: An Early to Midseason Golden Nematode Resistant Variety for Use as Chipstock or Tablestock," and R. L. Plaisted et al., "Pike: A Full Season Scab and Golden Nematode Resistant Chipstock Variety."

17. Why a formal seed program is not appropriate in the Andean highlands is analyzed in Graham Thiele, "Informal Seed Systems in the Andes: Why Are They Important and What Should We Do With Them?" For estimates of the average holdings of Bolivian potato farmers, see Sanabria, *Coca Boom*, pp. 132–33; and Thiele, "Informal Seed Systems," p. 84.

18. This quote and what follows are based on interviews with André Devaux, Sucre-PROINPA, April 3, 1990, pp. 1–3; and Cochabamba-PROINPA, April 6, 1990, pp. 24–29. Also see Thiele, "Informal Seed Systems," pp. 84–86.

19. Jorge Quiroga and Greta Watson, "Diagnósticos multidisciplinarios y su rol en el establecimiento de prioridades mediante investigación agrícola," pp. 466–76; Antonio Gandarillas and André Devaux, "PROINPA's Agroecological Approach for Potato Research in Bolivia," pp. 1–4.

20. PROINPA, *Informe Anual Compendio 1993–1994*, pp. i–iii.

21. Jeffrey Ronald Jones, "Technological Change and Market Organization in Cochabamba, Bolivia: Problems of Agricultural Development Among Potato Producing Small Farmers," p. 81.

22. Based on an interview with Nelson Estrada, Toralapa-PROINPA, April 6, 1990, pp. 32–33.

23. R. L. Plaisted et al., "Broadening the Range of Adaptation of Andigena (Neo-Tuberosum) Germplasm."

24. Huamán, Glomirzaie, and Amoros, "Potato," p. 23.

25. Heinz Brücher, *Useful Plants,* pp. 30–32. On andigena's pre-potato famine dominance, see Jonathan D. Sauer, *Historical Geography,* pp. 152–54.

26. PROINPA, *Informe Compendio del Programa de Investigación de la Papa 1996–1998,* p. 26.

27. André Devaux, Nelson Estrada, and Enrique Carrasco, "Frost Tolerance in Potatoes: A Challenge to Andean Biodiversity," p. 1.

28. Ibid., pp. 1–6. Also see Huamán, Glomirzaie, and Amoros, "Potato," p. 23.

29. M. Iwanaga and P. Schmiediche, "Using Wild Species to Improve Potato Cultivars."

30. Nelson Estrada Ramos, "Utilization of Wild and Cultivated Diploid Potato Species to Transfer Frost Resistance into the Tetraploid Common Potato, *Solanum Tuberosum L.*"

31. E. Carrasco et al., "Frost-Tolerant Varieties for the Andean Highlands," pp. 227, 231.

32. See the list in PROINPA, *Informe Anual Compendio 1993–1994,* p. 60.

33. Graham Thiele, G. Watson, and R. Torrez, "Cómo y dónde involucrar agricultores en la selección de variedades," pp. 1–22.

Chapter 10. Rustic Beds

1. See Guillermoprieto, *Heart,* pp. 179–80, 207.

2. Compiled from James W. Wilkie, Edwardo Alemán, and José Guadalupe Ortega, eds., *Statistical Abstract of Latin America,* vol. 35, pp. 12, 135.

3. Based on an interview with André Devaux, Cochabamba-PROINPA, February 29, 1996, pp. 2–3. All interviews and notes are on deposit at the Vanderbilt University Library, where they are organized by country. Pages for each country file are numbered sequentially.

4. What follows is based on interviews and site visits with Willman García, Toralapa Station-PROINPA, March 1, 1996, pp. 10–15.

5. PROINPA, *Informe Anual Compendio 1994–1995,* p. 32.

6. CIP, *Potatoes for the Developing World,* p. 63.

7. PROINPA, *Informe Compendio 1996–1998,* pp. 5, 26–27. Also see Carrasco et al., "Frost-Tolerant Varieties."

8. The acronym PRACIPA stands for Programa Andino Cooperativo de la Investicación de la Papa.

9. IBTA, *Boletín Técnico* 1 (May, 1995), pp. 2–7.

10. René Andrews, "Integrated Pest Management of Potato Tuber Moth in Cochabamba, Bolivia," pp. 8–9; PROINPA, *Informe Anual Compendio 1994–1995,* pp. 44–45; PROINPA, *Informe Compendio 1996–1998,* pp. 5, 34–35. On the tuber moth's life cycle, see K. V. Raman, "Potato Tuber Moth," pp. 1–14.

11. What follows is based on an interview with Gino Aguirre, Toralapa-PROINPA, March 1, 1996, pp. 8–10.

12. PROINPA, *Informe Anual Compendio 1994–1995,* pp. 64–66. The figures are updated based on PROINPA, *Informe Compendio 1996–1998,* p. 6.

13. See PROINPA, *Informe Anual Compendio 1994–1995,* pp. 61–62; and PROINPA, *Informe Compendio 1996–1998,* pp. 5, 42.

14. IBTA, *Boletín Técnico* 1 (May, 1995), p. 8.

15. Much of what follows is based on an interview with Jaime Herbas, Tarija-PROINPA, March 5, 1996, pp. 39–41.

16. See Thomas S. Walker, "Varietal Change," pp. 12–22.

17. "Investigaciones del IBTA: Retorna la variedad de papa 'Americana,'" *El Pais* (Tarija edition), December 11, 1995, p. 4.

18. Bolivia's national seed program is PROSEMPA, the acronym stands for Proyecto Semilla de la Papa.

19. PROINPA, *Informe Compendio 1996–1998*, pp. 14–18. Also see G. Aguirre et al., "Rustic Seedbeds: A Bridge Between Formal and Traditional Seed Systems in Bolivia," pp. 195–96.

20. This and much of what follows is based on an interview with Hernán Cordozo, Iscayachi-PROINPA, March 9, 1996, pp. 52–58.

21. PROINPA, *Producción de tubérculos-semilla de papa en cama protegida*, pp. 2–11. Also see the data in Aguirre et al., "Rustic Seedbeds," pp. 199–202.

22. Graham Thiele, André Devaux, and Carlos Soria, "Innovación tecnológica en la papa: de la oferta de tecnología al impacto macro-económico," p. 9; Aguirre et al., "Rustic Seedbeds," p. 202; and Thiele, "Informal Seed Systems," p. 95.

23. Interview with Javier Franco, Tarija-PROINPA, March 5, 1996, p. 47.

24. What follows is based on interviews with Javier Franco, Iscayachi-PROINPA, March 9, 1996, pp. 46–48, 50–52, 59–60. Also see PROINPA, *Informe Anual Compendio 1993–1994*, pp. 35–38; and *Informe Anual Compendio 1994–1995*, pp. 49–51.

25. On nematode control, see PROINPA, *Informe Compendio 1996–1998*, pp. 4, 36–37.

26. On quinoa and tarwi, see Vietmeyer, *Lost Crops*, pp. 149–61, 181–89.

27. PROINPA, *Informe Compendio 1996–1998*, pp. 14–18.

28. Aguirre et al., "Rustic Seedbeds," p. 202.

Chapter 11. Potato Weevils

1. See Wendell C. Bennett, "The Andean Highlands," pp. 46–47. Also see Gregory Knapp, *Ecología cultural prehispánica del Ecuador*.

2. Troll, "Cordilleras," p. 33.

3. Knapp, *Ecología cultural*, p. 25.

4. In general, see Gregory Knapp, *Andean Ecology: Adaptive Dynamics in Ecuador*, pp. 20–23.

5. Knapp, *Ecología cultural*, pp. 33–35.

6. Troll, "Cordilleras," pp. 20–21, 32–34.

7. Knapp, *Andean Ecology*, pp. 110–15, 119–45.

8. Troll, "Cordilleras," p. 33.

9. John V. Murra, "The Historic Tribes of Ecuador," pp. 792–93, 808–11. On Inca resettlement policies, see the testimony of Father Bernabé Cobo in Cobo, *History*, pp. 189–93.

10. Trotsky Guerrero Carrión, *Modernización agraria y pobreza rural en el Ecuador*, pp. 57, 71, and tables 17–19.

11. See Lang, *Inside Development*, pp. 146–48.

12. Economic Commission for Latin America (ECLA), *Statistical Yearbook for Latin America 1973*, p. 282.

13. United Nations (UN), *1979 Yearbook of World Energy Statistics*. The 7.32 rate used to convert metric tons to barrels of crude oil is taken from United Nations (UN), *1994 Energy Statistics Yearbook*, p. 1.

14. Guerrero Carrión, *Modernización*, pp. 17–18.

15. FAO, *Production Yearbook* 1998, p. 23. For Ecuador's population, see PRB, *World Population Data Sheet* 2000, p. 5.

16. Guerrero Carrión, *Modernización*, p. 65.

17. The acronym INIAP stands for Instituto Nacional Autónomo de Investigaciones Agropecuarias. What follows is based on an interview with Wilson Vásquez, Quito-INIAP, April 29, 1997, pp. 27–31. All notes and interviews are on deposit at the Vanderbilt University Library, where they are organized by country. Pages for each country file are numbered sequentially.

18. INIAP and FORTIPAPA (Fortalecimiento de la Investigación y Producción de Semilla

de Papa en el Ecuador), *Informe Anual 1995 Compendio*, p. 16; Charles C. Crissman, John M. Antle, and Susan M. Capalbo, eds., *Economic, Environmental, and Health Tradeoffs in Agriculture: Pesticides and the Sustainability of Andean Potato Production*, p. 88.

19. INIAP-FORTIPAPA, *Informe Anual 1996 Compendio*, p. 20.

20. The acronym FORTIPAPA stands for Fortalecimiento de la Investigacíon y Produccíon de Semilla de Papa en el Ecuador. The acronym for the Swiss Development Agency is COSUDE, which stands for Cooperacíon Suiza para el Desarrollo.

21. INIAP-FORTIPAPA, *Informe Anual 1998 Compendio*, pp. 31–33; INIAP-FORTIPAPA, *Informe Anual 1996*, p. 32.

22. On Phureja, see INIAP-FORTIPAPA, *Informe Anual 1996*, pp. 36–37; INIAP-FORTIPAPA, *Informe Anual 1998*, p. 14; and Vietmeyer, *Lost Crops*, pp. 96–97.

23. See Héctor Andrade et al., "The Role of the User in the Selection and Release of Potato Varieties in Ecuador."

24. INIAP-FORTIPAPA, *Informe Anual 1995*, p. 13; *Informe Anual 1996*, pp. 30–31. On the Frito-Lay project, see *Informe Anual 1998*, pp. 35–36.

25. Patricio Espinosa, Fabio Muños, and Jorge Carrillo, "Algunos aspectos del consumo doméstico de la papa en Quito, Guayaquil, y Cuenca," p. 11.

26. Ibid., pp 14–15.

27. Patricio Espinosa, "Pruebas de aceptabilidad de cuatro nuevas variedades mejoradas de papa del INIAP a nivel de consumidor final urbano en Quito," pp. 7–12.

28. Espinosa, Muños, and Carrillo, "Algunos aspectos," pp. 16–17. For a summary of the project's work with consumers, see INIAP-FORTIPAPA, *Informe Anual 1998*, pp. 80–81.

29. INIAP-FORTIPAPA, *Informe Anual 1995*, p. 16.

30. This and much of what follows is based on interviews with Charles Crissman, CIP-FORTIPAPA, May 5, 1997, pp. 54–55.

31. David R. Lee and Patricio Espinosa, "Economic Reforms and Changing Pesticide Policies in Ecuador and Colombia," pp. 121, 123, 125–26.

32. See Charles Crissman et al, "The Carchi Study Site: Physical, Health, and Potato Farming Systems in Carchi Province," pp. 112–17.

33. Compiled from table 5.9 in Crissman et al., "Carchi Study Site," p. 113.

34. Ibid., p. 113–14.

35. CIP, "Pesticides Threaten Ecuadorian Farmers," *International Potato Center Annual Report 1994*, p. 25.

36. On the sample, see Crissman et al., "Carchi Study Site," pp. 105–106; on carbofuran, see Crissman et al., "Carchi Study Site," p. 114. For the impact on farmer health, see Charles C. Crissman, Donald C. Cole, and Fernando Carpio, "Pesticide Use and Farm Worker Health in Ecuadorian Potato Production," pp. 595–96.

37. Donald C. Cole et al., "Health Impacts of Pesticide Use in Carchi Farm Populations," p. 225. Also see the interview with Crissman, CIP-FORTIPAPA, May 7, 1997, pp. 54–55.

38. United States Agency for International Development (USAID), Cooperative for American Relief to Everywhere (CARE), Peru, and CIP, "Ocho reglas de oro para el uso seguro de los plaguicidas.

39. Crissman et al., "Carchi Study Site," p. 112.

40. USAID-CARE-CIP, "Así vive el gorgojo de los Andes," pp. 1–10. Also see Lang, *Inside Development*, pp. 114–15.

41. Crissman et al., "Carchi Study Site," pp. 104–105.

42. This and much of what follows is based on interviews with Fausto López, FORTIPAPA-Montufar, May 6, 1997, pp. 59–63.

43. INIAP-FORTIPAPA, *Informe Anual 1998*, p. 14; *Informe Anual 1995*, p. 37. Conversions are based on an exchange rate of 3,860 sucres to one U.S. dollar.

44. For a study of the technology's adoption by farmers, see INIAP-FORTIPAPA, *Informe Anual 1998*, pp. 76–77.

45. Fausto Cisneros et al., "A Strategy for Developing and Implementing Integrated Pest Management."

46. See O. Ortiz et al., "Economic Impact of IPM Practices on the Andean Potato Weevil in Peru," p. 106, in general, pp. 95–110.

47. Crissman et al., "Carchi Study Site," p. 110.

48. Ibid., p. 113.

49. J. Revalo, S. Garcés, and J. Andrade, "Resistance of Commercial Potato Varieties to Attack by *Phytophthora infestans* in Ecuador."

50. Based on an interview with Doña Marta, Carchi Province, May 6, 1997, pp. 55–59.

51. For a breakdown of production costs, see Crissman et al., "Carchi Study Site," pp. 109–11, and table 5.6 on p. 110.

52. Crissman et al., "Carchi Study Site," p. 114. On small-scale cattle ranching, see Lang, *Feeding a Hungry Planet*, pp. 113–28.

Chapter 12. Watersheds

1. The Spanish acronym FUNDAGRO stands for Fundo para el Dessarollo Agropecuaria.

2. The Spanish acronym CIAT stands for Centro Internacional de Agricultura Tropical. For a discussion of CIAT and its research on rice and tropical pastures, see Lang, *Feeding a Hungry Planet*, pp. 100–29.

3. This and most of what follows are based on an interview with Oswaldo Paladines, FUNDAGRO-Quito, May 9, 1997. All notes and interviews are on deposit at the Vanderbilt University Library, where they are organized by country. Pages for each country file are numbered sequentially.

4. Also see Guerrero Carrión, *Modernización,* pp. 51–52.

5. The Spanish acronym CONDESAN stands for Consorcio para el Desarrollo Sostenible de la Ecoregion Andina.

6. For a concise statement of CONDESAN's mission and a list of project sites, see Joshua Posner, "Ecoregional Research: A Vision from the Andes."

7. CIP, "CONDESAN: An Ecoregional Approach to Research for the High Andes," in *International Potato Center Annual Report 1995*, pp. 27–30.

8. Knapp, *Andean Ecology,* pp. 68–72.

9. For a brief description of floury maize, see Christopher R. Dowswell, R. L. Paliwal, and Ronald P. Cantrell, *Maize in the Third World,* pp. 22–23, 112.

10. Taxo is a species of passion fruit. See Vietmeyer, *Lost Crops,* p. 293, footnote 15, in general, pp. 287–95. On guanabana, see Vietmeyer, *Lost Crops,* p. 238, in general, pp. 229–39.

Chapter 13. Multiplication

1. This and much of what follows is based on interviews with Fausto Merino, Riobamba-FORTIPAPA, May 7, 1997, pp. 74–83. All notes and interviews are on deposit at the Vanderbilt University Library, where they are organized by country. Pages for each country file are numbered sequentially.

2. Compiled from INIAP-FORTIPAPA, *Informe Anual 1996*, pp. 68–74.

3. INIAP-FORTIPAPA, *Informe Anual 1995*, p. 15.

4. The same seed analysis applies to Bolivia; see Aguirre et al., "Rustic Seedbeds," pp. 195–96.

5. This and much of what follows is based on an interview with Myriam Trujillo, Riobamba-FORTIPAPA, May 7, 1997, pp. 83–85.

6. Based on interviews at Guacoma, May 7, 1997, pp. 85–89.

7. This and what follows are based on interviews at Guabug and neighboring communities, May 8, 1997, pp. 95–102.

8. CIP, "Sprout Cuttings: A Rapid Multiplication Technique for Potatoes."

9. Diffused-light storage was pioneered at CIP. See Robert H. Booth, Roy L. Shaw, and CIP, *Principles of Potato Storage*, pp. 27–39; and CIP, *Potatoes for the Developing World*, pp. 94–96.

10. Comparing sprout multiplication to the traditional practice, INIAP put the rate of return on sprout multiplication at 500 percent; see INIAP-FORTIPAPA, *Informe Anual 1998*, p. 15.

11. Ibid., pp. 15–16. The calculation of the station's production presumes that seed tubers weighed an average of sixty-five grams each.

12. Ibid., pp. 77–79.

13. See Guerrero Carrión, *Modernizacion*, pp. 58–59.

14. Chile is an example; see Lang, *Feeding a Hungry Planet*, pp. 89–99.

15. Ibid., pp. 2, 13–14, 142–44.

Chapter 14. Potato Famine

1. Compiled from McDevitt, *World Population Profile: 1996*, p. A-3.

2. Michael Anderson, *Population Change in North-Western Europe, 1750–1850*, p. 23.

3. Zuckerman, *Potato*, pp. 115, 216.

4. Thomas Robert Malthus, *An Essay on the Principle of Population*, edited and with criticism by Philip Appleman, p. 20.

5. Thomas Robert Malthus, *An Essay on the Principle of Population*, selected and introduced by Donald Winch, p. 21.

6. Ibid., p. 230.

7. John Killen, ed., *The Famine Decade: Contemporary Accounts 1841–1851*, pp. 253–54.

8. These are standard estimates. See Cathal Póirtéir, introduction to *The Great Irish Famine*, p. 9; and Christine Kinealy, *A Death-Dealing Famine: The Great Hunger in Ireland*, pp. 190–91.

9. Killen, *Contemporary Accounts*, p. 253.

10. E. R. R. Green, "Agriculture," p. 123; figures converted to the metric system.

11. In August 1847, over three million people were being fed at government soup kitchens; see O'Neill, "Administration of Relief," p. 241.

12. Zuckerman, *Potato*, pp. 40–41.

13. See R. B. McDowell, "Ireland on the Eve of the Famine," pp. 55–71. On the attitudes of the rich towards the poor, see Zuckerman, *Potato*, pp. 42–45.

14. See Zuckerman, *Potato*, pp. 201–205; and Salaman, *History*, pp. 289–91.

15. Zuckerman, *Potato*, pp. 21–29.

16. Green, "Agriculture," pp. 101–104.

17. Converted to the metric system from figures and calculations in Zuckerman, *Potato*, p. 30.

18. Ibid., pp. 34–35; Green, "Agriculture," pp. 109–10.

19. Green, p. 89.

20. Zuckerman, *Potato*, p. 130.

21. Green, "Agriculture," pp. 91–95.

22. Killen, *Contemporary Accounts*, p. 29.

23. Ibid., pp. 31–32.

24. Kinealy, *Death-Dealing*, p. 5.

25. Oliver MacDonagh, "Irish Overseas Emigration during the Famine," p. 328; figures converted to the metric system.

26. Zuckerman, *Potato*, p. 155.

27. Ibid., pp. 149–58.

28. McDowell, "Eve of the Famine," p. 77.

29. Zuckerman, *Potato*, p. 156.

30. Green, "Agriculture," pp. 118–19.

31. Zuckerman, *Potato*, pp. 129, 149–50, 155.

32. Ibid., p. 43.

33. Malthus, *Essay*, selected and introduced by Winch, pp. xvii, 226–27.

34. Zuckerman, *Potato*, pp. 197–207.

35. Ibid., pp. 43–45, 203–10.

36. Salaman, *History*, pp. 290–91.

37. Ibid., p. 294. On grain exports from Ireland, see John Percival, *The Great Famine 1845–1851*, pp. 64–67.

38. On the contrast between the Tory and Whig administrations, see Kinealy, *Death-Dealing*, pp. 92–106; Salaman, *History*, pp. 294–303; and O'Neill, "Administration of Relief," pp. 212–34. For the numbers fed in the soup kitchens, see O'Neill, "Administration of Relief," p. 241.

39. See Sir William P. MacArthur, "Medical History of the Famine," pp. 263–89; and Laurence M. Geary, "Famine, Fever, and the Bloody Flux," pp. 82–83.

40. MacArthur, "Medical History," pp. 293–95. On the Kilrush workhouse, see Percival, *Great Famine*, pp. 106–107.

41. Compiled from the census report in Killen, *Contemporary Accounts*, p. 254.

42. MacArthur, "Medical History," pp. 279–81, 302–306.

43. Kinealy, *Death-Dealing*, p. 26.

44. On the Gregory Clause and its consequences, see Christine Kinealy, *This Great Calamity*, pp. 218–26; and Kinealy, *Death-Dealing*, pp. 123–25.

45. Percival, *Great Famine*, pp. 95–97, 101–105.

46. James S. Donnelly, Jr., "Mass Eviction and the Great Famine," pp. 155–56.

47. Percival, *Great Famine*, pp. 163–64, generally, pp. 162–80.

48. For the quote, see Kinealy, *Death-Dealing*, p. 145, generally, pp. 135–46. Also see Percival, *Great Famine*, pp. 163–71.

49. See MacDonagh, "Irish Immigration," p. 388; generally, pp. 319–88. Also see Kinealy, *Calamity*, p. 298, generally, pp. 297–304.

50. Kinealy, *Calamity*, p. 297.

51. Zuckerman, *Potato*, p. 191.

52. Kinealy, *Death-Dealing*, pp. 83–91.

53. Bergman, "Potato Blight," p. 291. The conversion rate used is 1 hectoliter = 3.53 cubic feet, or 2.84 bushels. There are 27.2 kilos to a bushel. Generally, see Bergman, "Potato Blight," pp. 390–93.

54. Ibid., p. 392.

55. Ibid., pp. 413–31.

56. Zuckerman, *Potato*, p. 212.

57. Kinealy, *Death-Dealing*, p. 2.

58. Zuckerman, *Potato*, p. 182.

59. Anderson, *Population Change*, p. 32.

60. Zuckerman, *Potato,* pp. 164–79.

61. Roger Price, "Poor Relief and Social Crisis in Mid-Nineteenth-Century France," p. 425, generally, pp. 424–29.

62. Price, "Poor Relief," pp. 430–33, 437. Also see Kinealy, *Death-Dealing,* p. 87.

63. Zuckerman, *Potato,* p. 183. French famine policies were not unlike strategies currently advocated; see Robert W. Solow, "How to Stop Hunger."

64. On the potato's role in the highlands, see Malcomb Gray, "The Highland Potato Famine of the 1840s," pp. 357–58.

65. T. M. Devine, *The Great Highland Famine: Hunger, Emigration, and the Scottish Highlands in the Nineteenth Century,* p. 91.

66. Ibid., pp. 273–77.

67. Green, "Agriculture," p. 126–27; figures converted to the metric system.

68. Percival, *Great Famine,* pp. 172–73.

69. Kinealy, *Death-Dealing,* pp. 151–55.

70. Anderson, *Population Change,* p. 25.

71. Kinealy, *Death-Dealing,* p. 123, generally, pp. 121–25. Also see Percival, *Great Famine,* pp. 179–80.

72. Kinealy, *Great Calamity,* p. 353.

Chapter 15. Hungry Ghosts

1. See the summary in William Chester Jordan, *The Great Famine: Northern Europe in the Early Fourteenth Century,* pp. 12–15.

2. Percival, *Great Famine 1845–1851,* pp. 90–94.

3. See Lawrence Goodwyn, *Democratic Promise: The Populist Moment in America,* pp. 26–31, generally, pp. 1–176, and Gavin Wright, *Old South New South: Revolutions in the Southern Economy Since the Civil War,* pp. 115–19.

4. This account follows Lang, *Inside Development,* pp. 146–48, generally, pp. 208–29.

5. Jasper Becker reviews the Soviet case; see *Hungry Ghosts: Mao's Secret Famine,* pp. 37–46. Also see Robert Conquest, *The Harvest of Sorrow: Soviet Collectivization and the Terror-Famine.*

6. Becker, *Hungry Ghosts,* pp. 47–54.

7. Ibid., p. 49, generally, pp. 54–57.

8. Ibid., p. 83, generally, pp. 58–82.

9. Ibid., p. 93, generally, pp. 83–111.

10. Dali L. Yang, *Calamity and Reform in China: State, Rural Society, and Institutional Reform Since the Great Leap Forward,* pp. 21–41, 64.

11. Becker, *Hungry Ghosts,* pp. 235–54, 261–62.

12. Ibid., p. 263, generally, pp. 260–65. For a sobering account of the terror behind the cultural revolution, see Liu Binyan, "An Unnatural Disaster." Deng's reforms and their impact are discussed in Frank Leeming, *The Changing Geography of China,* pp. 73–111.

13. See Ben Kiernan, *The Pol Pot Regime: Race, Power, and Genocide in Cambodia under the Khmer Rouge, 1975–1979,* pp. 456–60; and Patrick Heuveline, "'Between One and Three Million': Towards the Demographic Reconstruction of a Decade of Cambodian History," p. 60.

Chapter 16. Demography

1. Quinn, *Story of B,* pp. 298–301, generally, pp. 287–306.

2. Malthus, *Essay,* selected and introduced by Winch, pp. 14–15.

3. Kingsley Davis, "Population and Resources: Fact and Interpretation," p. 7, generally, pp. 2–8.

4. U.S. Bureau of the Census, *SAUS* 1999, p. 686.

5. McDevitt, *World Population Profile: 1996*, pp. A-27, A-30. The *Profile* has estimates for 2000.

6. For Russia's wheat imports, see U.S. Bureau of the Census, *SAUS* 1997, p. 853; for 1990 to 2000 population estimates, see McDevitt, *World Population Profile: 1996*, p. 10.

7. Ellen Jamison et al., *World Population Profile: 1994*, p. A-5; PRB, *World Population Data Sheet* 2000, pp. 6–7.

8. Compiled from IRRI, *Rice Statistics 1990*, p. 2; and FAO, *Quarterly Bulletin* 1999, 3/4, pp. 18–19.

9. Virginia Abernethy, "Optimism and Overpopulation," pp. 86–87.

10. U.S. Bureau of the Census, *SAUS* 1999, p. 278.

11. See John R. Gillis, Louise A. Tilly, and David Levine, "The Quiet Revolution."

12. Kingsley Davis, "Population and Resources," pp. 3–4.

13. Jordan, *Great Famine*, pp. 14–15, generally, pp. 7–39.

14. Anderson, *Population Change*, p. 23. The 1950 population is compiled from McDevitt, *World Population Profile: 1996*, p. A-9.

15. Amartya Sen, "Population, Delusion, and Reality," p. 63.

16. Anderson, *Population Change*, p. 23.

17. See Warren Thompson, "Population." Also see John R. Weeks, *Population: An Introduction to Concepts and Issues*, pp. 93–99.

18. Anderson, *Population Change*, pp. 30–31.

19. Ibid., pp. 40, 50–51, 82–83.

20. World Bank, *World Development Report 1984*, pp. 60–61.

21. Anderson, *Population Change*, pp. 76–77.

22. On school enrollments in England, see World Bank, *World Development Report 1984*, p. 61, in general, pp. 56–63. Also see the articles in Gillis, Tilly, and Levine, *Declining Fertility*. For the current total fertility rates, see PRB, *World Population Data Sheet* 2000, p. 8.

23. See Max Singer, "The Population Surprise," and Peter G. Peterson, "Gray Dawn: The Global Aging Crisis."

24. Anderson, *Population Change*, pp. 45–48, 71; World Bank, *World Development Report 1984*, p. 61.

25. World Bank, *World Development Report 1984*, p. 61.

26. Anderson, *Population Change*, pp. 30–31.

27. Compiled from McDevitt, *World Population Profile: 1996*, pp. A-6–A-10; also see Sen, "Population," p. 63.

28. McDevitt, *World Population Profile: 1996*, p. A-6.

29. See Kingsley Davis, *The Population of India and Pakistan*, p. 45.

30. See Anderson, *Population Change*, pp. 56–60.

31. Compiled from Kingsley Davis, *India and Pakistan*, pp. 34–35; and World Bank, *World Tables*, vol. 2, *Social Data*, p. 42.

32. Compiled from United Nations (UN), *Demographic Yearbook* 1954, pp. 593–94; and World Bank, *Social Data*, p. 75.

33. Compiled from UN, *Demographic Yearbook* 1954, p. 593–94; and World Bank, *Social Data*, p. 58.

34. The statistics on Ceylon's (Sri Lanka's) infant mortality are from UN, *Demographic Yearbook* 1951, pp. 331–33. For case studies of Sri Lanka, see N. K. Sarkar, *The Demography of Ceylon*, pp. 128–29, and generally, pp. 125–36. Also see S. A. Meegama, "Malaria Eradication

and Its Effects on Mortality Levels," pp. 207–208, and generally, pp. 207–37. Also see William Petersen, *Population*, pp. 560–76.

35. Anderson, *Population Change*, pp. 60–61.

36. World Bank, *World Development Report 1984*, pp. 52, 61.

37. The quotation and related figures are cited in Petersen, *Population*, pp. 10, 333. On the hazards of population estimates, see Ronald D. Lee, "Long-Run Global Population Forecasts: A Critical Appraisal."

38. U.S. Bureau of the Census, *SAUS* 1971, pp. 794–96; PRB, *World Population Data Sheet 1999*, p. 2.

39. Kingsley Davis, "Population and Resources," pp. 18–19; McDevitt, *World Population Profile: 1996*, p. A-27.

40. For the Kerala case, see Richard W. Franke and Barbara H. Chasin, "Development Without Growth: The Kerala Experiment"; Akash Kapur, "Poor but Prosperous"; B. A. Prakash, *Kerala's Economy*, pp. 43–60, 349–67; and Sen, "Population," pp. 70–71. Also see Richard W. Franke and Barbara H. Chasin, *Kerala: Radical Reform as Development in an Indian State*. The 2000 estimates for Indonesia and the United States are from Thomas McDevitt, *World Population Profile: 1998*, p. A-40, A-43.

41. T. V. Anthony, "The Family Planning Programme—Lessons from Tamil Nadu's Experience," pp. 321–27; Leela Visaria and Pavin Visaria, "India's Population in Transition," p. 22; PRB, *Success in a Challenging Environment: Fertility Decline in Bangladesh*, pp. 20–24. Also see Sen, "Population," p. 70.

42. McDevitt, *World Population Profile: 1996*, p. A-27; Sen, "Population," pp. 70–71; and PRB, *World Population Data Sheet* 1999, p. 6.

43. McDevitt, *World Population Profile: 1998*, p. A-40; World Bank, *Social Data*, pp. 42–43.

44. See Weeks, *Population*, pp. 7–10; McDevitt, *World Population Profile*, p. A-6.

45. World Bank, *Entering the 21st Century: World Development Report 1999/2000*, p. 251; Sen, "Population," pp. 65–66.

46. United Nations Development Program (UNDP), *Human Development Report 1998*, p. 184; "China Survey: The Titan Stirs," p. 3, and generally, pp. 3–18. Also see World Bank, *Development Report 1999/2000*, p. 250.

47. Sen, "Population," p. 66.

48. World Bank, *Development Report 1999/2000*, p. 245.

49. Lang, *Feeding a Hungry Planet*, p. 15.

50. See Ester Boserup, *Population and Technological Change: A Study of Long-Term Trends*.

51. McDevitt, *World Population Profile: 1998*, pp. A-3, A-39–A-43. Also see Ronald D. Lee, "Population Forecasts," pp. 58–61; and Bill McKibben, "A Special Moment in History."

52. If total fertility rates have dropped, why is the world's population still increasing? The answer has to do with what demographers call population momentum.

Consider a hypothetical country with 1000 people in the year 2000, 40 percent (400 people) of whom are nineteen or younger, which is about average for the less developed countries of Asia. Assume that everyone marries at age twenty, that they all have two children within the first five years of marriage, and that the crude death rate averages 10 people per thousand per year. How large will the population be in twenty years? The crude death rate will subtract out a total of 200 people over twenty years (10 people per year x 20 years). So in 2020, only 800 of the people alive in 2000 will still be alive. In the meantime, how many have we added? Consider just the 400 people who were nineteen or younger in 2000; assume an equal number at each age. By 2020, 300 of them (those born between 1980 and 1994) will have been married at least five years, long enough to have had the stipulated two children. Of the 300 parents, half are women. Hence, the increment by 2020 for the 1980–94 cohort will be 300 (150 women x 2 children each). Adding this 300 to the 800 people still alive in 2020 gives us 1,100, or a 10-percent increase over 2000. Of course, I have left out some of the fine points. The fertility of young couples married less than five years in 2000 and in 2020 is not added in. But the point is clear enough. Even with a total fertility rate at or below replacement, it takes time for a "young" country's overall growth rate to stabilize.

A country's age distribution—that is, the percentage of its population under twenty—has an impact all its own on population growth. So too do factors like the age at which people marry or how they space their children. Changes in such parameters can make a big difference very fast. In the example above, assume that only 20 percent of the population (200 people) is nineteen or younger in 2000, which is about right for western Europe. Change the age at which everyone marries to twenty-five. Leave everything else the same. What happens? In 2020, only 100 people, those born between 1980 and 1990, will be over thirty and hence old enough and married long enough to have had two children. For this cohort, the total additions by 2020 will add up to just 100 (50 wives x 2 children). In the meantime, 200 people will have died. Of course, we should add in the children born to couples married less than five years in 2000, plus those married long enough to have at least one child by 2020. Even so, such a population will grow very little, if at all.

For a brief discussion of population momentum, see Weeks, *Population,* p. 292.

53. Singer, "Population Surprise," pp. 24–25.

54. Kingsley Davis, "Population and Resources," pp. 16–17.

55. Steven Butler and Kevin Whitelaw, "Japan's Baby Bust"; PRB, *World Population Data Sheet* 2000, p. 7.

56. The figure for Japan is from Butler and Whitelaw, "Baby Bust," p. 43; the U.S. figure is compiled from the middle series estimates in U.S. Bureau of the Census, *SAUS* 1997, p. 17.

Chapter 17. Late Blight

1. CIP, "Taming the Late Blight Dragon," in *International Potato Center Annual Report 1996,* p. 18. Also see the interview with Edward French, CIP-Lima, July 12, 1988, pp. 35–36. All notes and interviews are on deposit at the Vanderbilt University Library, where they are organized by country. Pages for each country file are numbered sequentially.

2. Roger J. McHugh, "The Famine in Irish Oral Tradition," pp. 396–97.

3. Salaman, *History,* pp. 300–301.

4. Killen, *Contemporary Accounts,* pp. 33–35.

5. Zuckerman, *Potato,* pp. 207–208. Also see William F. Fry and Stephen B. Goodwin, "Resurgence of the Irish Potato Famine Fungus," p. 365.

6. See Leslie J. Dowley and Eugene O'Sullivan, *A Short History of the Potato, the Famine, Late Blight, and Irish Research on Phytophthora infestans,* p. 11. Also see Zuckerman, *Potato,* pp. 205–209; and Fry and Goodwin, "Resurgence," pp. 365–66.

7. Dowley and O'Sullivan, *Short History,* p. 13.

8. We may soon know for sure which type attacked Ireland's potato crop. See J. B. Ristaino, G. R. Parra, and C. Trout Groves, "PCR Amplification of *Phytophthora infestans* from 19th Century Herbarium Specimens."

9. Roxanne Nelson, "The Blight is Back," p. 20; Fry and Goodwin, "Resurgence," p. 365. For the quotation, see Jonathan D. Sauer, *Historical Geography,* p. 152.

10. McHugh, "Oral Tradition," p. 39; Killen, *Contemporary Accounts,* pp. 30–31.

11. This and much of what follows is based on an interview with Greg Forbes, CIP-Quito, April 28, 1997, pp. 2–7. In general, also see Jan W. Henfling, "Late Blight of Potato."

12. Randall C. Rowe, Sally A. Miller, and Richard M. Riedel, "Late Blight of Potato and Tomato."

13. Fry and Goodwin, "Resurgence," p. 364–65.

14. Willie Kirk and Jeffrey Stein, "Recommendations for Late Blight Control in Michigan for 2000."

15. On the difference between systemic and protectant fungicides, see Phil Nolte, "Potato Pointers."

16. Kirk and Stein, "Recommendations." On the factors involved in forecasting, also see R. J. Hijmans, G.A. Forbes, and T.S. Walker, "Estimating the Global Severity of Potato Late Blight with a GIS-Linked Disease Forecaster," p. 84.

17. CIP, "Late Blight: The Challenge Escalates," in *International Potato Center Annual Report 1993* (Lima: CIP, 1994), p. 12; CIP, "Scientists Develop New Potato Clones to Counter Late Blight, World's Worst Agricultural Disease," p. 1.

18. CIP, "Challenge Escalates," p. 12.

19. PROINPA and IBTA, *Informe Compendio 1996–1998*, p. 11; R. Torrez et al., "Implementing IPM for Late Blight in the Andes." Also see the interview with Oscar Navia, CIP-PROINPA, March 4, 1996, p. 30.

20. Spraying at the rate of once every four to five days adds up quickly; see George R. Mackay, "Resistance: The Foundation of Integrated Pathogen Management of Late Blight *(Phytophthora infestans)*," p. 3.

21. Nova Scotia Department of Agriculture and Marketing, "Potato Late Blight Forecast."

22. Oregon State University, "Disease Forecasting."

23. See Michigan State University, "2000 Potato Disease Weather Monitoring."

24. Martin A. Draper et al., "Leaf Blight Diseases of Potato"; Fry and Goodwin, "Resurgence," pp. 366–67.

25. See Richard Horton, "Infection: The Global Threat," p. 26.

26. Ellen Ruppel Shell, "Resurgence of a Deadly Disease," pp. 56–59.

27. CIP, "Research Needed to Halt Late Blight Strains," in *International Potato Center Annual Report 1995*, pp. 10–11; Fry and Goodwin, "Resurgence," pp. 368–70.

28. CIP, "Challenge Escalates," p. 10; Doris Stanley, "Potatoes Once Again Under Attack," pp. 10, 12.

29. On the role infected seed tuber played, see Stephen B. Goodwin et al., "Genetic Changes Within Populations of *Phytophthora infestans* in the United States and Canada During 1994–1996: Role of Migration and Recombination," pp. 944–48; and Mary L. Powelson and Debra A. Inglis, "Seed Piece Treatment Key to Protecting Against Late Blight."

30. Pat Mooney, "The Hidden 'Hot Zone,'" p. 5.

31. Fry and Goodwin, "Resurgence," p. 367–68.

32. Mooney, "'Hot Zone,'" p. 5.

33. See University of Idaho, "Potato Late Blight,"; Kirk and Stein, "Recommendations"; and Goodwin et al., "Genetic Changes," pp. 939–40. For the statistics on fungicide use, see United States Department of Agriculture (USDA), National Agricultural Statistics Service, *Agricultural Chemical Usage 1999 Field Crop Summary*, p. 57.

34. See the data in Scott, Rosegrant, and Ringler, *Roots and Tubers*, p. 15, 32–33. The total value of the 1996 potato crop in developing countries is estimated by multiplying 108.1 million metric tons by the 1993 price of $160 per metric ton.

35. CIP, "Late Blight Project Initiated Early," in *International Potato Center Annual Report 1997*, pp. 16–17. Also see the data in O. Ortiz et al., "Understanding Farmers' Response of Late Blight: Evidence from Peru, Bolivia, Ecuador, and Uganda," pp. 102–106.

36. See Kirk and Stein, "Recommendations." For the results of tests on new fungicides, see Powelson and Inglis, "Featured Research." Oregon State University's Melheur Experiment Station lists some thirty-eight registered fungicides for use on potatoes, plus four that

are restricted to emergency use only. See the station's website at www.cropinfo.net/Potatoblight.htm.

37. See Torrez et al., "Late Blight in the Andes."

38. CIP, "The Development of Durable Resistance to Late Blight: A Global Initiative," pp. 5–8.

39. Mackay, "Resistance," p. 3.

40. C. Fonseca et al., "Economic Impact of the High-Yielding Late-Blight-Resistant Variety Canchán-INIAA in Peru," p. 58.

41. J. L. Rueda et al., "Economic Impact of High-Yielding, Late-Blight Resistant Varieties in the Eastern and Central African Highlands," pp. 27–28.

42. Bill Hardy, B. Trognitz, and G. Forbes, "Late Blight Breeding at CIP: Progress to Date"; CIP, "Research Needed," pp. 11–12.

43. CIP, "Late Blight Dragon," pp. 18–20.

44. See Marc Chislain and Bodo Trognitz, "Molecular Breeding for Late Blight Resistance in Potato," pp. 11–15; and Marc Chislain et al., "Identification of QTLs for Late Blight Resistance in a Cross Between *S. phureja* and a Dihaploid *S.tuberosum* and Association with a Plant Defense Gene," pp. 67–70.

45. See Bodo Trognitz, G. Forbes, and B. Hardy, "Resistance to Late Blight of Potato from Wild Species," pp. 6–9; and Bodo Trognitz et al., "Breeding Potatoes with Durable Resistance to Late Blight."

46. CIP, "Research Needed," p. 12; CIP, "Project Initiated Early," p. 18.

47. Trognitz et al., "Durable Resistance," p. 6.

48. Global Initiative on Late Blight (GILB), "Up Date on the G x E (Genotype by Environment) Study," pp. 2–4.

49. GILB, "Standard International Field Trials (SIFT) Are Established in Developing Countries"; CIP, "New Potato Clones," p. 3.

50. CIP, "New Potato Clones," pp. 2–3.

51. GILB, "Focus on Partners: Farmers Groups, Extension Organizations, and Research Institutes Team Up to Tackle Late Blight in Seven Countries," pp. 1–2; CIP, *International Potato Center Annual Report 1999*, pp. 9–10.

52. Torrez et al., "Late Blight in the Andes," pp. 94–97.

Chapter 18. Tuber Moth

1. USDA, National Agricultural Statistics Service, *Agricultural Chemical Usage,* p.57.

2. K. Fuglie et al., "Economic Impact of IPM Practices on the Potato Tuber Moth in Tunisia," p. 70; G. M. Moawad et al., "Biological Control of the Potato Tuber Moth, *Phthorimaea operculella* (Zeller) in Potato Fields and Storage," pp. 1–2.

3. Fuglie et al., "IPM Practices," p. 68; CIP, "The Tunisia Story: Turning Tables on Chemical Use," pp. 21–22.

4. Based on interviews with Aziz Lagnaoui, CIP-Egypt, July 9, 1996, pp. 26–27. All notes and interviews are on deposit at the Vanderbilt University Library, where they are organized by country. Pages for each country file are numbered sequentially. Also see Richard T. Roush and Ward M. Tingey, "Strategies for the Management of Insect Resistance to Synthetic and Microbial Insecticides."

5. CIP, "Biotechnology: A Tool for Building Alternatives," *International Potato Center Annual Report 1992*, pp. 23–24.

6. CIP, "Integrated Pest Management," *International Potato Center Program Report 1993–1994*, pp. 87–88; K. V. Raman, "Integrated Insect Pest Management for Potatoes in Developing Countries," pp. 2–3.

7. Fuglie et al., "IPM Practices," pp. 68–69. On tuber moth species and sex pheromones, see Raman, "Pest Management," pp. 1–3.

8. Much of the above is based on interviews with Aziz Lagnaoui, CIP-Tunisia, July 3, 1996, pp. 29–37. Also see CIP, *Major Potato Diseases, Insects, and Nematodes,* p. 87; CIP-CARE-USAID, "Asi vive la polilla de la papa"; and Raman, "Potato Tuber Moth."

9. Fuglie et al., "IPM Practices," pp. 68–73.

10. On the steps involved in an IPM strategy, see Cisneros et al., "Strategy."

11. Fuglie et al., "IPM Practices," pp. 70–77.

12. McDevitt, *World Population Profile: 1998,* pp. A-7, A-40.

13. The 1968–70 data is from Douglas Horton and A. Monares, "A Small, Effective Seed Multiplication Program: Tunisia," p. 7; for 1997–99, see FAO, *Quarterly Bulletin* 1999, 3/4, p. 38.

14. This and much of what follows is based on field trips and interviews with Mohammed Fahem et al., CIP-Tunisia, June 28, 1996, pp. 1–13; June 29, 1996, pp. 13–18; and July 1, 1996, pp. 26–27.

15. The principles for storage under straw are discussed in Lisińska and Leszczyński, *Potato Science and Technology,* pp. 156–58.

16. Based on a field trip and interview with Aziz Lagnaoui, CIP-Tunisia, June 28, 1996, pp. 11–12; and a visit to INRAT's entomology lab, INRAT-Tunis, July 3, 1996, pp. 28–35.

17. See Michèle Blanchard-Lemée et al., *Mosaics of Roman Africa: Floor Mosaics from Tunisia.*

18. See Albert Hourani, *A History of the Arab Peoples,* pp. 1–58.

Chapter 19. Natural Enemies

1. Lang, *Feeding a Hungry Planet,* pp. 39–41.

2. Raman, "Pest Management," p. 4.

3. PRB, *World Population Data Sheet* 2000, p. 2. The combined population of Algeria, Morocco, Lybia, and Tunisia is 75 million.

4. FAO, *Production Yearbook* 1998, p. 19.

5. CIP-FAO, *Potatoes in the 1990s,* p. 29; FAO, *Quarterly Bulletin* 1999, 3/4, p. 38.

6. For per-capita consumption in 1961–63, see CIP-FAO, *Potatoes in the 1990s,* p. 31. For current per-capita consumption, see G. M. Moawad et al., "Large-Scale Implementation of Integrated Pest Management in Egypt," p. 8.

7. A. Lagnaoui and R. El-Bedewy, "An Integrated Pest Management Strategy for Controlling Tuber Moth in Egypt."

8. FAO, *Production Yearbook* 1998, pp. 3, 5, 20, 22.

9. This and much of what follows is based on field trips and interviews with Ramzy El-Bedewy, CIP-Egypt, July 9, 1996, pp. 2–13, 18–20. All notes and interviews are on deposit at the Vanderbilt University Library, where they are organized by country. Pages for each country file are numbered sequentially. Also see CIP, "Potatoes for Egypt: An IPM Success," *International Potato Center Annual Report 1996,* pp. 16–17. I wrote this short article for CIP's *Annual Report.* This chapter and the CIP article are both based on the same field notes, but they are not duplicates.

10. Interview with Galal M. Moawad, Plant Protection Research Institute-Cairo, July 9, 1996, pp. 3–7.

11. This description is based on the GV production unit at both the Plant Protection Research Institute and CIP's Kafr El-Zayat Research Station. See El-Bedewy, interview, CIP-Egypt, pp. 6, 27–28. Also see K. V. Raman and J. Alcázar, "Biological Control of Potato Tuber Moth Using Phthorimaea Baculovirus."

12. See Moawad et al., "Implementation," pp. 8–9; and Moawad et al., "Biological Control," p. 3.

13. The story is told in Logan, *Dirt,* pp. 151–54.

14. See Phyllis A.W. Martin, "An Iconoclastic View of *Bacillus thuringiensis* Ecology."

15. Moawad et al., "Implementation," p. 9.

16. This and what follows are based on an interview with Magdy A. Madkour, Agricultural and Genetic Engineering Institute-Cairo, July 9, 1996, pp. 8–13.

17. For details, see Stephen Nottingham, *Eat Your Genes: How Genetically Modified Food Is Entering Our Diet,* pp. 18–19, 48–49.

18. Anita Manning, "Insects Penetrate Genetically Engineered Cotton."

19. Based on interviews with Aziz Lagnaoui, CIP-Egypt, July 10, 1996, pp. 26–27.

20. Moawad et al., "Implementation," p. 9, generally, pp. 10–12. On the Bt, GV, and neem combination tested, see Moawad et al., "Biological Control," pp. 11–13.

21. This and what follows are based on interviews with Ramzy El-Bedewy, CIP-Egypt, July 9, 1996, pp. 7–8; and July 10, 1996, pp. 20–21.

22. These thresholds are noted in Lagnaoui and El-Bedewy, "Integrated Pest Management," p. 6.

23. Much of what follows is based on interviews with Ramzy El-Bedewy and on-farm visits, CIP-Egypt, July 10, 1996, pp. 23–27.

24. Ibid., p. 24.

Chapter 20. Freezing a French Fry

1. This is not the company's real name.

2. The description that follows is based on field notes from Farm Fry-Egypt, July 11, 1996, pp. 31–38. All notes and interviews are on deposit at the Vanderbilt University Library, organized by country. Pages for each country file are numbered sequentially. For an excellent description of the steps in making both potato chips and french fries, see Lisińska and Leszczyński, *Potato Science and Technology,* pp. 166–232.

3. For confirmation of these figures, see Lisińska and Leszczyński, *Potato Science and Technology,* pp. 146–47.

4. Interview with Ramzy El-Bedewy, CIP-Egypt, July 11, 1996, pp. 37–38.

Chapter 21. Mother Plants

1. Based on an interview with Steven Tindimubona, Kabale-Uganda, December 17, 1997, p. 96. All notes and interviews are on deposit at the Vanderbilt University Library, organized by country. Pages for each country file are numbered sequentially.

2. On Uganda's civil strife, see Thomas P. Ofcansky, *Uganda: Tarnished Pearl of Africa,* pp. 26–29, 33–68.

3. On the domestication of crops in Africa, including sorghum, see Harlan, *Crops and Man,* pp. 140–46.

4. Louise Sperling, "Farmer Participation and the Development of Bean Varieties in Rwanda," p. 97.

5. Susuti seems much like Brazil's chuchu or Mexico's chajote; the latter can be found in the produce section of the supermarket. See Harlan, *Crops and Man,* p. 79.

6. The Rift Valley stretches some twenty-four hundred kilometers, from Lake Manyara in Tanzania to the Red Sea. See the maps and photographs in Colin Willock, *Africa's Rift Valley,* pp. 14–15.

7. This and what follows are based on an interview with Mohammed Yusuf, Africare-Kabale, December 16, 1997, pp. 88–89.

8. CIP, "Integrated Control of Bacterial Wilt Paying Off in East Africa," *International Potato Center Annual Report 1997*, p. 14. I wrote this article for CIP's *Annual Report* and have drawn on it freely. Haile Kidane-Mariam estimated the Kabale District's annual production at between 20,000 and 25,000 hectares; yields varied from 6 to 12 metric tons. The high and low estimates noted here use an average of 6 and 8 metric tons, respectively. See the interview with Haile Kidane-Mariam, CIP-Kabale, December 18, 1997, p. 110.

9. What follows is based on an interview with Berga Lemaga, CIP-Kabale, December 17, 1997, pp. 90–94.

10. See Berga Lemaga, "Integrated Control of Bacterial Wilt: Literature Review and Work Plan, 1995–1997." pp. 1–3.

11. Berga Lemaga et al., "Integrated Control of Potato Bacterial Wilt in Southwestern Uganda," pp. 4–6. The average yield increase is calculated from pp. 9, 11, tables 2, 5, and 6. Also see CIP, "Bacterial Wilt," p. 14.

12. On field control generally, see E. R. French, "Integrated Control of Bacterial Wilt of Potatoes," pp. 8–10. Also see E. R. French, ed., *Bacterial Wilt Manual,* sections 2–4.

13. See Lemaga, interview, CIP-Kabale, December 17, 1997; and CIP, "Control of Bacterial Wilt," p. 15.

14. This and much of what follows are based on Kidane-Mariam, interview, CIP-Kabale, December 18, 1997, pp. 100–10.

15. Haile Kidane-Mariam estimated Uganda's total acreage in good years at 75,000 hectares, Kenya's at 100,000 hectares, and Ethiopia's at 40,000 hectares for a total of 215,000; the calculation of total production presumes an average yield of six metric tons per hectare.

16. See Shell, "Resurgence," pp. 46–49.

17. Tindimubona, interview, Kabale-Uganda, December 17, 1997, pp. 94–96.

18. Based on an interview with Edith Rubereti, Kabale-Uganda, December 17, 1997, pp. 97–100.

19. See the interview with Mr. Sentaro, Kabale-Uganda, December 18, 1997, pp. 100–101.

20. What follows is based on interviews and demonstrations at the Kalengyere Research Station, December 1997, pp. 100–10. Also see the interview with Francis Alacho, Kabale-Uganda, December 19, 1997, pp. 110–12.

21. Peter Ewell, CIP's Director in Kenya, estimated that each Maguga seedling tuber sold for approximately twenty cents apiece. Presuming an eighty-kilo bag is filled with seed tubers weighing an average of sixty grams apiece, then each bag holds about 1,333 tubers, which adds up to a cost of $266.60 per bag (1,333 tubers x 20 cents).

22. On the origins of Cruza-148, see CIP, "Challenge Escalates," *International Potato Center Annual Report 1993,* p. 13.

23. Kidane-Mariam, interview, Kalengyere-Uganda, December 17, 1997, p. 103.

24. Compiled from tables 2, 4, and 5 in Haile M. Kidane-Mariam, "Trip Report on Visits to Uganda, Ethiopia, 19–28 November, 1997: Update on the Farmer-Based Seed System in the Target NARS," pp. 7, 9, 10. For the ten farmers of Kabale District in Kidane-Mariam's table 2, I estimated initial Kalengyere seed stocks by assuming a multiplication rate of 4.2. See Kidane-Mariam, p. 6, table 1.

25. Ibid., pp. 2, 3, 4.

Chapter 22. Tissue Culture

1. Consultative Group on International Agricultural Research (CGIAR), Technical Advisory Committee (TAC), and CGIAR Secretariat, *Report of the Fourth External Program and Management Review of the Centro Internacional de la Papa (CIP)*, pp. 48–49.

2. What follows is based on a field trip to Kenya's Plant Quarantine Station with Peter Ewell, CIP director for East Africa, and Haile Kidane-Mariam, CIP-Maguga, December 23, 1997, pp. 19–23. Also, see the interview with Peter Ewell, January 2, 1998, pp. 23–26. All notes and interviews are on deposit at the Vanderbilt University Library, organized by country. Pages for each country file are numbered sequentially.

Also see John H. Dodds and Lorin W. Roberts, *Experiments in Plant Tissue Culture*, pp. 113–21.

Chapter 23. The Potato Revolution

1. What follows is based on an interview with Noël Pallais, CIP-Lima, July 14, 1988, pp. 33–34. All notes and interviews are on deposit at the Vanderbilt University Library, organized by country. Pages for each country file are numbered sequentially.

2. Noël Pallais, "True Potato Seed Quality," pp. 784–92.

3. CIP, "Think Globally Act Locally: The Key to Success in India's True Potato Seed Program," *International Potato Center Annual Report 1996*, p. 26; CIP, *Proyecto Chacasina*, p. 15, illustration 64. Also see the qualifications in footnote 39, chapter 4.

4. Some twenty-eight viruses can infect potato plants. Of these, only potato spindle tuber viroid (PSTVd) and potato virus T (PVT) can be transmitted through true seed or pollen. In both cases, the transmission occurred under experimental conditions in which researchers wanted to spread the virus. On these points, see Luis F. Salazar, *Potato Viruses and Their Control*, pp. 7, 58–60.

5. Noël Pallais, interview, CIP-Lima, March 12, 1996, p. 72.

6. C. F. Quirós et al., "Increase of Potato Genetic Resources in Their Center of Diversity: The Role of Natural Outcrossings and Selection by the Andean Farmer," pp. 107–13; T. Johns and S. L. Keens, "Ongoing Evolution of the Potato on the Altiplano of Western Bolivia," pp. 409–24; and Salaman, *History*, pp. 10, 38, 54, 159–60.

7. Quirós et al., "Potato Genetic Resources," p. 112; Johns and Keen, "Evolution of the Potato," p. 421; Salaman, *History*, pp. 159–60; CIP, "Andean Farmers Use TPS to Renew Their Crops," *International Potato Center Annual Report 1992*, p. 41; and Ali M. Golmirzaie and Humberto A. Mendoza, "Breeding Strategies for True Potato Seed Production," p. 1.

8. Salaman, *History*, pp. 160–61, generally, pp. 159–87.

9. Pallais, interview, CIP-Lima, July 14, 1988, pp. 33–34.

10. Golmirzaie and Mendoza, "Breeding Strategies," pp. 2–6.

11. Pallais, "Seed Quality," pp. 788–89.

12. See Lang, *Feeding a Hungry Planet*.

13. What follows is based on an interview with Ramzy El-Bedewy, CIP-Egypt, July 10, 1996, pp. 17–19, 29.

14. See Noël Pallais, "True Potato Seed: A Global Perspective"; also see related articles in *CIP Circular* (March 1994), pp. 1–11.

15. CIP, "A Breakthrough in TPS Use," *International Potato Center Annual Report 1992*, pp. 39–40.

16. Depending on germination and transplanting rates, it takes between 20 and 50 grams of TPS to produce one hectare of potatoes. In India's case, the hypothetical acreage that 680

kilos can cover assumes that 20 grams suffice. Thus, a kilo (1000 grams) of TPS will seed 50 hectares; 680 kilos will seed 34,000 hecatres (680 x 50 = 34,000).

Chapter 24. True Potato Seed

1. FAO, *Quarterly Bulletin* 1999, 3/4, p. 38.

2. CIP-FAO, *Potatoes in the 1990s*, p. 29.

3. The figures were 77 percent in 1976–77 and 82 percent in 1990–91. See the tables in S. S. Sangwan, *Production and Marketing of Potato in India: A Case Study of Uttar Pradesh*, pp. 212–16; and P. C. Gaur and S. K. Pandey, "TPS Production and Adoption of Technology in India," p. 27.

4. See K. L. Chadha and J. S. Grewal, "Potato Research in India—History, Infrastructure, and Achievements," pp. 2–3; and CIP, "New Wheat and Rice Varieties Ignite Asian Production," *International Potato Center Annual Report 1995*, pp. 18–19. On the potato's role in crop rotations, see Roy et al., "Intensification of Potato Production," pp. 205–10. On the green revolution in rice, see Lang, *Feeding a Hungry Planet*, pp. x–xii, 1–54.

5. Chada and Grewal, "Potato Research," pp. 10–14; V. S. Khatana et al., "Economic Impact of True Potato Seed on Potato Production in Eastern and Northeastern India," pp. 140–42.

6. Mukhtar Singh, "The Potato in Retrospect and Prospect in India," p. 7; Khatana et al., "Economic Impact," pp. 140–42.

7. For seed costs, see T. R. Sharma, B. M. Goydani, and R. C. Sharma, "True Potato Seed: A Cheaper and Better Alternative to Seed Tubers," p. 142. For the survey results, see Luis A. Maldonado, Julia E. Wright, and Gregory J. Scott, "Constraints to Production and Use of Potato in Asia," pp. 74–76.

8. This and much of what follows is based on interviews with Dr. S. K. Bardhan Roy, Midnapore-West Bengal, December 26, 1996, pp. 66–67. All notes and interviews are on deposit at the Vanderbilt University Library, organized by country. Pages for each country file are numbered sequentially.

9. Compiled from 1990–91 census data reported in Directorate of Agriculture, Government of West Bengal, *Estimates of Area and Production of Principal Crops in West Bengal 1992–1993*, pp. 8–9.

10. See West Bengal Agricultural Research Service Association (WBARSA), "Farm Focus: Potato," p. 2; Khatana et al., "Economic Impact," p. 10; and Roy et al., "Intensification of Potato Production," p. 209.

11. Temperature data is from Directorate of Agriculture, Government of West Bengal, *Annual Report for 1995–1996: Section of Economic Botanist-III (Anandapore Farm)*, p. 12.

12. CIP, "Think Globally," *International Potato Center Annual Report 1996*, p. 25. I wrote the original report and have drawn on it freely; even so, this chapter and the CIP report vary considerably. For some of the basic research done in India, see M. D. Upadhya, ed., *True Potato Seed (TPS) in South and South East Asia: Proceedings of the Regional Workshop for Researchers on True Potato Seed, Jointly Organized by the Indian Council of Agricultural Research and the International Potato Center, New Delhi, India, 4–8 January 1989*.

13. Based on an interview with Khyal C. Thakur, CIP-Delhi, December 20, 1996, pp. 38–41.

14. Based on interviews at Mahendra Farm, December 17, 1996, pp. 2–16.

15. Based on interviews at Modipuram, December 19, 1996, pp. 26–27; K. C. Thakur et al., "Bulk pollen extraction procedures and the potency of the extracted pollen"; and Nisha Bhargava et al., "An Efficient Potato Pollen Extractor for Bulk Pollen Collection."

16. Directorate of Agriculture, *Annual Report 1995–1996*, pp. 16–17. Also see Roy, interview, Midapore-West Bengal, December 26, 1996, pp. 68–69.

17. CIP, "1995 Field Report," *International Potato Center Annual Report 1995*, p. 16.

18. See the interviews at Modipuram, December 19, 1996, pp. 26–28.

19. Directorate of Agriculture, *Annual Report 1995–1996*, pp. 17, 19; CIP, "Think Locally," p. 25. For the cost-benefit analysis, see Sharma, Goydani, and Sharma, "Cheaper and Better," pp. 119–24.

20. "From 1 square meter of seed bed about 4 kilos of minitubers (C1) are available. One gram of TPS covers 5 square meters of seed bed"; see WBARSA, "Farm Focus," p. 5. For the number of seeds per gram of TPS and the number of tubers harvested per square meter, see Directorate of Agriculture, *Annual Report 1995–1996*, pp. 17, 19.

21. CIP, "Think Locally," p. 26. These are conservative estimates. Dr. Roy claimed that at the Anandapore Station ten grams were sufficient to produced four hundred kilos of minitubers; see Roy, interview, Midnapore-West Bengal, December 26, 1996, pp. 68–72.

22. WBARSA, "Farm Focus," p. 5. In prior tests, the minituber had also demonstrated a yield advantage of between a fifth and a third over Kufri Jyoti; see S. K. Bardhan Roy, A. K. Chakravorty, and A. K. Roy, "Farmers' Participatory On-Farm Trials with TPS Seedling Tubers in the Rice-Based Cropping Systems of West Bengal."

23. For a summary of farmers' evaluations, see Directorate of Agriculture, *Annual Report 1995–1996*, pp. 32–34. Also see WBARSA, "Farm Focus," p. 5. Roy estimated farmer satisfaction at 90 percent.

24. "Opium Permit Triggers Row in MP," *Telegraph* (Calcutta), December 26, 1996, p. 6.

25. What follows is based on a field trip and interviews with Dr. S. K. Bardhan Roy, Midnapore, December 27, 1996, pp. 74–86. For a description of prevailing rotations, see Roy et al., "Intensification of Potato Production," pp. 208–11.

26. Rainfall data for Midnapore can be found in Directorate of Agriculture, *Area Production*, p. 10.

27. See the interview with Mr. Ghosh, Midnapore, December 27, 1996, pp. 82–85.

28. See the interview at the Garhbeta I block office, Midnapore, December 27, 1996, pp. 76–78.

29. CIP, "Think Locally," p. 26.

30. Khatana et al., "Economic Impact," pp. 148, 152.

Chapter 25. Super Seed

1. Based on interviews with Mahesh Upadhya, CIP-Lima, March 12, 1996, pp. 75–77; and CIP-Lima, May 14, 1997, pp. 12–16. All notes and interviews are on deposit at the Vanderbilt University Library, organized by country. Pages for each country file are numbered sequentially.

2. See Salazar, *Potato Viruses*, pp. 7, 12–14, 23–25, 47, in general, pp. 23–43, 47–60. Note that viruses rarely infect cells in the meristematic tissue such as roots and shoots—hence, the utility of meristem multiplication for generating virus-free potato stock.

3. See CIP, *Proyecto Chacasina*, p. 63; and CIP, "Think Globally," p. 26.

4. "TPS transmits few potato viruses. The only two agents that it transmits efficiently are PVT and PSTVd; however, these can be detected easily"; see Salazar, *Potato Viruses*, p. 184.

5. It takes between 20 and 50 grams of TPS to plant a hectare; see CIP, "TPS Hybrid May Eliminate Need for Seed Tubers," *International Potato Center Annual Report 1997*, p. 29, and CIP, "Think Globally," p. 26.

6. Based on an interview with Noël Pallais, CIP-Lima, May 14, 1997, pp. 16–19.

7. CIP, "TPS May Eliminate Seed Tubers," p. 28.

8. Peter Schmiediche, "Report on a Trip to Vietnam from 13 to 18 January 1997 to Attend a Workshop on the ADB-Financed TPS Project," p. 5; CIP, "TPS May Eliminate Seed Tubers," p. 29.

9. Schmiediche, "Report," p. 3; CIP, "TPS May Eliminate Seed Tubers," p. 29.

10. For the guide, see CIP, *Proyecto Chacasina*.

11. CIP, "Chacasina: True Seed in the Andes," *International Potato Center Annual Report 1996* pp. 27–28; Pallais, interview, CIP-Lima, May 14, 1997, pp. 16–19. Also see Catherine Elton, "Peru's Potatoes Saved by Science."

12. A. Chilver et al., "On-Farm Profitability of TPS Utilization Technologies."

13. Cloning may be a dead end. A cloned sheep like the famous Dolly is simply not as robust as the original; see R.C. Lewontin, "The Confusion Over Cloning."

Chapter 26. Starch

1. CIP, "Sweet-Potato Facts"; B. Bashaasha et al., *Sweetpotato in the Farming and Food Systems of Uganda: A Farm Survey Report*, p. 1.

2. Piperno and Pearsall, *Lowland Neotropics*, pp. 164–65, 274, 312; Hawkes, "Domestication," pp. 487–89; and Harlan, *Crops and Man*, p. 228.

3. Carl O. Sauer, *Agricultural Origins*, p. 47.

4. Salaman, *History*, pp. 131–32.

5. See Jennifer A. Woolfe, *Sweet Potato: An Untapped Food Source*, pp. 16–17; and Salaman, *History*, pp. 132–33.

6. Hawkes, "Domestication," p. 488; and Jonathan D. Sauer, *Historical Geography*, pp. 38–41. On the genetic evidence, see Dapeng Zhang et al., "RADP Variation in Sweetpotato (*Ipomoea batatas* (L) Lam) Cultivars from South America and Papua New Guinea." For excellent articles on Polynesian crops and their spread, see Paul Alan Cox and Sandra Anne Banack, eds., *Islands, Plants, and Polynesians: An Introduction to Polynesian Ethnobotany*.

7. At least one yam species was domesticated in the Americas. Today, however, species from Africa have largely displaced the indigenous crop; see Heiser, *Seed to Civilization*, pp. 146–48. Also see Harlan, *Crops and Man*, pp. 182–83.

8. Based on an interview with Edward (Ted) Carey, CIP-Nairobi, December 9, 1997, p. 1. All notes and interviews are on deposit at the Vanderbilt University Library, organized by country. Pages for each country file are numbered sequentially. For a brief summary of the origin and diffusion of the sweet potato, see Huamán and Zhang, "Sweetpotato," pp. 29–32.

9. Woolfe, *Sweet Potato*, p. 27.

10. Ibid., p. 10.

11. For a summary of cultural practices, see Bashaasha et al., *Sweetpotato in Uganda*, pp. 2–46. On the use of sweet-potato greens, see Woolfe, *Sweet Potato*, pp. 92–93, and generally, pp. 92–105.

12. On sweets and jams, see Woolfe, *Sweet Potato*, pp. 330–33.

13. On sweet potato flour, see ibid., pp. 294–95, 344–45; on alcoholic beverages, see pp. 379–83.

14. On starch production, see ibid., pp. 366–70; on noodles, see pp. 371–73; on sweet potato's use as a fuel, see pp. 48–49.

15. See Burton, *Potato*, p. 146, table 21, note 1.

16. Woolfe, *Sweet Potato*, p. 41.

17. See Woolfe, *Human Diet*, pp. 24–25.

18. Woolfe, *Sweet Potato*, p. 145, and generally, pp. 145–58.

19. Ibid., pp. 411–14, 417–36.

20. The percentages and much of what follows are based on the Carey interview, CIP-Nairobi, December 9, 1997, pp. 1–9. On the potential for increasing the dry-matter content in sweet potatoes, including orange-fleshed varieties, see C. Brabet et al., "Starch Content and Properties of 106 Sweetpotato Clones from the World Germplasm Collection Held at CIP, Peru."

21. In general, see E. E. Carey et al., "Collaborative Sweetpotato Breeding in Eastern, Central, and Southern Africa."

22. See CIP, "The Promise of Vitamin A," *International Potato Center Annual Report 1998*, 13–14.

23. Bashaasha et al., *Sweetpotato in Uganda*, p. 25.

24. Kenya Agricultural Research Institute (KARI) and CIP, "Orange Fleshed Sweetpotato Varieties." Also see V. Hagenimana, L .M. K'osambo, and E. E. Carey, "Potential of Sweetpotato in Reducing Vitamin A Deficiency in Africa."

25. See Jill E. Wilson et al., *Agro-Facts: Sweet Potato Breeding*, pp. 8–12.

26. Based on Carey, interviews, CIP-Nairobi, December 9, 1997, pp. 6–9; and January 7, 1998, pp. 28–30.

27. See the notes on the visit to the Kakamega Research Institute, KARI-Kenya, January 6, 1998, p. 128. Also see Zósimo Huamán, *Systematic Botany and Morphology of the Sweetpotato Plant*, pp. 12–14.

Chapter 27. Food Security

1. I wrote a shorter version of this article for CIP and have drawn on it freely; see CIP, "Food Secuity in East Africa: A Battle on Many Fronts," *International Potato Center Annual Report 1998*, pp. 10–12. For the figures on population, see Gregory J. Scott et al., "Sweetpotato in Ugandan Food Systems: Enhancing Food Security and Alleviating Poverty," p. 337. On consumption and the role of women in production, see Bashaasha et al., *Sweetpotato in Uganda*, pp. 2, 7–8, 35–36.

2. This and much of what follows is based on an interview with G. W. Otim-Nape, NARO-Namulonge, December 14, 1997, pp. 73–76. All notes and interviews are on deposit at the Vanderbilt University Library, organized by country. Pages for each country file are numbered sequentially.

3. Woolfe, *Human Diet*, pp. 24–25, Johnathan D. Sauer, *Historical Geography*, pp. 55–62. Manioc is also called cassava.

4. The figures are from Scott et al., "Food Security," p. 339.

5. G. W. Otim-Nape et al., *Cassava Mosaic Virus Disease in Uganda: The Current Pandemic and Approaches to Control*, pp. 1–12. For a technical discussion of the virus and how it spreads, see G. W. Otim-Nape, J. M. Thresh, and D. Fargette, "*Bemisia tabaci* and Cassava Mosaic Virus Disease in Africa."

6. Scott et al., "Food Security," p. 342.

7. Otim-Nape et al., *Virus Disease*, pp. 45–53.

8. Based on interviews with Philip Ndolo, Alupe Research Station, KARI-Kenya, January 5, 1998, pp. 119–20.

9. For an overview of how rural families produce, store, and consume sweet potatoes in Uganda, see Bashaasha et al., *Sweetpotato in Uganda*, pp. 10–38.

10. Based on an interview with Nicole Smit, CIP-Kampala, December 10, 1997, pp. 33–38.

11. On Uganda's sweet potato weevil, see N. Smit and B. Odongo, "Integrated Management for Sweetpotato in East Africa." Also see CIP, "Uganda: The Sweetpotato Option," *International Potato Center Annual Report 1996*, pp. 8–9.

12. On traditional storage, see Bashaasha et al., *Sweetpotato in Uganda*, pp. 30–31.

13. What follows is partly based on an interview with Robert Odeu, Dokolo-Uganda, December 13, 1997, pp. 64–67.

14. On storage and processing, see the interviews at the Kawanda Research Institute, NARO-Uganda, December 11, 1997, pp. 39–42; and the interviews at Serere District Farm Institute, December 12, 1997, pp. 46–51. Also see Bashaasha et al., *Sweetpotato in Uganda,* p. 31; and V. Hagenimana and C. Owori, "Feasibility, Acceptability, and Production of Sweetpotato-Based Products in Uganda," p. 276.

15. In 1998, the rains held up through January.

16. See the Kampala to Soroti trip notes, Uganda, December 11, 1997, pp. 43–45; and the Kabale to Kampala trip notes, Uganda, December 19, 1997, pp. 112–14.

17. See Dai Peters and Christopher Wheatley, "Small Scale Agro-Enterprises Provide Opportunities for Income Generation: Sweetpotato Flour in East Java, Indonesia."

18. What follows is based on interviews with Dai Peters, Serere District Farm Institute, December 12, 1997, pp. 46–51.

19. See Hagenimana and Owori, "Sweetpotato-Based Products," pp. 279–80.

20. This and what follows are based on interviews at Awoja, December 12, 1997, pp. 55–60.

21. On pig production in Asia, see Woolfe, *Sweet Potato,* pp. 417–34.

22. On cultural practices, see Bashaasha et al., *Sweetpotato in Uganda,* pp. 21–29.

23. Based on interviews with Philip Ndolo, Alupe Research Station-Kenya, January 5, 1998, pp. 116–18.

24. Smit, interview, CIP-Kampala, December 10, 1997, p. 36.

25. See the interview with Mrs. Mwanzi, Alupe-Kenya, January 5, 1998, pp. 123–24.

26. See Hagenimana, K'osambo, and Carey, "Potential of Sweetpotato," and CIP, "Eradicating Childhood Blindness in Africa: The Promise of Orange-Flesh Sweetpotatoes," *International Potato Center Annual Report 1999,* p. 11.

Chapter 28. Farmer Field Schools

1. See Elske van de Fliert, *Integrated Pest Management: Farmer Field Schools Generate Sustainable Practices: A Case Study of Central Java.*

2. IRRI, *Rice Almanac,* pp. 58–60, 105.

3. See van de Fliert, *Field Schools,* pp. 14, 26–28; and Lang, *Feeding a Hungry Planet.* pp. 39–41. For the statistics on rice production, see IRRI, *Rice Statistics 1990,* p. 2.

4. Ann R. Braun et al., "Improving Profits from Sweetpotato by Linking IPM with Diversification of Markets," p. 10; van de Fliert, *Field Schools,* pp. 26–28.

5. This is based on 1997–99 averages. See FAO, *Quarterly Bulletin* 1999, 3/4, p. 40.

6. This and much of what follows is based on an interview with Ann Braun, CIP-Bogor, October 21, 1995, pp. 29–31. All notes and interviews are on deposit at the Vanderbilt University Library, organized by country. Pages for each country file are numbered sequentially. Also see Elske van de Fliert and Rini Asmunati, "Identification of IPM and IPM Training Needs for Sweetpotato in East and Central Java," pp. 7, 9–11.

7. Yields ranged from 10 to 30 metric tons per hectare; see van de Fliert and Asmunati, "IPM Training Needs," p. 11. The 1997–99 FAO averages are about 9.6 metric tons for Indonesia and 6 for Vietnam; see FAO, *Quarterly Bulletin* 1999, 3/4, pp. 40–41.

8. The above and much of what follows is based on interviews and field notes, Java-Bendunganjati, November 5, 1995, pp. 84–86.

9. Braun et al., "Profits from Sweetpotato," p. 9.

10. Based on an interview with Wiyanto, Java-Bendunganjati, November 5, 1995, pp. 87–88.

11. Rini Asmunati's contributions are noted in Braun et al., "Profits from Sweetpotato," p. 15.

12. See Elske van de Fliert et al., "From Basic Approach to Tailored Curriculum: Participatory Development of a Farmer Field School Model for Sweetpotato," pp. 1–5. On training, see van de Fliert, *Field Schools*, pp. 31–33. CIP is a long-time advocate of farmer-to-farmer training; see CIP, *Potatoes for the Developing World*, pp. 5, 109–13.

13. Van de Fliert, *Field Schools*, pp. 285–91; Braun et al., "Farmer Field Schools for Sweetpotato in Indonesia," pp. 201–202.

14. Based on interviews at Bendunganjati, November 8, 1995, pp. 105–106.

15. Based on interviews at Bendunganjati, November 5, 1995, pp. 93–96; and an interview with Ann Braun, CIP-Bogor, November 16, 1995, pp. 131–35. Also see J. A. Sutherland, "A Review of the Biology of the Sweetpotato Weevil *Cylas formicarius*." For the distribution of weevil species, see G. William Wolfe, "The Origin and Dispersal of the Pest Species *Cylas* with a Key to the Pest Species Groups of the World," p. 15.

16. See Braun et al., "Farmer Field Schools," pp. 198–99.

17. Braun et al., "Profits from Sweetpotato," pp. 12–13.

18. Based on field notes taken at the Magetan market, November 10, 1995, pp. 118–20.

19. Based on an interview with Christopher Wheatley, CIP-Bogor, November 13, 1995, pp. 127–28. Also see Memed Gunawan, Christopher Wheatley, and Irfansyah, "Current Status and Prospect for Sweetpotato in Indonesia," pp. 14, 21, 29.

20. Dai Peters presented the results of this project in Uganda at Namulonge Research Station, December 14, 1997, pp. 78–80. Also see Peters and Wheatley, "Small Scale Agro-Enterprises."

21. See Braun et al., "Farmer Field Schools," p. 199.

22. Based on an interview with Merle Shepherd, Bogor-Indonesia, November 17, 1995, pp. 136–40.

23. CIP, "Widening the Circle," *International Potato Center Annual Report 1998*, pp. 31–32.

Chapter 29. Feeding China

1. Wheatley, interview, CIP-Bogor, November 13, 1995, pp. 124–27. All notes and interviews are on deposit at the Vanderbilt University Library, organized by country. Pages for each country file are numbered sequentially.

2. See Lester R. Brown, *Who Will Feed China? Wake-Up Call for a Small Planet.*

3. Charles S. Gitomer, *Potato and Sweetpotato in China: Systems, Constraints, and Potential*, pp. 58–59. For case studies of sweet potato utilization worldwide, including China, see Woolfe, *Sweet Potato*, pp. 486–503. Also see David Machin and Solveig Nyvold, eds. *Roots, Tubers, Plantains, and Bananas in Animal Feeding.*

4. CIP, "Agrarian Transformation Underground: Potato and Sweetpotato in China," *International Potato Center Annual Report 1998*, p. 17.

5. Ibid., p. 17. For a look at China's expanding market for fries, see L. Zhang et al., "U.S. Opportunities in China's Frozen French Fry Market."

6. FAO, *Quarterly Bulletin* 1999, 3/4, p. 40.

7. CIP, "Sweet-Potato Facts," p. 4.

8. Lester R. Brown, *Who Will Feed China?* pp. 46–47. In general, also see Vaclav Smil, "Is There Enough Chinese Food?"

9. FAO, *Quarterly Bulletin* 1999, 3/4, p. 38.

10. CIP, "Agrarian Transformation," p. 17.

11. McDevitt, *World Population Profile: 1996*, pp. A-6, A-7.

12. Vaclav Smil, *China's Environmental Crisis: An Inquiry into the Limits of National Development*, pp. 42–49, 56–57, 140–50.

13. FAO, *Quarterly Bulletin* 1999, 3/4, pp. 42, 44, 45.

14. For an account of the delegation's visit to CIP, see Kenneth J. Brown, *Roots and Tubers Galore*, pp. 86–91. Also see CIP, *CIP Circular* (June, 1984), pp. 1–2; and CIP, *CIP Circular* (Sept., 1985), p. 7.

15. For percentages of sweet potato production by province, see Gitomer, *Potato and Sweetpotato*, p. 25. The production estimate is from CIP and the Chinese Academy of Sciences (CAAS), "Virus Cleanup Boosts Chinese Sweet Potato Production," p. 1.

16. The Center for Integrated Agricultural Development (CIAD) and the Sichuan Academy of Agricultural Sciences (SAAS), "Potential for Market Diversification of Sweet Potato Starch and Flour in Sichuan Province," pp. 5, 27.

17. Wheatley, interview, CIP-Bogor, November 13, 1995, pp. 124–27; and CIAD-SAAS, "Market Diversification," pp. 5–6.

18. CIAD-SAAS, "Market Diversification," pp. 4–5.

19. Christopher Wheatley, Lin Liping, and Song Bofu, "Enhancing the Role of Small-Scale Sweetpotato Starch Enterprises in Sichuan, China," p. 270.

20. Ibid., pp. 270–73.

21. CIAD-SAAS, "Market Diversification," pp. 2, 17–18.

22. CIAD-SAAS, "Social and Economic Aspects of the Small-Scale, Sweetpotato Based Pig Production System in Suining Municipality, Sichuan Province," pp. 5, 12–13, 17, 24–25, 28, 31, 36–41. On the inefficiency of pork production, also see Smil, "Is There Enough Chinese Food?" p. 34.

23. The figures are 1997–99 averages; see FAO, *Quarterly Bulletin* 1999, 3/4, p. 40.

24. The ELISA kit description is based on an interview with Lee Tsai-hsia, Asian Vegetable Research and Development Center-Taiwan, May 25, 1993, pp. 29–31. On ELISA technology, also see CGIAR, *1984 Annual Report*, pp. 37–38.

25. CIP-CAAS, "Virus Cleanup," p. 1.

26. On multiplication, see ibid., p. 2. On productivity increases, see Thomas S. Walker, "Trip Report to Vietnam and China: 18 October–2 November 1997," p. 7. For the 1998 totals, see K. Fuglie et al., "Economic Impact of Virus-Free Sweetpotato Planting Material in Shandong Province, China," pp. 250–52. In CIP-CAAS, "Virus Cleanup," p. 3, it is estimated that when 55 percent of the acreage was planted to virus-free stock some 2.4 million farm families benefited; hence, I expect that an 80-percent coverage rate would benefit 25 percent more, or a total of 3.1 million farm families. For estimates of the impact on household income, see Fuglie et al., "Sweetpotato Planting Material," pp. 252–53.

27. Thomas S. Walker, "Trip Report," p. 6.

28. For uses of sweet potato starch, see Gitomer, *Potato and Sweetpotato*, p. 59.

29. CIP, "Dry Matter Counts," *International Potato Center Annual Report 1996*, p. 10. The starch-recovery rate in China is estimated at between 12 and 15 percent; see Woolfe, *Sweet Potato*, p. 367. Prospects for adding to dry-matter content are good; see Brabet et al.,"Starch Content and Properties."

30. Gitomer, *Potato and Sweetpotato*, pp. 48–56.

31. CIP, "International Cooperation," *International Potato Center Annual Report 1992*, p. 50.

32. Kenneth J. Brown, *Roots and Tubers Galore*, p. 88.

33. CIP,"CIP-24: A Well-Traveled Potato," *International Potato Center Annual Report 1994*, p. 15; Song Bofu et al., "Economic Impact of CIP-24 in China," pp. 35–37, 43–44.

34. Thomas S. Walker, "Trip Report," pp. 2–3.

35. Gitomer, *Potato and Sweetpotato*, pp. 25, 93–94, 103, 110.

36. CIP, "Upward Project Reduces Poverty in China," *International Potato Center Annual Report 1996*, pp. 30–31.

37. Based on interviews with Jocelyn Tsao and Mr. Hsieh, Taichung-Taiwan, May 31, 1993, pp. 37–43.

38. CIP, "True Potato Seed: A Piece of the Puzzle," *International Potato Center Annual Report 1995*, p. 17; *CIP Circular* (June, 1984), p. 7.

39. Gitomer, *Potato and Sweetpotato*, p. 110.

40. Scott, Rosegrant, and Ringler, *Roots and Tubers*, pp. 15, 37–38. For current yield data, see FAO, *Quarterly Bulletin* 1999, 3/4, pp. 38–39. For China's potato acreage in the early 1980s, see FAO, *Production Yearbook 1982*.

41. Gitomer, *Potato and Sweetpotato*, pp. 159–60.

42. Scott, Rosegrant, and Ringler, *Roots and Tubers*, pp. 15, 37–38.

43. Ibid., p. 37. The conversion rate used here is 22 percent for the potato harvest and 30 percent for the sweet potato harvest; see Woolfe, *Sweet Potato*, p. 122.

44. See Roy L. Prosterman, Tim Hanstad, and Li Ping, "Can China Feed Itself?"

45. Scott, Rosegrant, and Ringler, *Roots and Tubers*, p. 43.

Chapter 30. Potato Lessons

1. Liu Binyan and Perry Link, "A Great Leap Backward," p. 19.

2. Lang, *Inside Development*, p. 145.

3. Mainland China's population for 2000 and 2015 is estimated at 1.253 and 1.413 billion, respectively; see McDevitt, *World Population Profile: 1996*, p. A.7. The 2015 estimate is taken as midway between 2010 and 2020.

4. For the 2000 estimate, see U.S. Bureau of the Census, *SAUS* 1998, p. 9.

5. For rice-production statistics, see FAO, *Quarterly Bulletin* 1999, 3/4, pp. 18–19. The data on U.S. and world rice exports is for 1996–98; see U.S. Bureau of the Census, *SAUS* 1998, pp. 677–78; and *SAUS* 1999, pp. 684–85.

6. Vaclav Smil, "Who Will Feed China?" pp. 802–803. For 1998 global grain exports, see U.S. Bureau of the Census, *SAUS* 1999, p. 685; also see Vaclav Smil, *Feeding the World: A Challenge for the Twenty-First Century*, p. 299.

7. Odin Tunali, "A Billion Cars: The Road Ahead," p. 24; U.S. Bureau of the Census *SAUS* 1998, p. 633.

8. Tunali, "Billion Cars," pp. 24, 26, 29; William Greider, *One World Ready or Not: The Manic Logic of Global Capitalism*, pp. 148–49, 164–67, 444–45.

9. Mark Hertsgaard, "Our Real China Problem," p. 106.

10. Ibid., pp. 99–100, 102, 109.

11. Binyan and Link, "Leap Backward," p. 21.

12. The material that follows is from CIAD-SAAS, "Market Diversification," and CIAD-SAAS, "Social and Economic Aspects."

13. Thomas P. Lyons, "Feeding Fujian: Grain Production and Trade 1986–1996," pp. 512–13. For a quick visual survey of China's demographics, agriculture, and overall development, see Robert Benewick and Stephanie Donald, *The State of China Atlas*.

14. Smil, *China's Environmental Crisis*, pp. 168, 171.

15. Lyons, "Feeding Fujian," pp. 521–30.

16. Ibid., p. 544; Smil, *China's Environmental Crisis*, p. 199.

17. On China's prospects, see Smil, *China's Environmental Crisis*, pp. 192–97; and Smil, *Feeding the World*, pp. 291–315. In the end, China may take the big-development road to disaster; see Audrey R. Topping, "Ecological Roulette: Damming the Yangtze."

Chapter 31. The Patented Potato

1. Mark Lappé and Britt Bailey, *Against the Grain: Biotechnology and the Corporate Takeover of Your Food*, p. 64; and Michael Pollan, "Playing God in the Garden," pp. 48–50.

2. Moawad et al., "Biological Control," p. 2.

3. On Bt in organic gardening, see Lappé and Bailey, *Against the Grain*, pp. 65–66; and Nottingham, *Eat Your Genes*, pp. 48–49, 55–56.

4. Rachel Carson, *Silent Spring*, pp. 289–90.

5. Pollan, "Playing God," pp. 51, 62.

6. Nottingham, *Eat Your Genes*, pp. 47–49, 54–55, 94.

7. Pollan, "Playing God," pp. 51, 62–63, 82.

8. For cautions with respect to Bt transgenic plants, see Mark E. Whalon, Utami Rahardja, and Patchara Verakalasa, "Selection and Management of *Bacillus Thuringiensis*–Resistant Colorado Potato Beetle," pp. 309, 317–19.

9. Pollan, "Playing God," p. 63. On "resistance monitoring protocol," see George G. Kennedy and Ned M. French, "Monitoring Resistance in Colorado Potato Beetle Populations;" in general, see Roush and Tingey, "Strategies for the Management of Insect Resistance."

10. Lappé and Bailey, *Against the Grain*, pp. 64–70; Nottingham. *Eat Your Genes,* pp. 48–49, 55–58.

11. Nottingham, *Eat Your Genes*, p. 47.

12. See "Possible Threat to Monarch Butterfly Posed by Bt Corn."

13. Lappé and Bailey, *Against the Grain*, pp. 56–71.

14. Pollan, "Playing God," pp. 44–47, 62–63, 82.

15. Nottingham, *Eat Your Genes*, pp. 39–40, 43–45. On Buctril and Round Up, see Lappé and Bailey, *Against the Grain*, pp. 41–43 and 50–61, respectively.

16. Nottingham, *Eat Your Genes*, pp. 43–45.

17. Based on a report aired on "Morning Edition," National Public Radio, December 13, 1999.

18. Lappé and Bailey, *Against the Grain*, p. 114.

19. Nottingham, *Eat Your Genes*, pp. 86–88; Jane Rissler and Margaret Mellon, *The Ecological Risks of Engineered Crops*, pp. 9–70.

20. Lappé and Bailey, *Against the Grain*, pp. 41–45, 115–17; Nottingham, *Eat Your Genes*, pp. 42–44.

21. Nottingham, *Eat Your Genes*, p. 161.

22. CGIAR, TAC, and CGIAR Secretariat, *Report of the Fourth External Programme*, p. 4.

23. See CIP, *Tissue Culture Propagation of Potato*.

24. See CIP, "Sprout Cuttings," and CIP, "Stem Cuttings: A Rapid Multiplication Technique for Potatoes."

25. See CIP, "Potato Farming Grew Out of U.S.-Vietnam Peace Talks," and CIP, *Potatoes for the Developing World*, pp. 115–19.

26. For example, see A. C. de Avila et al., "Boosting Tuber-Seed Production in Brazil: Serological Techniques and Antiserum Production." In general, see Salazar, *Potato Viruses*, pp. 121–35.

27. CIP, "An Improved Method for Fighting Bacterial Wilt: NCM-ELISA Detection Kit and Training Materials"; CIP, "Milestone Reached in Bacterial Wilt Research," *International Potato Center Annual Report 1997*, p. 13.

28. The story is told in CIP, *Potatoes for the Developing World*, pp. 109–13, 135–40.

29. Compiled from the list in CIP, *International Potato Center Annual Report 1997*, pp. 53–55.

30. Kenneth J. Brown, *Roots and Tubers Galore*, pp. 4–5.

31. Compiled from CIP, "Staff," *International Potato Center Annual Report 1998*, pp. 59–61.

32. CIP, "Training Highlights," *International Potato Center Annual Report 1998*, pp. 52–53. For a more detailed discussion of CIP's approach to training, see CIP, "Training," pp. 433–38.

33. Compiled from CIP, "Donor Contributions," *International Potato Center Annual Report 1997*, p. 33; CIP, "Donor Contributions," *International Potato Center Annual Report 1998*, p. 38; and CIP, "Donor Contributions," *International Potato Center Annual Report 1999*, p. 15.

34. CIP, *Medium-Term Plan 1998–2000*.

35. Kenneth J. Brown, *Roots and Tubers Galore*, p. 3.

36. On Chata Roja, see CIP, "From the Lab to the Land," *International Potato Center Annual Report 1999*, p. 10. On the "hairy" potato, see CIP, "Biotechnology," *International Potato Center Annual Report 1992*, p. 27. Finally, see CIP, "Breeding Deep-Rooted Potatoes," *International Potato Center Annual Report 1997*, p. 30.

37. Bofu et al., "Economic Impact of CIP-24 in China," pp. 31–33.

38. CIP, "Biotechnology Speeds New Potato Development," *International Potato Center Annual Report 1997*, p. 21.

39. Ibid., pp. 21–23; CIP, "Back to the Molecular Future," *International Potato Center Annual Report 1998*, pp. 28–30.

40. My thanks to Peter Schmiediche for this observation.

41. V. Cañedo et al., "Assessing Bt-Transformed Potatoes for Potato Tuber Moth, *Phthorimaea operculella* (Zeller), Management," pp. 161–62, 166–67.

42. Cynthia Gawron-Burke and Timothy B. Johnson, "Development of *Bacillus Thuringiensis*–Based Pesticides for the Control of Potato Insect Pests."

43. CIP, "Back to the Molecular Future," p. 30.

44. See "Vegetable Vaccines," and Nottingham, *Eat Your Genes*, p. 78.

45. Nottingham, *Eat Your Genes*, pp. 64–79.

46. For an overview of CIP's pest-management strategy, see K.V. Raman, "Potato Pest Management in Developing Countries."

Chapter 32. The Macro Level

1. T. Scarlett Epstein, "Viewpoints: Are Babies Made at the Macro-Level?" Also see T. Scarlett Epstein, *A Manual for Culturally-Adapted Market Research in the Development Process;* and T. Scarlett Epstein, Janet Gruber, and Graham Mytton, *A Training Manual for Development Market Research Investigators*.

2. Douglas Horton, "Farming Systems Research: Twelve Lessons from the Mantaro Valley Project," pp. 96–97.

3. Ibid., pp. 100–101, 105.

4. Norman Borlaug, "Challenges for Global Food and Fiber Production," pp. 53–54.

5. In 1998, the world's "agricultural population," defined as "all persons actively engaged in agriculture and their non-working dependants," totaled 2.6 billion; hence 2 billion villagers is a reasonable estimate. See FAO, *Production Yearbook* 1998, pp. viii, 19.

6. Measured in current dollars, median household income rose from $29,943 to $37,005; for the figures on GDP, median income, and income distribution, see U.S. Bureau of the Census, *SAUS* 1999. pp. 459, 474, and 479, respectively.

7. Jeff Madrick, "In the Shadow of Prosperity," p. 40, 42.

8. In Mexico, the top quintile earns 55.3 percent of the income, compared to 37.1 percent in Germany or 39.8 percent in the United Kingdom; see World Bank, *World Development Report 1998–1999: Knowledge for Development,* pp. 198–99.

9. U.S. Bureau of the Census, *SAUS* 1999, pp. 479, 483.

10. Madrick, "Shadow," p. 41.

11. See Andrew Hacker, "War Over the Family," and Michael Massing, "Crime and Drugs: The New Myths."

12. Herman E. Daly, *Ecological Economics and the Ecology of Economies: Essays in Criticism,* pp. 3–7.

13. Clifford Cobb, Ted Halstead, and Jonathan Rowe, "If the GDP is Up, Why is America Down?" pp. 64–65.

14. Daly, *Ecological Economics,* pp. 8–9.

15. Cobb, Halstead, and Rowe, "Why is America Down?" pp. 70, 72.

16. Willy Brandt, *North-South: A Program for Survival. Report of the Independent Commission on International Development Issues.*

17. Solow, "How to Stop Hunger."

18. Cobb, Halstead, and Rowe, "Why is America Down?" p. 68.

19. James Fallows, "Why Americans Hate the Media." Also see Robert Hughes, "Why Watch It Anyway."

20. Fallows, "Americans Hate the Media," pp. 56–57, 60.

21. Robert W. McChesney, *Rich Media, Poor Democracy: Communication Politics in Dubious Times,* pp. 81–91.

22. Jeffrey W. Bentley, "Facts, Fantasies, and Failures of Farmer Participatory Research."

23. Muhammad Yunas with Alan Jolis, *Banker to the Poor,* pp. 73, 298.

24. Abu N. M. Wahid, *The Grameen Bank: Poverty Relief in Bangladesh,* pp. 12–13, 32–38, 52–57.

25. Yunas and Jolis, *Banker,* pp. 291–301.

26. Ibid., pp. 240–41.

27. On how the S&Ls loaned money, see S. C. Gwynne, "Adventures in the Loan Trade." On the collapse and the policies that led up to it, see L. J. Davis, "Chronicle of a Debacle Foretold: How Deregulation Begat the S&L Scandal"; and Robert Sherrill, "S&Ls, Big Banks, and Other Triumphs of Capitalism." The estimated cost of the clean-up ranges from $500 billion to $1.4 trillion; see Michael M. Thomas, "The Greatest American Shambles."

28. See Leon Levy and Jeff Madrick, "Hedge Fund Mysteries."

29. On the quality of the media's economic coverage, see David J. Rothkopf, "The Disinformation Age." On the global media and its corporate interests, see McChesney, *Rich Media, Poor Democracy,* pp. 78–119.

30. U.S. Bureau of the Census, *SAUS* 1999, p. 479.

31. Greider, *One World,* pp. 21, 57–80.

32. For the labor force and production statistics, see Grieder, *One World,* pp. 110–11, 113, 216; for Motorola, see p. 91. Generally, see pp. 83–93.

33. Ibid., pp. 116–21.

34. Ibid., pp. 328–29.

35. Ibid., pp. 22–23, 243–45.

36. James Fallows, "Hurry Up Please It's Time," p. 32.

37. Charles Dickens, "A Christmas Carol," p. 107.

BIBLIOGRAPHY

Abernethy, Virginia. "Optimism and Overpopulation." *Atlantic Monthly,* December, 1994, pp. 84–91.

Aguirre, G., J. Calderón, D. Buitrago, V. Iriarte, J. Ramos, J. Blajos, G. Thiele, and A. Devaux. "Rustic Seedbeds: A Bridge Between Formal and Traditional Seed Systems in Bolivia." In *Impact on a Changing World: International Potato Center Program Report 1997–1998,* pp. 195–203. Lima: CIP, 1999.

Ahmad, Kamal Uddin, and Mumtaj Kamal. *Wealth from the Potato.* Dacca, Bangladesh: Dacca Press, 1980.

Anderson, Michael. *Population Change in North-Western Europe, 1750–1850.* London: Macmillan, 1988.

Andrade, Héctor, X. Cuesta, E. Carrera, F. Yumisaca, W. Escobar, and E. Yáñes. "The Role of the User in the Selection and Release of Potato Varieties in Ecuador." *CIP Circular* (April, 1997): 21–22.

Andrews, Jean. "The Peripatetic Chili Pepper: Diffusion of the Domesticated Capsicums Since Columbus." In *Chilies to Chocolate: Food the Americas Gave the World,* edited by Nelson Foster and Linda S. Cordell, pp. 81–93. Tucson: University of Arizona Press, 1996.

Andrews, René. "Integrated Pest Management of Potato Tuber Moth in Cochabamba, Bolivia." *CIP Circular* (August, 1992): 8–9

Anthony, T. V. "The Family Planning Programme—Lessons from Tamil Nadu's Experience." *Indian Journal of Social Science* 5 (1992): 319–27.

Argüelles, José. *The Mayan Factor: Path Beyond Technology.* Santa Fe: Bear and Company, 1987.

Arbizu, Carlos, Z. Huamán, and A. Golmirzaie. "Other Andean Roots and Tubers." In *Biodiversity in Trust: Conservation and Use of Plant Genetic Resources in CGIAR Centers,* edited by Dominic Fuccillo, Linda Sears, and Paul Stapleton, pp. 39–56. Cambridge: Cambridge University Press, 1997.

Aveni, Anthony F. *Empires of Time: Calendars, Clocks, and Cultures.* New York: Basic Books, 1989.

———. *Stairways to the Stars: Skywatching in Three Great Ancient Cultures.* New York: John Wiley and Sons, 1997.

Aveni, Anthony F., and Helaine Silverman. "Between the Lines: Reading the Nazca Markings as Rituals Writ Large." *Sciences* 31 (July–August, 1991): 36–42.

Bahn, Paul G. *The Cambridge Illustrated History of Prehistoric Art.* Cambridge: Cambridge University Press, 1998.

Bahn, Paul G., and Jean Vertut. *Journey through the Ice Age.* 2nd ed., rev. Los Angeles: University of California Press, 1997.

Baribeau, B. "The Tuber-Unit Seed Plot in Quebec." *American Potato Journal* 12 (1935): 62–64.

Bashaasha, B., R. O. M. Mwanga, C. Ocitti p'Obwoya, and P. T. Ewell. *Sweetpotato in the*

Farming and Food Systems of Uganda: A Farm Survey Report. Nairobi: CIP-NARO, 1995.

Becker, Jasper. *Hungry Ghosts: Mao's Secret Famine.* New York: Free Press, 1996.

Bellomo, Randy V., and William F. Kean. "Evidence of hominid-controlled fire at the FxJj 20 site complex, Karari Escarpment." In *Koobi Fora Research Project,* edited by Glynn Ll. Isaac and Barbara Isaac. Vol. 5, *Plio-Pleistocene Archeology,* pp. 224–33. Oxford: Clarendon Press, 1997.

Benewick, Robert, and Stephanie Donald. *The State of China Atlas.* New York: Penguin, 1999.

Bennett, Wendell C. "The Andean Highlands: An Introduction." In *Handbook of South American Indians,* edited by Julian H. Steward. Vol. 2., *Andean Civilizations,* pp. 1–60. Washington, D.C.: U.S. Government Printing Office, 1946.

Bentley, Jeffrey W. "Facts, Fantasies, and Failures of Farmer Participatory Research." *Agriculture and Human Values* 11 (spring–summer, 1994): 140–50.

Bergman, M. "The Potato Blight in the Netherlands and Its Social Consequences (1844–1847)." *International Review of Social History* 1 (1967): 390–431.

Bethell, Leslie. "A Note on the Native American Population on the Eve of the European Invasions." In *The Cambridge History of Latin America,* edited by Leslie Bethell. Vol. 1, *Colonial Latin America,* pp. 145–46. London: Cambridge University Press, 1984.

———, ed. *The Cambridge History of Latin America.* 11 vols. London: Cambridge University Press, 1984–1995.

Bhargava, Nisha, Arvind Bhargava, S. N. Bhargava, K. C. Thakur, and M. D. Upadhya. "An Efficient Potato Pollen Extractor for Bulk Pollen Collection." *American Potato Journal* 68 (1991): 581–84.

Binyan, Liu. "An Unnatural Disaster." *New York Review,* April 8, 1998, pp. 3–6.

Binyan, Liu, and Perry Link. "A Great Leap Backward." *New York Review,* October 8, 1998, pp. 19–23.

Blanchard-Lemée, Michèle, Mongi Ennaifer, Hèdi Slim, and Lafita Slim. *Mosaics of Roman Africa: Floor Mosaics from Tunisia.* New York: George Braziller, 1996.

Bofu, Song, Tian Weiming, Wang Jimin, Wang Chunlin, Yan Zhengui, Wang Shengwu, and H. Huarte. "Economic Impact of CIP-24 in China." In *Case Studies of the Economic Impact of CIP-Related Technology,* edited by Thomas Walker and Charles Crissman, pp. 31–49. Lima: CIP, 1996.

Booth, Robert H., Roy L. Shaw, and International Potato Center (CIP). *Principles of Potato Storage.* Lima: CIP, 1981.

Borlaug, Norman. "Challenges for Global Food and Fiber Production." *K. Skogs-o. Lantbr.akad. Tidskr.* Supplement 21 (1988): 15–55.

Boserup, Ester. *Population and Technological Change: A Study of Long-Term Trends.* Chicago: University of Chicago Press, 1981.

Brabet, C., D. Reynoso, D. Dufour, C. Mestres, J. Arredondo, and G. Scott. "Starch Content and Properties of 106 Sweetpotato Clones from the World Germplasm Collection Held at CIP, Peru." In *Impact on a Changing World: International Potato Center Program Report 1997–1998,* pp. 279–86. Lima: CIP, 1999.

Bradley, Fern Marshall. *Rodale's Garden Answers.* Emmaus, Pennsylvania: Rodale Press, 1995.

Brandt, Willy. *North-South: A Programme for Survival. Report of the Independent Commission on International Development Issues.* Cambridge, Massachusetts: MIT, 1980.

Braudel, Fernand. *The Structure of Everyday Life. The Limits of the Possible.* Vol. 1, *Civilization and Capitalism: 15th–18th Centuries.* Translated by Siân Reynolds. London: Collins, 1981.

Braun, Ann R., Elske van de Fliert, Christopher Wheatley, Gordon Prain, and Yudi Widodo. "Improving Profits from Sweetpotato by Linking IPM with Diversification of Markets." *CIP Circular* (December, 1995): 8–15.

Braun, Ann R., E. Priatna, R. Asmunati, Wiyanto, Y. Widodo, and E. van de Fliert. "Farmer Field Schools for Sweetpotato in Indonesia." In *International Potato Center Program Report 1995–1996*, pp. 198–204. Lima: CIP, 1997.

Bright, Christopher. "Invasive Species: Pathogens of Globalization." *Foreign Policy* (fall, 1999): 50–64.

Brooks, Alison S., David M. Helgren, Jon S. Cramer, Alan Franklin, William Hornyak, Jody M. Keating, Richard G. Klein, William J. Rink, Henry Schwarcz, J. N. Leith Smith, Kathlyn Stewart, Nancy E. Todd, Jacques Verniers, and John E. Yellen. "Dating and Context of Three Middle Stone Age Sites with Bone Points in the Upper Semliki Valley, Zaire." *Science* 268 (April 28, 1995): 549–53.

Brooks, Jerome E. *The Mighty Leaf.* Boston: Little, Brown and Company, 1952.

Brown, Kenneth J. *Roots and Tubers Galore: The Story of CIP's Global Research Program and the People Who Shaped It.* Lima: CIP, 1993.

Brown, Lester R. *Who Will Feed China? Wake-Up Call for a Small Planet.* New York: W. W. Norton, 1995.

Brown, Lester R., and Brian Halweil. "The Drying of China." *World Watch* (July/August, 1998): 10–21.

Brücher, Heinz. *Useful Plants of Neotropical Origin and Their Wild Relatives.* Berlin: Springer-Verlag, 1989.

Burton, William G. *The Potato: A Survey of Its History and of the Factors Influencing Its Yield, Nutritive Value, Quality and Storage.* 2nd ed., rev. Wageningen: H. Veenman and Zonen N. V., 1966.

Butler, Steven, and Kevin Whitelaw. "Japan's Baby Bust." *U.S. News and World Report,* October 5, 1998, pp. 42–44.

Cañedo, V., J. Benavides, A. Golmirzaie, F. Cisneros, M. Chislain, and A. Lagnaoui. "Assessing Bt-Transformed Potatoes for Potato Tuber Moth, *Phthorimaea operculella* (Zeller), Management." In *Impact on a Changing World: International Potato Center Program Report 1997–1998*, pp. 161–69. Lima: CIP, 1999.

Cárdenas, Martín. *Manual de plantas económicas de Bolivia.* 2nd ed. La Paz: Amigos del Libro, 1989.

Cardich, Augusto. "Native Agriculture in the Highlands of the Peruvian Andes." *National Geographic Research* 3 (1987): 22–39.

Carey, E. E., S. T. Gichuki, P. J. Ndolo, G. Turyamureeba, R. Kapinga, N. B. Lutaladio, and J. M. Teri. "Collaborative Sweetpotato Breeding in Eastern, Central, and Southern Africa." In *International Potato Center Program Report 1995–1996*, pp. 49–57. Lima: CIP, 1997.

Carrasco, E., A. Devaux, W. García, and R. Esprella. "Frost-Tolerant Varieties for the Andean Highlands." In *International Potato Center Program Report 1995–1996*, pp. 227–32. Lima: CIP, 1997.

Carson, Rachel. *Silent Spring.* Boston: Houghton Mifflin, 1962.

Center for Integrated Agricultural Development (CIAD) and the Sichuan Academy of Agricultural Sciences (SAAS). "Potential for Market Diversification of Sweetpotato Starch and Flour in Sichuan Province." Project report. Beijing: Beijing Agricultural University, 1995.

———. "Social and Economic Aspects of the Small-Scale, Sweetpotato Based Pig Production System in Suining Municipality, Sichuan Province." Project report. Beijing: Beijing Agricultural University, 1995.

Chadha, K. L., and J. S. Grewal. "Potato Research in India—History, Infrastructure, and Achievements." *Advances in Horticulture* 7 (1993): 1–20.

Chauvet, Jean-Marie, Éliette Brunel Deschamps, and Christine Hillaire. *Dawn of Art: The Chauvet Cave.* New York: Harry N. Abrams, 1996.

Chiappelli, Fredi, ed. *First Images of the Americas: The Impact of the New World on the Old.* 2 vols. Los Angeles: University of California Press, 1976.

Chilver, A., T. Walker, V. Khatana, H. Fano, R. Suherman, and A. Rizk. "On-Farm Profitability of TPS Utilization Technologies." In *Impact on a Changing World: International Potato Center Program Report 1997–1998,* pp. 213–19. Lima: CIP, 1999.

"China Survey: The Titan Stirs." *Economist,* November 28, 1992, pp. 3–18.

Chislain, Marc, and Bodo Trognits. "Molecular Breeding for Late Blight Resistance in Potato." *CIP Circular* (September, 1996): 10–15.

Chislain, Marc., B. Trognitz, R. Nelson, Ma. del R. Herrera, L. Portal, M. Orillo, and F. Trognits. "Identification of QTLs for Late Blight Resistance in a Cross Between *S. phureja* and a Dihaploid *S. tuberosum* and Association with a Plant Defense Gene." In *Impact on a Changing World: International Potato Center Program Report 1997–1998,* pp. 67–71. Lima: CIP, 1999.

Cieza de León, Pedro de. *The Discovery and Conquest of Peru.* Edited and translated by Alexandra Parma Cook and Noble David Cook. Durham: Duke University Press, 1998.

Cisneros, Fausto, Jesús Alcázar, María Palacios, and Oscar Ortiz. "A Strategy for Developing and Implementing Integrated Pest Management." *CIP Circular* (December, 1995): 2–7.

Clement, Charles R. "A Center of Crop Genetic Diversity in Western Amazonia: A New Hypothesis of Indigenous Fruit-Crop Distribution." *BioScience* 39 (October, 1989): 624–31.

Clottes, Jean, and Jean Courtin. *The Cave Beneath the Sea: Paleolithic Images at Cosquer.* New York: Harry N. Abrams, 1996.

Cobb, Clifford, Ted Halstead, and Jonathan Rowe. "If the GDP is Up, Why is America Down?" *Atlantic Monthly,* October, 1995, pp. 59–78.

Cobo, Bernabé. *A History of the Inca Empire.* Edited and translated by Roland Hamilton. Austin: University of Texas Press, 1979.

Coe, Michael D. *Mexico: From the Olmecs to the Aztecs.* 4th ed., rev. Singapore: Thames and Hudson, 1994.

Coe, Sophie D., and Michael D. Coe. *The True History of Chocolate.* London: Thames and Hudson, 1996.

Cole, Donald C., Fernando Carpio, Jim A. Julian, and Nifa León. "Health Impacts of Pesticide Use in Carchi Farm Populations." In *Economic, Environmental, and Health Tradeoffs in Agriculture: Pesticides and the Sustainability of Andean Potato Production,* edited by Charles C. Crissman, John M. Antle, and Susan M. Capalbo, pp. 209–30. Boston: Kluwer, 1998.

Conquest, Robert. *The Harvest of Sorrow: Soviet Collectivization and the Terror-Famine.* New York: Oxford University Press, 1986.

Conrad, Geoffrey W., and Arthur A. Demarest. *Religion and Empire: The Dynamics of Aztec and Inca Expansionism.* Cambridge: Cambridge University Press, 1984.

Consultative Group on International Agricultural Research (CGIAR). *1984 Annual Report.* Washington, D.C.: CGIAR, 1984.

Consultative Group on International Agricultural Research (CGIAR), Technical Advisory Committee (TAC), and CGIAR Secretariat. *Report of the Fourth External Program and Management Review of the Centro Internacional de la Papa (CIP).* Washington, D.C.: TAC Secretariat, Food and Agricultural Organization of the United Nations, 1995.

Cook, David Noble. *Demographic Collapse: Indian Peru, 1520 to 1620.* Cambridge: Cambridge University Press, 1981.

Cook, Sherburne F., and Woodrow Borah. *Essays in Population History: Mexico and the Caribbean.* Los Angeles: University of California Press, 1971.

Cowan, Wesley C., and Patty Jo Watson, eds. *The Origins of Agriculture: An International Perspective.* Washington, D.C.: Smithsonian Institution Press, 1992.

Cox, Paul Alan, and Sandra Anne Banack, eds. *Islands, Plants, and Polynesians: An Introduction to Polynesian Ethnobotany.* Portland, Oregon: Dioscorides Press, 1991.

Crissman, Charles C., Donald C. Cole, and Fernando Carpio. "Pesticide Use and Farm Worker Health in Ecuadorian Potato Production." *American Journal of Agricultural Economics* 76 (August, 1994): 593–97.

Crissman, Charles C., John M. Antle, and Susan M. Capalbo, eds. *Economic, Environmental, and Health Tradeoffs in Agriculture: Pesticides and the Sustainability of Andean Potato Production.* Boston: Kluwer, 1998.

Crissman, Charles C., Patricio Espinosa, Cecile E. H. Ducrot, Donald C. Cole, and Fernando Carpio. "The Carchi Study Site: Physical, Health, and Potato Farming Systems in Carchi Province." In *Economic, Environmental, and Health Tradeoffs in Agriculture: Pesticides and the Sustainability of Andean Potato Production,* edited by Charles C. Crissman, John M. Antle, and Susan M. Capalbo, pp. 85–120. Boston: Kluwer, 1998.

Critchfield, Richard. *The Villagers: Changed Values, Altered Lives.* New York: Anchor, 1994.

Crosby, Alfred W., Jr. *The Columbian Exchange: Biological and Cultural Consequences of 1492.* Westport: Greenwood, 1972.

———. "Metamorphosis of the Americas." In *Seeds of Change,* edited by Herman J. Viola and Carolyn Margolis, pp. 70–89. Washington, D.C.: Smithsonian Institution Press, 1991.

Cutler, Hugh C., and Martín Cárdenas. "Chicha, a Native South American Beer." *Botanical Museum Leaflets* 13 (1947): 33–60.

Daly, Herman E. *Ecological Economics and the Ecology of Economies: Essays in Criticism.* New York: Elgar, 1999.

Davidson, Alan. "Europeans' Wary Encounter with Tomatoes, Potatoes, and Other New World Foods." In *Chilies to Chocolate: Food the Americas Gave the World,* edited by Nelson Foster and Linda S. Cordell, pp. 1–14. Tucson: University of Arizona Press, 1996.

Davis, Kingsley. *The Population of India and Pakistan.* Princeton: Princeton University Press, 1951.

———. "Population and Resources: Fact and Interpretation." In *Resources, Environment, and Population: Present Knowledge, Future Options,* edited by Kingsley Davis and Mikhail S. Bernstam, pp. 1–21. New York: Oxford University Press, 1991.

Davis, Kingsley, and Mikhail S. Bernstam, eds. *Resources, Environment, and Population: Present Knowledge, Future Options.* New York: Oxford University Press, 1991.

Davis, L. J. "Chronicle of a Debacle Foretold: How Deregulation Begat the S&L Scandal." *Harpers,* September, 1990, pp. 50–66.

Dean, Bill B. *Managing the Potato Production System.* Binghampton, New York: Food Products Press, 1994.

Dean, Daniel. "Our Changing Potato Industry." *American Potato Journal* 10 (1933): 108–14.

de Ávila, A. C., L. F. Salazar, O. A. Hidalgo, J. Nakashima, and A. N. Dusi. "Boosting Tuber-Seed Production in Brazil: Serological Techniques and Antiserum Production." *CIP Circular* (March, 1989): 1–6.

de la Vega, Garcilaso. *The Royal Commentaries of the Inca Garcilaso de la Vega.* Translated by María Jolas and edited by Alain Gheerbrant. London: Cassell and Company, 1963.

Denevan, William M. "The Aboriginal Population of Amazonia." In *The Native Population of the Americas in 1492,* edited by William M. Denevan, pp. 205–34. Madison: University of Wisconsin Press, 1976.

———. *The Native Population of the Americas in 1492.* Madison: University of Wisconsin Press, 1976.

de Santillana, Giorgio, and Hertha von Dechend. *Hamlet's Mill: An Essay on Myth and the Frame of Time.* Boston: Godine, 1977.

Devaux, André, Nelson Estrada, and Enrique Carrasco. "Frost Tolerance in Potatoes: A Challenge to Andean Biodiversity." Cochabamba: PROINPA, 1995. Mimeographed.

Devine, T. M. *The Great Highland Famine: Hunger, Emigration, and the Scottish Highlands in the Nineteenth Century.* Edinburgh: John Donald Publishers, 1988.

Diamond, Jared. *Guns, Germs, and Steel: The Fate of Human Societies.* New York: W. W. Norton, 1997.

Dickens, Charles. "A Christmas Carol." In *Christmas Stories.* Illustrated by Arthur Rackham. Norwalk, Connecticut: Easton Press, 1992.

Directorate of Agriculture, Government of West Bengal. *Annual Report for 1995–1996: Section of Economic Botanist-III (Anandapore Farm).* Midnapore: Directorate of Agriculture, 1995.

———. *Estimates of Area and Production of Principal Crops in West Bengal 1992–1993.* Calcutta: Government of West Bengal, 1995.

Dodds, John H., and Lorin W. Roberts. *Experiments in Plant Tissue Culture.* 2nd ed. Cambridge: Cambridge University Press, 1985.

Donnelly, James S. "Mass Eviction and the Great Famine." In *The Great Irish Famine,* edited by Cathal Póirtéir, pp. 155–73. Dublin: Mercier Press, 1995.

Dowley, Leslie J., and Eugene O'Sullivan. *A Short History of the Potato, the Famine, Late Blight, and Irish Research on Phytophthora infestans.* Carfow, Ireland: Oak Park Research Center, 1995.

Dowswell, Christopher R., R. L. Paliwal, and Ronald P. Cantrell. *Maize in the Third World.* Boulder, Colorado: Westview Press, 1996.

Draper, Martin A., Gary A. Secor, Neil C. Gudmestad, H. Arthur Lamey, and Duane Preston. "Leaf Blight Diseases of Potato." North Dakota State University Extension Service, PP-1084, July, 1994. Available at www.ext.nodak.edu/extpubs/plantsci/hortcrop/pp1084w.htm.

Duncan, David Ewing. *The Calendar.* London: Fourth Estate, 1998.

Economic Commission for Latin America (ECLA). *Statistical Yearbook for Latin America 1973.* New York: United Nations, 1974.

Economy, Elizabeth. "Painting China Green." *Foreign Affairs,* March/April, 1999, pp. 14–18.

Elton, Catherine. "Peru's Potatoes Saved by Science." *Christian Science Monitor,* March 15, 2000.

Epstein, T. Scarlett. *A Manual for Culturally-Adapted Market Research in the Development Process.* London: RWAL: 1988.

———. "Viewpoints: Are Babies Made at the Macro-Level?" *Populi* (June, 1992): 17.

Epstein, T. Scarlett, Janet Gruber, and Graham Mytton. *A Training Manual for Development Market Research Investigators.* London: BBC, 1991.

Erickson, Clark L. "Archaeological Methods for the Study of Ancient Landscapes of the Llanos de Mojos in the Bolivian Amazon." In *Archaeology in the Lowland American Tropics,* edited by Peter W. Stahl, pp. 66–95. Cambridge: Cambridge University Press, 1995.

Espinosa, Patricio. "Pruebas de aceptabilidad de cuatro nuevas variedades mejoradas de papa del INIAP a nivel de consumidor final urbano en Quito." Quito: FORTIPAPA, 1996. Mimeographed.

———, ed. *Recetario de las raíces y tubérculos andinos.* Quito: Ediciones ABYA-YALA, 1997.

Espinosa, Patricio, Fabio Muños, and Jorge Carrillo. "Algunos aspectos del consumo doméstico de la papa en Quito, Guayaquil, y Cuenca." Quito: FORTIPAPA, 1995. Mimeographed.

Espinosa, Patricio, and Crissman, Charles C. "Aspectos del consumo urbano de las raíces y tubérculos andinos y actitud del consumidor en Ecuador." Working paper. Quito: CIP, 1996.

Espinosa, Patricio, Rocío Vaca, Jorge Abad, and Charles Crissman. *Raíces y tubérculos andinos: Cultivos marginados en el Ecuador.* Quito: CIP, 1996.

Estrada Ramos, Nelson. "Utilization of Wild and Cultivated Diploid Potato Species to Transfer Frost Resistance into the Tetraploid Common Potato, *Solanum Tuberosum L.*" In *Plant Cold Hardiness,* edited by Paul H. Li, pp. 339–53. New York: Alan R. Liss, 1987.

Fallows, James. "Why Americans Hate the Media." *Atlantic Monthly,* February, 1996, pp. 45–64.

———. "Hurry Up Please It's Time." *New York Review,* September 23, 1999, pp. 29–34.

Fonseca, C., R. Labarta, A. Mendoza, J. Landeo, and T. S. Walker. "Economic Impact of the High-Yielding Late-Blight-Resistant Variety Canchán-INIAA in Peru." In *Case Studies of the Economic Impact of CIP-Related Technology,* edited by Thomas Walker and Charles Crissman, pp. 51–63. Lima: CIP, 1996.

Food and Agricultural Organization of the United Nations (FAO). *FAO Production Yearbook.* Rome: FAO, various years.

———. *FAO Quarterly Bulletin of Statistics.* Rome: FAO, various years.

Foote, Timothy. *The World of Bruegel 1525–1569.* New York: Time-Life, 1968.

Foster, Nelson, and Linda S. Cordell. *Chilies to Chocolate: Food the Americas Gave the World.* Tucson: University of Arizona Press, 1992.

Franke, Richard W., and Barbara H. Chasin. *Kerala: Radical Reform as Development in an Indian State.* San Francisco: Institute for Food and Development Policy, 1989.

———. "Development Without Growth: The Kerala Experiment." *Technology Review* (April, 1990): 43–51.

Freidel, David, Linda Schele, and Joy Parker. *Maya Cosmos: Three Thousand Years on the Shaman's Path.* New York: Quill, 1993.

French, E. R. "Integrated Control of Bacterial Wilt of Potatoes." *CIP Circular* (June, 1994): 8–11.

———, ed. *Bacterial Wilt Manual.* Lima: CIP, 1996.

Fry, William F., and Stephen B. Goodwin. "Resurgence of Irish Potato Famine Fungus." *BioScience* 47 (June, 1997): 363–71.

Fuccillo, Dominic, Linda Sears, and Paul Stapleton, eds. *Biodiversity in Trust: Conservation and Use of Plant Genetic Resources in CGIAR Centers.* Cambridge: Cambridge University Press, 1997.

Fuglie, K., H. Ben Zalah, M. Essamet, A. Ben Temime, and A. Fahmouni. "Economic Impact of IPM Practices on the Potato Tuber Moth in Tunisia." In *Case Studies of the Economic Impact of CIP-Related Technology,* edited by Thomas S. Walker and Charles C. Crissman, pp. 67–81. Lima: CIP, 1996.

Fuglie, K., L. Zhang, L. Salazar, and T. Walker. "Economic Impact of Virus-Free Sweetpotato Planting Material in Shandong Province, China." In *Impact on a Changing World: International Potato Center Program Report 1997–1998,* pp. 249–54. Lima: CIP, 1999.

Galinat, Walton C. "Maize: Gift from America's First People." In *Chilies to Chocolate: Food the Americas Gave the World* , edited by Nelson Foster and Linda S. Cordell, pp. 47–60. Tucson: University of Arizona Press, 1992.

———. "The Origin of Maize: Grain of Humanity." *Economic Botany* 49 (1995): 3–12.

Gandarillas, Antonio, and André Devaux. "PROINPA's Agroecological Approach for Potato Research in Bolivia." *CIP Circular* (August, 1992): 1–9.

Garrett, Laurie. *The Coming Plague: Newly Emerging Diseases in a World Out of Balance.* New York: Farrar, Straus, and Giroux, 1994.

Gaur, P. C., and S. K. Pandey. "TPS Production and Adoption of Technology in India." In *Production and Utilization of True Potato Seed in Asia,* edited by M. D. Upadhya, B. Hardy, P. C. Gaur, and S. G. Ilangantileke, pp. 25–35. Lima: CIP, 1996.

Gawron-Burke, Cynthia, and Timothy B. Johnson. "Development of *Bacillus Thuringiensis*-Based Pesticides for the Control of Potato Insect Pests." In *Advances in Potato Pest Biology and Management,* edited by G. W. Zehnder, M. L. Powelson, R. K. Jansson, and

K. V. Raman, pp. 522–34. St. Paul, Minnesota: American Phytopathological Society, 1994.

Geary, Laurence M. "Famine, Fever, and the Bloody Flux." In *The Great Irish Famine*, edited by Cathal Póirtéir, pp. 74–85. Dublin: Mercier Press, 1995.

Gerling, D., and R. T. Meyer, eds. *Bemisia 1995: Taxonomy, Biology, Damage, Control and Management*. Hants, U.K.: Intercept, 1996.

Gillis, John R., Louise A. Tilly, and David Levine. "The Quiet Revolution." In *The European Experience of Declining Fertility, 1850–1979: The Quiet Revolution*, edited by John R. Gillis, Louise A. Tilly, and David Levine, pp. 1–9. Cambridge: Blackwell, 1992.

———, eds. *The European Experience of Declining Fertility, 1850–1979: The Quiet Revolution*. Cambridge: Blackwell, 1992.

Gitomer, Charles S. *Potato and Sweetpotato in China: Systems, Constraints, and Potential*. CIP: Lima, 1996.

Global Initiative on Late Blight (GILB). "Standard International Field Trials (SIFT) Are Established in Developing Countries." *GILB Newsletter* (August, 1999): 4–5.

———. "Up Date on the G x E (Genotype by Environment) Study." *GILB Newsletter* (August, 1999): 2–4.

———. "Focus on Partners: Farmers Groups, Extension Organizations, and Research Institutes Team Up to Tackle Late Blight in Seven Countries." *GILB Newsletter* (April, 2000): 1–4.

Golmirzaie, Ali M., and Humberto A. Mendoza. "Breeding Strategies for True Potato Seed Production." *CIP Circular* (December, 1988): 1–8.

Goodwin, Stephen B., Christine D. Smart, Robert W. Sandrock, Kenneth L. Deahl, Zamir K. Punja, and William E. Fry. "Genetic Changes Within Populations of *Phytophthora infestans* in the United States and Canada During 1994–1996: Role of Migration and Recombination." *Phytopathology* 88 (1998): 939–49.

Goodwyn, Lawrence. *Democratic Promise: The Populist Moment in America*. New York: Oxford University Press, 1976.

Gore, Rick. "Ancient Americans." *National Geographic,* October, 1997, pp. 93–99.

Gould, Stephen Jay. *The Panda's Thumb: More Reflections in Natural History*. New York: W. W. Norton, 1980.

———. *The Flamingo's Smile: Reflections in Natural History*. New York: W. W. Norton, 1985.

Goulding, Michael, Nigel J. H. Smith, and Dennis J. Mahar. *Floods of Fortune: Ecology and Economy Along the Amazon*. New York: Columbia University Press, 1996.

Govinda, Lama Anagarika. *The Inner Structure of the I Ching*. San Francisco: Wheelwright Press, 1981.

Grau, Alfredo, and Julio Rea. "Yacon." In *Roots and Tubers: Ahipa, Arracacha, Maca, and Yacon,* edited by Michael Hermann and J. Heller, pp. 199–242. Rome: International Plant Genetics Resources Institute, 1997.

Gray, Malcomb. "The Highland Potato Famine of the 1840s." *Economic History Review* 7 (April, 1955): 357–68.

Gray, Roger W., Verson L. Sorenson, and Willard W. Cochrane. *An Economic Analysis of the Impact of Government Programs on the Potato Industry of the United States*. University of Minnesota, North Central Regional Publication No. 42, 1954.

Green, E. R. R. "Agriculture." In *The Great Famine: Studies in Irish History, 1845–1852,* edited by R. Dudley Edwards and T. Desmond Williams, pp. 89–128. 1956. New ed. with an introduction by Cormac Ó. Gráda. Dublin: Lilliput Press, 1994.

Greider, William. *One World Ready or Not: The Manic Logic of Global Capitalism*. New York: Touchstone: 1997.

Groube, Les. "The Taming of Rain Forests: A Model for Late Pleistocene Forest Exploitation in New Guinea." In *Foraging and Farming: The Evolution of Plant Exploitation,* edited by David R. Harris and Gordon C. Hillman, pp. 292–304. London: Unwin Hyman, 1989.

Groube, Les, Jay Chappell, J. Muke, and D. Price. "A 40,000 Year-Old Human Occupation Site At Huon Peninsula, Papau New Guinea." *Nature* 324 (1986): 453–55.

Grun, Paul. "The Evolution of Cultivated Potatoes." *Economic Botany* 44 (1990): 39–55.

Guenthner, Joseph, Biing-Hwang Lin, and Annette E. Levi. "The Influence of Microwave Ovens On the Demand for Fresh and Frozen Potatoes." *Journal of Food Distribution Research* (September, 1991): 45–52.

Guerrero Carrión, Trotsky. *Modernización agraria y pobreza rural en el Ecuador.* Loja: Editorial Universitaria, 1992.

Guillermoprieto, Alma. *The Heart that Bleeds.* New York: Vintage, 1995.

Guillet, David. "Terracing and Irrigation in the Peruvian Highlands." *Current Anthropology* 28 (1987): 409–30.

Gunawan, Memed, Christopher Wheatley, and Irfansyah. "Current Status and Prospect for Sweet Potato in Indonesia." Working paper. Bogor: CIP, 1995.

Gwynne, S. C. "Adventures in the Loan Trade." *Harpers,* September, 1983, pp. 22–26.

Hacker, Andrew. "War Over the Family." *New York Review,* December 4, 1997, pp. 34–38.

Hagenimana, V., and C. Owori. "Feasibility, Acceptability, and Production Costs of Sweetpotato-Based Products in Uganda." In *International Potato Center Program Report 1995–1996,* pp. 276–81. Lima: CIP, 1997.

Hagenimana, V., L. M. K'osambo, and E. E. Carey. "Potential of Sweetpotato in Reducing Vitamin A Deficiency in Africa." In *Impact on a Changing World: Program Report 1997–1998,* pp. 287–94. Lima: CIP, 1999.

Hamilton, Earl J. "What the New World Gave the Economy of the Old." In *First Images of the Americas,* edited by Fredi Chiappelli, 2 vols., pp. 853–84. Los Angeles: University of California Press, 1976.

Hardy, Bill. "Carlos Ochoa—Potato Prize Winner." *CIP Circular* (September, 1994): 8–9.

Hardy, Bill., B. Trognitz, and G. Forbes. "Late Blight Breeding at CIP: Progress to Date." *CIP Circular* (April, 1995): 2–5.

Harlan, Jack R. *Crops and Man.* 2nd ed., rev. Madison, Wisconsin: American Society of Agronomy, 1992.

Harris, David R. "The Origins of Agriculture in the Tropics." *American Scientist* 60 (March–April, 1972): 181–93.

———. "Alternative Pathways Toward Agriculture." In *Origins of Agriculture,* edited by Charles A. Reed, pp. 179–243. The Hague: Mouton Publishers, 1977.

———. "An Evolutionary Continuum of Plant-People Interaction." In *Foraging and Farming: The Evolution of Plant Exploitation,* edited by David R. Harris and Gordon C. Hillman, pp. 11–26. London: Unwin Hyman, 1989.

Harris, David R., and Gordon C. Hillman, eds. *Foraging and Farming: The Evolution of Plant Exploitation.* London: Unwin Hyman, 1989.

Hawkes, J. G. Introduction to *The History and Social Influence of the Potato,* by Redcliffe N. Salaman. Revised with an introduction by J. G. Hawkes. 1949. Cambridge: Cambridge University Press, 1985.

———. "The Domestication of Roots and Tubers in the American Tropics." In *Foraging and Farming: The Evolution of Plant Exploitation,* edited by David R. Harris and Gordon C. Hillman, pp. 481–503. London: Unwin Hyman, 1989.

———. *The Potato: Evolution, Biodiversity, and Genetic Resources.* Washington, D.C.: Smithsonian Institution Press, 1990.

Hawkes, J. G., and J. Francisco-Ortega. "The Potato in Spain During the Late 16th Century." *Economic Botany* 46 (1992): 86–97.

Hawkins, Arthur. "Highlights of a Half-Century in Potato Production." *American Potato Journal* 34 (1957): 25–29.

Hecht, Susana, and Alexander Cockburn. *The Fate of the Forest: Developers, Destroyers, and Defenders of the Amazon.* New York: Harpers, 1990.

Heiser, Charles B., Jr. *Nightshades: The Paradoxal Plants.* San Francisco: W. H. Freeman and Company, 1969.

————. *Of Plants and People.* Norman: University of Oklahoma Press, 1985.

————. *Seed to Civilization: The Story of Food.* 2nd ed., rev. Cambridge, Massachusetts: Harvard University Press, 1990.

Henfling, Jan W. "Late Blight of Potato." CIP technical bulletin no. 4. Lima: CIP, 1987.

Hermann, Michael. *Andean Roots and Tubers: Research Priorities for a Neglected Food Resource.* Lima: CIP, 1992.

————. "Arracacha." In *Roots and Tubers: Ahipa, Arracacha, Maca, and Yacon,* edited by Michael Hermann and J. Heller, pp. 75–172. Rome: International Plant Genetics Resources Institute, 1997.

Hermann, Michael, and J. Heller. "Andean Roots and Tubers at the Crossroads." In *Andean Roots and Tubers: Ahipa, Arracacha, Maca, and Yacon,* edited by Michael Hermann and J. Heller, pp. 5–11. Rome: International Plant Genetics Resources Institute, 1997.

————, eds. *Andean Roots and Tubers: Ahipa, Arracacha, Maca, and Yacon.* Rome: International Plant Genetic Resources Institute, 1997.

Hermann, Michael, N. K. Quynh, and D. Peters. "Reappraisal of Edible Canna as a High-Value Starch Crop in Vietnam." In *Impact on a Changing World: International Potato Center Program Report 1997–1998,* pp. 415–24. Lima: CIP, 1999.

Hertsgaard, Mark. "Our Real China Problem." *Atlantic Monthly,* November, 1997, pp. 97–114.

Heuveline, Patrick. "'Between One and Three Million': Towards the Demographic Reconstruction of a Decade of Cambodian History." *Demographic Studies* 52 (1998): 49–65.

Hijmans, R. J., G. A. Forbes, and T. S. Walker. "Estimating the Global Severity of Potato Late Blight with a GIS-Linked Disease Forecaster." In *Impact on a Changing World: International Potato Center Program Report 1997–1998,* pp. 83–90. Lima: CIP, 1999.

Horton, Douglas. "Farming Systems Research: Twelve Lessons from the Mantaro Valley Project," *Agricultural Administration* 23 (1986): 93–107.

————. *Potatoes: Production, Marketing, and Programs for Developing Countries.* Boulder, Colorado: Westview Press, 1987.

Horton, Douglas, and A. Monares. "A Small, Effective Seed Multiplication Program: Tunisia." Working paper. Lima: CIP, 1985.

Horton, Richard. "Infection: The Global Threat." *New York Review,* April 6, 1995, pp. 24–28.

Hourani, Albert. *A History of the Arab Peoples.* Cambridge, Massachusetts: Belknap, 1991.

"How Poor Is China?" *Economist,* October 12, 1996, pp. 35–36.

Huamán, Zósimo. "Conservation of Potato Genetic Resources at CIP." *CIP Circular* (June, 1986): 1–7.

————. "Systematic Botany and Morphology of the Sweetpotato Plant." Technical information bulletin. Lima: CIP, 1992.

————. "Ex situ Conservation of Potato Genetic Resources at CIP." *CIP Circular* (September, 1994): 2–7.

Huamán, Zósimo, A. Glomirzaie, and W. Amoros, "The Potato." In *Biodiversity in Trust: Conservation and Use of Plant Genetic Resources in CGIAR Centers,* edited by Dominic Fuccillo, Linda Sears, and Paul Stapleton, pp. 21–28. Cambridge: Cambridge University Press, 1997.

Huamán, Zósimo, and D. P. Zhang. "Sweetpotato." In *Biodiversity in Trust: Conservation and Use of Plant Genetic Resources in CGIAR Centers,* edited by Dominic Fuccillo, Linda Sears, and Paul Stapleton, pp. 29–38. Cambridge: Cambridge University Press, 1997.

Hughes, Robert. "Behold the Stone Age." *Time,* February 13, 1995, pp. 52–62.

Hughes, Robert. "Why Watch It Anyway." *New York Review,* February 16, 1995, pp. 37–42.

Hyslop, John. *The Inca Road System.* New York: Academic Press, 1984.

IBTA (Instituto Boliviano de Tecnología Agropecuaria). *Boletín Técnico* 1 (May, 1995).

INIAP (Instituto Nacional Autónomo de Investigaciones Agropecuarias) and FORTIPAPA (Fortalecimiento de la Investigación y Producción de Semilla de Papa en el Ecuador). *Informe Anual 1995 Compendio.* Quito: FORTIPAPA, 1996.

———. *Informe Anual 1996 Compendio.* Quito: FORTIPAPA, 1997.

———. *Informe Anual 1998 Compendio.* Quito: FORTIPAPA, 1999.

International Potato Center (CIP). *International Potato Center Annual Report.* Lima: CIP, various years.

———. "Sprout Cuttings: A Rapid Multiplication Technique for Potatoes." CIP training guide. Lima: CIP, 1981.

———. "Stem Cuttings: A Rapid Multiplication Technique for Potatoes." CIP training guide. Lima: CIP, 1981.

———. *Potatoes for the Developing World.* Lima: CIP, 1984.

———. *Tissue Culture Propagation of Potato.* CIP slide training guide. Lima: CIP, 1986.

———. *Exploration, Maintenance, and Utilization of Sweet Potato Genetic Resources.* Lima: CIP, 1988.

———. "Potato Farming Grew Out of U.S.-Vietnam Peace Talks." *Backgrounder Press Release.* May, 1991, pp. 1–4.

———. "Yacón in Hokkaido, Japan." *CIP Circular* (June, 1994): 11.

———. *International Potato Center Program Report 1993–1994.* Lima: CIP, 1995.

———. "The Development of Durable Resistance to Late Blight: A Global Initiative." Pamphlet. Lima: CIP, 1996.

———. *Major Potato Diseases, Insects, and Nematodes.* Lima: CIP, 1996.

———. *Proyecto Chacasina: Produzcamos papa con semilla sexual.* Lima: CIP, 1996.

———. "Sweet-Potato Facts." Pamphlet. Lima: CIP, 1996.

———. *Medium-Term Plan 1998–2000.* Lima: CIP, 1997.

———. *Pocket Guide to Nine Exotic Andean Roots and Tubers.* Lima: CIP, 1997.

———. "Scientists Develop New Potato Clones to Counter Late Blight, World's Worst Agricultural Disease." CIP press release. May 24, 1998, pp. 1–4.

———. *Impact on a Changing World: International Potato Center Program Report: 1997–1998.* Lima: CIP, 1999.

———. "An Improved Method for Fighting Bacterial Wilt: NCM-ELISA Detection Kit and Training Materials." In *Impact on a Changing World: International Potato Center Program Report 1997–1998,* p. 122. Lima: CIP, 1999.

———. "Training." In *Impact on a Changing World: International Potato Center Program Report 1997–1998,* pp. 433–38. Lima: CIP, 1999.

International Potato Center (CIP) and the Chinese Academy of Sciences (CAAS). "Virus Cleanup Boosts Chinese Sweet Potato Production." Working paper. Lima: CIP, 1998.

International Potato Center (CIP) and the Food and Agricultural Organization of the United Nations (FAO). *Potatoes in the 1990s: Situation and Prospects of the World Potato Economy.* Rome: FAO, 1995.

International Rice Research Institute (IRRI). *World Rice Statistics 1990.* Manila: IRRI, 1991.

———. *IRRI Rice Almanac, 1993–1995.* Manila: IRRI, 1993.

Isaac, Glynn Ll. Introduction to *Koobi Fora Research Project,* edited by Glynn Ll. Isaac and Barbara Isaac. Vol. 5, *Plio-Pleistocene Archeology,* pp. 1–11. Oxford: Clarendon Press, 1997.

Isaac, Glynn Ll., and Barbara Isaac, eds. *Koobi Fora Research Project.* Vol. 5, *Plio-Pleistocene Archeology.* Oxford: Clarendon Press, 1997.

Iwanaga, M., and P. Schmiediche. "Using Wild Species to Improve Potato Cultivars." *CIP Circular* (June, 1989): 1–7.

Jamison, Ellen, Frank Hobbs, Peter O. Way, and Ellen Staneck, eds. *World Population Profile 1994.* Washington, D.C.: U.S. Government Printing Office, 1994.

Jansson, Richard K., and Kandukuri V. Raman, eds. *Sweet Potato Pest Management: A Global Perspective.* Boulder, Colorado: Westview, 1991.

Johns, T., and S. L. Keens. "Ongoing Evolution of the Potato on the Altiplano of Western Bolivia." *Economic Botany* 40 (1986): 409–24.

Jones, Jeffrey Ronald. "Technological Change and Market Organization in Cochabamba, Bolivia: Problems of Agricultural Development Among Potato Producing Small Farmers." Ph.D. diss., University of California at Los Angeles, 1980.

Jordan, William Chester. *The Great Famine: Northern Europe in the Early Fourteenth Century.* Princeton, New Jersey: Princeton University Press, 1996.

Kahn, E. J., Jr. *The Staffs of Life.* Boston: Little, Brown and Company, 1985.

Kaplan, Lawrence, and Lucille N. Kaplan. "Beans of the Americas." In *Chilies to Chocolate: Food the Americas Gave the World,* edited by Nelson Foster and Linda S. Cordell, pp. 61–79. Tucson: University of Arizona Press, 1996.

Kapur, Akash. "Poor but Prosperous." *Atlantic Monthly,* September, 1998, pp. 40–45.

Kelley, Jonathan, and Herbert S. Klein. *Revolution and the Rebirth of Inequality: A Theory Applied to the National Revolution in Bolivia.* Los Angeles: University of California Press, 1981.

Kennedy, George G., and Ned M. French, "Monitoring Resistance in Colorado Potato Beetle." In *Advances in Potato Pest Biology and Management,* edited by G. W. Zehnder, M. L. Powelson, R. K. Jansson, and K. V. Ramon, pp. 278–93. St. Paul, Minnesota: American Phytopathological Society, 1994.

Kenya Agricultural Research Institute (KARI) and International Potato Center (CIP). "Orange Fleshed Sweetpotato Varieties." Pamphlet. Nairobi: KARI-CIP, 1997.

Khatana, V. S., M. D. Upadhya, A. Chilver, and C. Crissman. "Economic Impact of True Potato Seed on Potato Production in Eastern and Northeastern India." In *Case Studies of the Economic Impact of CIP-Related Technologies,* edited by Thomas S. Walker and Charles C. Crissman, pp. 139–56. Lima: CIP, 1996.

Kidane-Mariam, Haile M. "Trip Report on Visits to Uganda, Ethiopia, 19–28 November, 1997: Update on the Farmer-Based Seed System in the Target NARS." CIP: Nairobi, 1997. Mimeographed.

Kiernan, Ben. *The Pol Pot Regime: Race, Power, and Genocide in Cambodia under the Khmer Rouge, 1975–1979.* New Haven: Yale University Press, 1996.

Killen, John, ed. *The Famine Decade: Contemporary Accounts 1841–1851.* Belfast: Blackstaff Press, 1995.

Killion, Thomas, ed. *Gardens of Prehistory: The Archaeology of Settlement Agriculture in Greater Mesoamerica.* Tuscaloosa: University of Alabama Press, 1992.

Kinealy, Christine. *This Great Calamity.* Dublin: Gill and Macmillan, 1994.

———. *A Death-Dealing Famine: The Great Hunger in Ireland.* Chicago: Pluto Press, 1997.

Kirk, Willie, and Jeffrey Stein. "Recommendations for Late Blight Control in Michigan for 2000." Michigan State University, Potato Web Portal. Available at www.potato.msu.edu.

Klein, Herbert S. *Bolivia: The Evolution of a Multi-Ethnic Society.* 2nd ed., rev. New York: Oxford University Press, 1992.

Knaggs, H. Valentine. *Potatoes as Food and Medicine.* London: C. W. Daniel Company, 1932.

Knapp, Gregory. *Ecología cultural prehispánica del Ecuador.* Quito: Banco Central del Ecuador, 1988.

———. *Andean Ecology: Adaptive Dynamics in Ecuador.* Boulder, Colorado: Westview Press, 1991.

Kolata, Alan L., ed. *Tiwanaku and Its Hinterland: Archaeology and Paleoecology of an Andean Civilization.* Washington, D.C.: Smithsonian Institution Press, 1996.

Kolata, Alan L., and Charles R. Ortloff. "Tiwanaku Raised-Field Agriculture in the Lake Titicaca Basin." In *Tiwanaku and Its Hinterland: Archaeology and Paleoecology of an Andean Civilization,* edited by Alan L. Kolata, pp. 109–51. Washington, D.C.: Smithsonian Institution Press, 1996.

Kosok, Paul. *Life, Land, and Water in Ancient Peru.* New York: Long Island University Press, 1965.

Krupp, Edwin C. *Echoes of the Ancient Skies: The Astronomy of Lost Civilizations.* New York: Oxford University Press, 1983.

Lagnaoui, A., and R. El-Bedewy. "An Integrated Pest Management Strategy for Controlling Tuber Moth in Egypt." *CIP Circular* (April, 1997): 6–7.

Lang, James. *Conquest and Commerce: Spain and England in the Americas.* New York: Academic Press, 1975.

———. *Portuguese Brazil: The King's Plantation.* New York: Academic Press, 1979.

———. *Inside Development in Latin America: A Report from the Dominican Republic, Colombia, and Brazil.* Chapel Hill: University of North Carolina Press, 1988.

———. *Feeding a Hungry Planet: Rice, Research, and Development in Asia and Latin America.* Chapel Hill: University of North Carolina Press, 1996.

Lappé, Mark, and Britt Bailey. *Against the Grain: Biotechnology and the Corporate Takeover of Your Food.* Monroe, Maine: Common Courage Press, 1998.

Lechtman, Heather, and Ana María Soldi. *La tecnología en el mundo andino.* México: Universidad Nacional Autónoma de México, 1981.

Lee, David R., and Patricio Espinosa. "Economic Reforms and Changing Pesticide Policies in Ecuador and Colombia." In *Economic, Environmental, and Health Tradeoffs in Agriculture: Pesticides and the Sustainability of Andean Potato Production,* edited by Charles C. Crissman, John M. Antle, and Susan M. Capalbo, pp. 121–42. Boston: Kluwer, 1998.

Lee, Ronald D. "Long-Run Global Population Forecasts: A Critical Appraisal." In *Resources, Environment, and Population: Present Knowledge, Future Options,* edited by Kingsley Davis and Mikhail S. Bernstam, pp. 44–71. New York: Oxford University Press, 1991.

Leeming, Frank. *The Changing Geography of China.* Cambridge: Blackwell, 1993.

Lemaga, Berga. "Integrated Control of Bacterial Wilt: Literature Review and Work Plan, 1995–1997." *Technical Report Series No.3.* Nairobi: African Highland Initiative, 1997.

Lemaga, Berga, J. J. Hakiza, F. O. Alacho, and R. Kakuhenzire. "Integrated Control of Potato Bacterial Wilt in Southwestern Uganda." Working paper. Kabale: CIP, 1997.

Lemonik, Michael D. "Ancient Odysseys." *Time,* February 13, 1995, pp. 64–67.

Levenstein, Harry. *Paradox of Plenty: A Social History of Eating in America.* New York: Oxford University Press, 1993.

Levy, Leon, and Jeff Madrick. "Hedge Fund Mysteries." *New York Review,* December 17, 1998, pp. 73–77.

Lewontin, R. C. "The Confusion Over Cloning." *New York Review,* October 23, 1997, pp. 18–23.

Li, Paul H., ed. *Plant Cold Hardiness.* New York: Alan R. Liss, 1987.

Lisińska, G., and W. Leszczyński, *Potato Science and Technology.* New York: Elsevier, 1989.

Logan, William Bryant. *Dirt: The Ecstatic Skin of the Earth.* New York: Riverhead, 1995.

Lynch, John. *Spain Under the Hapsburgs.* 2 vols. Oxford: Basil Blackwell, 1964–1969.

Lyons, Thomas P. "Feeding Fujian: Grain Production and Trade 1986–1996." *China Quarterly* 155 (September, 1998): 513–45.

MacArthur, Sir William P. "Medical History of the Famine." In *The Great Famine: Studies in Irish History, 1845–1852*, edited by R. Dudley Edwards and T. Desmond Williams, pp. 263–315. 1956. New ed. with an introduction by Cormac Ó. Gráda. Dublin: Lilliput Press, 1994.

MacDonagh, Oliver. "Irish Overseas Emigration during the Famine." In *The Great Famine: Studies in Irish History, 1845–1852*, edited by R. Dudley Edwards and T. Desmond Williams, pp. 319–88. 1956. New ed. with an introduction by Cormac Ó. Gráda. Dublin: Lilliput Press, 1994.

Machin, David, and Solveig Nyvold, eds. *Roots, Tubers, Plantains, and Bananas in Animal Feeding*. Rome: FAO, 1992.

Mackay, George R. "Resistance: The Foundation of Integrated Pathogen Management of Late Blight (*Phytophthora infestans*)." *CIP Circular* (June, 1996): 2–5.

Madrick, Jeff. "In the Shadow of Prosperity." *New York Review*, August 14, 1997, pp. 40–44.

Maldonado, Luis A., Julia E. Wright, and Gregory J. Scott. "Constraints to Production and Use of Potato in Asia." *American Journal of Potato Research* 75 (March–April, 1998): 71–79.

Malthus, Thomas Robert. *An Essay on the Principle of Population*. Edited and with criticism by Philip Appleman. New York: W. W. Norton, 1976.

Malthus, Thomas Robert. *An Essay on the Principle of Population*. Selected and introduced by Donald Winch. Cambridge: Cambridge University Press, 1992.

Mamani, Mauricio. "El *chuño*: preparación, uso, almacenamiento." In *La tecnología en el mundo andino*, edited by Heather Lechtman and Ana María Soldi, pp. 235–46. México: Universidad Nacional Autónoma de México, 1981.

Mann, Jim, and A. Stewart Truswell, eds. *Essentials of Human Nutrition*. Oxford: Oxford University Press, 1998.

Manning, Anita. "Insects Penetrate Genetically Engineered Cotton." *USA Today*, July 23, 1996, p. 10-b.

Marshack, Alexander. *The Roots of Civilization*. Rev. ed. Mount Kisco, New York: Moyer Bell, 1991.

Martin, Phyllis A. W. "An Iconoclastic View of *Bacillus thuringiensis* Ecology." *American Entomologist* (summer, 1994): 85–90.

Massing, Michael. "Crime and Drugs: The New Myths." *New York Review*, February 1, 1996, pp. 16–20.

Masuda, Shozo, Izumi Shimada, and Craig Morris, eds. *Andean Ecology and Civilization: An Interdisciplinary Perspective on Andean Ecological Complementarity*. Tokyo: University of Tokyo Press, 1985.

McChesney, Robert W. *Rich Media, Poor Democracy: Communication Politics in Dubious Times*. Urbana: University of Illinois Press, 1999.

McDevitt, Thomas M., ed. *World Population Profile: 1996*. Washington, D.C.: U.S. Government Printing Office, 1996.

———. *World Population Profile: 1998*. Washington, D.C.: U.S. Government Printing Office, 1999.

McDowell, R. B. "Ireland on the Eve of the Famine." In *The Great Famine: Studies in Irish History, 1845–1852*, edited by R. Dudley Edwards and T. Desmond Williams, pp. 3–86. 1956. New ed. with an introduction by Cormac Ó. Gráda. Dublin: Lilliput Press, 1994.

McHugh, Roger J. "The Famine in Irish Oral Tradition." In *The Great Famine: Studies in Irish History, 1845–1852*, edited by R. Dudley Edwards and T. Desmond Williams, pp. 391–436. 1956. New ed. with an introduction by Cormac Ó. Gráda. Dublin: Lilliput Press, 1994.

McKibben, Bill. "A Special Moment in History." *Atlantic Monthly*, May, 1998, pp. 55–78.

McMullen, Neill. *Seeds and Agricultural Progress.* Washington, D.C.: National Planning Association, 1987.

McNeill, William H. "American Food Crops in the Old World." In *Seeds of Change,* edited by Herman J. Viola and Carlyn Margolis, pp. 43–59. Washington, D.C.: Smithsonian Institution Press, 1991.

Meegama, S. A. "Malaria Eradication and Its Effects on Mortality Levels." *Population Studies* 21 (November, 1967): 207–37.

Michigan State University. "2000 Potato Disease Weather Monitoring." Michigan State University, Potato Web Portal. Available at www.potato.msu.edu.

Moawad, G. M., R. El-Bedewy, H. K. M. Bekheit, and A. Lagnaoui. "Biological Control of the Potato Tuber Moth, *Phthorimaea operculella* (Zeller) in Potato Fields and Storage." Working paper. Lima: CIP, 1996.

Moawad, G. M., R. El-Bedewy, S. Abd El-Halim, H. K. M. Bekheit, A. Mabrouk, A. Farghaly, and A. Lagnaoui. "Large-Scale Implementation of Integrated Pest Management in Egypt." *CIP Circular* (April, 1997): 8–13.

Mooch, Joyce Lewinger, and Robert E. Rhoades. *Diversity, Farmer Knowledge, and Sustainability.* Ithaca: Cornell University Press, 1992.

Mooney, Patrick. "The Hidden 'Hot Zone': An Epidemic in Two Parts." Rural Advancement Foundation International (RAFI), *Occasional Paper Series* 2 (August, 1995): 1–12.

Moore, H. C. "Evidence that Certified Seed is Improved Seed." *Proceedings of the Annual Meetings of the Potato Association of America* 11 (1924): 26–27.

Mujica, Elías. "Terrace Culture and Pre-Hispanic Traditions." *CIP Circular* (August, 1995): 11–18.

Murra, John V. "The Historic Tribes of Ecuador." In *Handbook of South American Indians,* edited by Julian H. Steward. Vol. 2. *Andean Societies,* pp. 785–822. Washington, D.C.: U.S. Government Printing Office, 1946.

———. "El control vertical de un máximo de pisos ecológicos en la economía de las sociedades andinas." In *Formaciones económicas y políticas del mundo andino,* edited by John V. Murra, pp. 59–115. Lima: Instituto de Estudios Peruanos, 1975.

———. *The Economic Organization of the Inka State.* Greenwich, Connecticut: JAI Press, 1980.

———. "Andean Societies Before 1532." In *The Cambridge History of Latin America,* edited by Leslie Bethell. Vol 1., *Colonial Latin America,* pp. 59–91. London: Cambridge University Press, 1984.

———. "'El Archipiélago Vertical' Revisited." In *Andean Ecology and Civilization: An Interdisciplinary Perspective on Andean Ecological Complementarity,* edited by Shozo Masuda, Izumi Shimada, and Craig Morris, pp. 3–13. Tokyo: University of Tokyo Press, 1985.

———, ed. *Formaciones económicas y políticas del mundo andino.* Lima: Instituto de Estudios Peruanos, 1975.

Nabhan, Gary Paul. *The Desert Smells Like Rain.* San Francisco: North Point Press, 1982.

———. *Gathering the Desert.* Tucson: University of Arizona Press, 1985.

———. "Native Crops of the Americas: Passing Novelties or Lasting Contributions to Diversity?" In *Chilies to Chocolate: Food the Americas Gave the World,* edited by Nelson Foster and Linda S. Cordell, pp. 143–61. Tucson: University of Arizona Press, 1992.

National Potato Council (NPC). *Potato Statistical Yearbook.* Englewood, Colorado: NPC, various years.

Nelson, Roxanne. "The Blight is Back." *Scientific American,* June, 1998, pp. 20, 26.

Nolte, Phil. "Potato Pointers." *GILB Newsletter* (December, 1998): 2–3.

North, John. *Stonehenge: Neolithic Man and the Cosmos.* London: HarperCollins, 1996.

Nottingham, Stephen. *Eat Your Genes: How Genetically Modified Food Is Entering Our Diet.* New York: Zed Books, 1998.

Nova Scotia Department of Agriculture and Marketing. "Potato Late Blight Forecast." September 2, 1998. Available at http://agri.gov.ns.ca/pt/hort/updates/2potatobl.htm.

Novy, Richard G., G. A. Secor, B. L. Farnsworth, J. H. Lorenzen, E. T. Holm, D. A. Preston, N. C. Gudmestad, and J. R. Sowokinos. "Nor-Valley: A White-Skinned Chipping Cultivar With Cold-Sweetening Resistance," *American Journal of Potato Research* 75 (March–April 1998): 101–105.

Nowlan, Kevin B. "The Political Background." In *The Great Famine: Studies in Irish History 1845–1852,* edited by R. Dudley Edwards and T. Desmond Williams, pp. 131–206. 1956. New ed. with an introduction by Cormac Ó. Gráda. Dublin: Lilliput Press, 1994.

Ochoa, Carlos M. *The Potatoes of South America: Bolivia.* Translated by Donald Ugent. Cambridge: Cambridge University Press, 1990.

———. "The Andes, Cradle of the Potato." *Diversity* 7 (1991): 45–47.

———. *Las Papas de Sudamérica: Perú.* Lima: CIP, 1999.

Ochoa, Laurie. "Potato: Saving Peru's Dwindling Diversity." *Los Angeles Times,* August 26, 1993, pp. H10–H11.

Ofcansky, Thomas P. *Uganda: Tarnished Pearl of Africa.* Boulder, Colorado: Westview Press, 1996.

O'Neill, Thomas P. "The Organization and Administration of Relief, 1845–1852." In *The Great Famine: Studies in Irish History 1845–1852,* edited by R. Dudley Edwards and T. Desmond Williams, pp. 209–59. 1956. New ed. with an introduction by Cormac Ó. Gráda. Dublin: Lilliput Press, 1994.

Onuki, Yoshiro. "The 'Yunga' Zone in the Prehistory of the Central Andes: Vertical and Horizontal Dimensions in Andean Ecological and Cultural Processes." In *Andean Ecology and Civilization: An Interdisciplinary Perspective on Andean Ecological Complementarity,* edited by Shozo Masuda, Izumi Shimada, and Craig Morris, pp. 339–56. Tokyo: University of Tokyo Press, 1985.

Oregon State University. "Disease Forecasting." Available at www.bcc.orst.edu/lateblight/forecasting.htm.

Ortiz, O., J. Alcázar, W. Catalán, W. Villano, V. Cerna, H. Fano, and T. S. Walker. "Economic Impact of IPM Practices on the Andean Potato Weevil in Peru." In *Case Studies of the Economic Impact of CIP-Related Technologies,* edited by Thomas Walker and Charles Crissman, pp. 95–110. Lima: CIP, 1996.

Ortiz, O., P. Winters, H. Fano, G. Thiele, S. Guamán, R. Torrez, V. Barrera, J. Unda, and J. Hakiza. "Understanding Farmers' Response to Late Blight: Evidence from Peru, Bolivia, Ecuador, and Uganda." In *Impact on a Changing World: International Potato Center Program Report 1997–1998,* pp. 101–109. Lima: CIP, 1999.

Ortiz de Montellano, Bernard R. *Aztec Medicine, Health, and Nutrition.* New Brunswick: Rutgers University Press, 1990.

Ortloff, C. R. "La ingeniería hidráulica chimú." In *La tecnología en el mundo andino,* edited by Heather Lechtman and Ana María Soldi, pp. 91–134. Mexico City: Universidad Nacional Autónoma de México, 1981.

Otim-Nape, G. W., J. M. Thresh, and D. Fargette. "*Bemisia tabaci* and Cassava Mosaic Virus Disease in Africa." In *Bemisia 1995: Taxonomy, Biology, Damage, Control and Management,* edited by D. Gerling and R. T. Meyer, pp. 319–50. Hants, U.K.: Intercept, 1996.

Otim-Nape, G. W., A. Bua, J. M. Thresh, Y. Baguma, S. Ogwal, G. N. Semakula, G. Acola, B. Byabakama, and A. Martin. *Cassava Mosaic Virus Disease in Uganda. The Current Pandemic of and Approaches to Control.* Chatham, U.K.: Natural Resource Institute, 1997.

Pallais, Noël. "True Potato Seed Quality." *Theoretical and Applied Genetics* 73 (1987): 784–92.

———. "Origin of the International Potato Center." *HortScience* 26 (1991): 230, 323.

———. "True Potato Seed: A Global Perspective. *CIP Circular* (March, 1994): 2–4.

Pearsall, Deborah M. "The Origins of Plant Cultivation in South America." In *The Origins of Agriculture: An International Perspective,* edited by C. Wesley Cowan and Patty Jo Watson, pp. 173–205. Washington, D.C.: Smithsonian Institution Press, 1992.

Percival, John. *The Great Famine 1845–1851.* London: BBC Books, 1995.

Peters, Dai, and Christopher Wheatley. "Small Scale Agro-Enterprises Provide Opportunities for Income Generation: Sweetpotato Flour in East Java, Indonesia." *Quarterly Journal of International Agriculture* 36 (1997): 331–52.

Petersen, William. *Population.* 2nd ed., rev. New York: Macmillan, 1967.

Peterson, Peter G. "Gray Dawn: The Global Aging Crisis." *Foreign Affairs,* January/February 1999, pp. 42–55.

Peterson, Roger Tory. *Field Guide to Wildflowers of Northeastern and North-Central North America.* Boston: Houghton Mifflin, 1968.

Petit, Charles W. "Rediscovering America." *U.S. News and World Report,* October 12, 1998, pp. 56–64.

Piperno, Dolores R., and Deborah M. Pearsall. *The Origins of Agriculture in the Lowland Neotropics.* New York: Academic Press, 1998.

Plaisted, R. L., H. A. Mendoza, H. D. Thurston, E. E. Ewing, B. B. Brodie, and W. M. Tingey. "Broadening the Range of Adaptation of Andigena (Neo-Tuberosum) Germplasm, *CIP Circular* (June, 1987): 1–5.

Plaisted, R. L., D. E. Halseth, B. B. Brodie, S. A. Slack, J. B. Sieczka, B. J. Christ, K. M. Paddock, and M. W. Peck. "Andover: An Early to Midseason Golden Nematode Resistant Variety for Use as Chipstock or Tablestock." *American Journal of Potato Research* 75 (May–June, 1998): 113–16.

———. "Pike: A Full Season Scab and Golden Nematode Resistant Chipstock Variety." *American Journal of Potato Research* 75 (May–June 1998): 117–20.

Plutarch, *The Lives of the Noble Grecians and Romans.* Translated by James Amyot and Thomas North. 2 vols. Norwalk: Easton Press, 1940.

Póirtéir, Cathal. Introduction to *The Great Irish Famine,* edited by Cathal Póirtéir, pp. 9–17. Dublin: Mercier Press, 1995.

———, ed. *The Great Irish Famine.* Dublin: Mercier Press, 1995.

Pollan, Michael. "Playing God in the Garden." *New York Times Magazine,* October 25, 1998, pp. 44–51, 62–63, 82, 92–93.

Population Reference Bureau (PRB). *World Population Data Sheet.* Pamphlet. Washington, D.C.: PRB, various years.

———. *Success in a Challenging Environment: Fertility Decline in Bangladesh.* Washington, D.C.: PRB (1993).

Posner, Joshua. "Ecoregional Research: A Vision from the Andes." In *Impact on a Changing World: International Potato Center Program Report 1997–1998,* pp. 409–12. Lima: CIP, 1999.

"Possible Threat to Monarch Butterfly Posed by Bt Corn." *Diversity* 15 (1999): 17–18.

Powelson, Mary R., and Debra A. Inglis. "Seed Piece Treatment Key to Protecting Against Late Blight." Oregon State University. Available at www.bcc.orst.edu/lateblight/press-release.htm.

Prakash, B.A. *Kerala's Economy.* New Delhi: Sage, 1994.

Preston, Richard. *The Hot Zone.* New York: Random House, 1993.

Price, Roger. "Poor Relief and Social Crisis in Mid-Nineteenth-Century France." *European Studies Review* 13 (October, 1983): 423–54.

PROINPA (Programa de Investigación de la Papa). *Producción de tubérculos en cama protegida.* Technical manual. Cochabamba: PROINPA-IBTA, 1995.

PROINPA and IBTA (Instituto Boliviano de Tecnologia Agropecuaria). *Informe Anual Compendio 1993–1994.* Cochabamba: PROINPA, 1994.

————. *Informe Anual Compendio 1994–1995.* Cochabamba: PROINPA, 1996.

————. *Informe Compendio del Programa de Investigación de la Papa 1996–1998.* Cochabamba: PROINPA, 1998.

Prosterman, Roy L., Tim Hanstad, and Li Ping. "Can China Feed Itself?" *Scientific American,* November, 1996, pp. 70–76.

Protzen, Jean-Pierre. "Inca Quarrying and Stonecutting." *Nampa Pacha* 21 (1983): 183–214.

Pulgar Vidal, Javier. *Geografía del Perú: Las Ocho Regiones Naturales.* 9th ed. Lima: Editorial Inca, 1987.

Quinn, Daniel. *The Story of B.* New York: Bantam, 1996.

Quiroga, Jorge, and Greta Watson. "Diagnósticos multidisciplinarios y su rol en el establecimiento de prioridades mediante investigación agrícola." In *Memorias del simposio Latinoamericano sobre investigación y extensión y sistemas agropequarios,* pp. 466–76. Quito: FUNAGRO, 1993.

Quirós, C. F., R. Ortega, L. van Raamsdonk, M. Herrera-Montoya, P. Cisneros, E. Schmidt, and S. A. Bush. "Increase of Potato Genetic Resources in Their Center of Diversity: The Role of Natural Outcrossings and Selection by the Andean Farmer." *Genetic Resources and Crop Evolution* 39 (1992): 107–13.

Quirós, C. F., and R. Aliaga Cárdenas. "Maca." In. *Andean Roots and Tubers: Ahipa, Arracacha, Maca, and Yacon,* edited by M. Hermann and J. Heller, pp. 173–197. Rome: International Plant Genetic Resources Institute, 1997.

Raman, K. V. "Potato Tuber Moth." Technical information bulletin no. 3. Lima: CIP, 1980.

————. "Integrated Insect Pest Management for Potatoes in Developing Countries." *CIP Circular* (March, 1988): 1–8.

————. "Potato Pest Management in Developing Countries." In *Advances in Potato Pest Biology and Management,* edited by G. W. Zehnder, M. L. Powelson, R. K. Jansson, and K. V. Raman, pp. 583–96. St. Paul, Minnesota: American Phytopathological Society, 1994.

Raman, K. V. and J. Alcázar. "Biological Control of Potato Tuber Moth Using Phthorimaea Baculovirus." Training bulletin. Lima: CIP, 1992.

Rastovski, A., and A. van Es, eds. *Storage of Potatoes: Post-Harvest Behavior, Store Design, Storage Practice, and Handling.* Wageningen: Center for Agricultural Publishing and Documentation, 1987.

Reed, Charles A., ed. *Origins of Agriculture.* The Hague: Mouton Publishers, 1977.

Revalo, J., S. Garcés, and J. Andrade. "Resistance of Commercial Potato Varieties to Attack by *Phytophthora infestans* in Ecuador." *CIP Circular* (April, 1997): 20–21.

Rhoades, Robert E. "The Incredible Potato." *National Geographic,* May, 1982, pp. 668–94.

Rhoades, Robert E., and Douglas E. Horton. "Past Civilizations, Present World Needs, and Future Potential: Root Crop Agriculture Across the Ages." In *Proceedings of the Eighth Symposium of the International Society for Tropical Root Crops,* edited by CIP, pp. 8–19. Lima: CIP, 1989.

Ringle, Ken. "Raider of the Lost Spud." *Washington Post,* December 1, 1992.

Rissler, Jane, and Margaret Mellon. *The Ecological Risks of Engineered Crops.* Cambridge, Massachusetts: MIT Press, 1996.

Ristaino, J. B., G. R. Parra, and C. Trout Groves. "PCR Amplification of *Phytophthora infestans* from 19th Century Herbarium Specimens." *GILB Newsletter* (April, 2000): 4.

Roosevelt, Anna C. *Moundbuilders of the Amazon: Geophysical Archaeology on Marajo Island, Brazil.* New York: Academic Press, 1991.

Roosevelt, Anna C., M. Lima da Costa, C.Lopes Machado, M. Michab, N. Mercier, H. Valladas, J. Feathers, W. Barnett, M. Imazio da Silveira, A. Henderson, J. Sliva, B. Chernoff, D. S. Reese, J. A. Holman, N. Toth, and K. Schick. "Paleoindian Cave Dwellers in the Amazon: The Peopling of the Americas." *Science* 272 (April 19, 1996): 373–84.

Rosenblat, Ángel. "The Population of Hispaniola at the Time of Columbus." In *The Native Population of the Americas in 1492*, edited by William M. Denevan, pp. 43–66. Madison: University of Wisconsin Press, 1976.

Rothkopf, David J. "The Disinformation Age." *Foreign Policy* (spring, 1999): 83–96.

Roush, Richard T., and Ward M. Tingey. "Strategies for the Management of Insect Resistance to Synthetic and Microbial Insecticides." In *Advances in Pest Biology and Management*, edited by G. W. Zehnder, M. L. Powelson, R. K. Jansson, and K. V. Raman, pp. 237–54. St. Paul, Minnesota: American Phytopathological Society, 1994.

Rowe, John Howland. "Inca Culture at the Time of the Spanish Conquest." In *Handbook of South American Indians*, edited by Julian H. Steward. Vol. 2., *Andean Civilizations*, pp. 183–330. Washington, D.C.: U.S. Government Printing Office, 1946.

Rowe, Randall C., Sally A. Miller, and Richard Riedel. "Late Blight of Potato and Tomato." Ohio State University Factsheet HYG-3102-95. Available from www.ag.ohio-state.edu.

Roy, S. K. Bardhan, A. K. Chakravorty, and A. K. Roy. "Farmers' Participatory On-Farm Trials with TPS Seedling Tubers in the Rice-Based Cropping Systems of West Bengal." In *Production and Utilization of True Potato Seed in Asia*, edited by M. D. Upadhya et al., pp. 187–89. Lima: CIP, 1996.

Roy, S. K. Bardhan, T. Walker, V. S. Khatana, N. K. Saha, V. S. Verma, M. S. Kadian, A. J. Haverkort, and W. Bowen. "Intensification of Potato Production in Rice-Based Cropping Systems: A Rapid Rural Appraisal in West Bengal." In *Impact on a Changing World: International Potato Center Program Report 1997–1998*, pp. 205–12. Lima: CIP, 1999.

Rozin, Elizabeth. *The Primal Cheeseburger*. New York: Penguin, 1994.

Rudgley, Richard. *Essential Substances: A Cultural History of Intoxicants in Society.* Tokyo: Kodansha International, 1993.

———. *The Lost Civilizations of the Stone Age*. New York: Free Press, 1999.

Rueda, J. L., P. T. Ewell, T. S. Walker, M. Soto, M. Bicamumpaka, and D. Berríos. "Economic Impact of High-Yielding, Late-Blight Resistant Varieties in the Eastern and Central African Highlands." In *Case Studies of the Economic Impact of CIP-Related Technology*, edited by Thomas S. Walker and Charles C. Crissman, pp. 15–30. Lima: CIP, 1996.

Ruspoli, Mario. *The Cave of Lascaux*. New York: Harry N. Abrams, 1987.

Salaman, Redcliffe. *The History and Social Influence of the Potato*. Revised with an introduction by J. G. Hawkes. 1949. Cambridge: Cambridge University Press, 1986.

Salazar, Luis F. *Potato Viruses and Their Control*. Lima: CIP, 1996.

Salunkhe, D. K., S. S. Kadam, and S. J. Jadhav. *Potato: Production, Processing, and Products.* Boca Raton, Florida: CRC Press, 1991.

Sanabria, Harry. *The Coca Boom and Rural Social Change in Bolivia*. Ann Arbor: University of Michigan Press, 1993.

Sangwan, S. S. *Production and Marketing of Potato in India: A Case Study of Uttar Pradesh*. New Delhi: Mittal Publications, 1991.

Sarkar, N. K. *The Demography of Ceylon*. Colombo: Ceylon Government Press, 1957.

Sattaur, Omar. "The Lost Art of Waru Waru," *New Scientist* 118 (May 12, 1988): 50–51.

Sauer, Carl O. "Geography and Plant and Animal Resources." In *Handbook of South American Indians*, edited by Julian H. Steward. Vol. 6, *Physical Anthropology, Linguistics, and Cultural Geography of South American Indians*, pp. 487–543. Washington, D.C.: U.S. Government Printing Office, 1950.

———. *Agricultural Origins and Dispersals*. New York: American Geographical Society, 1952.

Sauer, Jonathan D. "Changing Perception and Exploitation of New World Plants in Europe, 1492–1800." In *First Images of the Americas: The Impact of the New World on the Old*, edited by Fredi Chiappelli, 2 vols., pp. 813–32. Los Angeles: University of California Press, 1976.

———. *Historical Geography of Crop Plants: A Select Roster*. Ann Arbor, Mich.: CRC Press, 1993.

Schmiediche, Peter. "Report on a Trip to Vietnam from 13 to 18 January 1997 to Attend a Workshop on the ADB-Financed TPS Project." CIP memorandum, April 4, 1997.

Schultes, Richard Evans. "Amazonian Cultigens and Their Northward and Westward Migration in Pre-Columbian Times." In *Pre-Columbian Plant Migrations*, edited by Doris Stone, pp. 17–37. Cambridge, Massachusetts: Harvard University Press, 1984.

Scott, Gregory J., J. Otieno, S. B. Ferris, A. K. Muganga, and L. Maldonado. "Sweetpotato in Ugandan Food Systems: Enhancing Food Security and Alleviating Poverty." In *Impact on a Changing World: International Potato Center Program Report 1997–1998*, pp. 337–47. Lima: CIP, 1999.

Scott, Gregory J., Mark W. Rosegrant, and Claudia Ringler, *Roots and Tubers for the 21st Century: Trends, Projections, and Policy Options*. Washington, D.C.: International Food Policy Research Institute, 2000.

Sellers, Jane B. *The Death of Gods in Ancient Egypt*. London: Penguin, 1992.

Sen, Amartya. "Population, Delusion, and Reality." *New York Review*, September 22, 1994, pp. 62–71.

Sharma, T. R., B. M. Goydani, and R. C. Sharma. "True Potato Seed: A Cheaper and Better Alternative to Seed Tubers." In *Production and Utilization of True Potato Seed in Asia*, edited by M. D. Upadhya et al., pp. 119–25. Lima: CIP, 1996.

Shell, Ellen Ruppel. "Resurgence of a Deadly Disease." *Atlantic Monthly*, April, 1997, pp. 45–60.

Sherrill, Robert. "S & Ls, Big Banks, and Other Triumphs of Capitalism." *Nation*, November 19, 1990, pp. 589–623.

Singer, Max. "The Population Surprise." *Atlantic Monthly*, August, 1999, pp. 22–25.

Singh, Mukhtar. "The Potato in Retrospect and Prospect in India." *Journal of the Indian Potato Association* 1 (1974): 6–10.

Smil, Vaclav. *China's Environmental Crisis: An Inquiry into the Limits of National Development*. London: M. E. Sharpe, 1993.

———. "Who Will Feed China?" *The China Quarterly* 143 (September, 1995): 801–14.

———. "Is There Enough Chinese Food?" *New York Review*, February 1, 1996, pp. 32–34.

———. *Feeding the World: A Challenge for the Twenty-First Century*. Cambridge, Massachusetts: MIT Press, 2000.

Smit, Nicole, and B. Odongo. "Integrated Management for Sweetpotato in East Africa." In *International Potato Center Program Report 1995–1996*, pp. 191–97. Lima: CIP, 1997.

Smith, Bruce D. "The Initial Domestication of *Cucurbita pepo* in the Americas 10,000 Years Ago." *Science* 276 (May 9, 1997): 932–34.

Snack Food Association (SFA). *50 Years: A Foundation for the Future*. Alexandria, Virginia: SFA, 1987.

Solow, Robert M. "How to Stop Hunger." *New York Review*, December 5, 1991, pp. 22–24.

Sørensen, Martin. *Yam Bean*. Rome: International Plant Genetics Resources Institute, 1996.

Sørensen, Martin, Wolfgang J. Grüneberg, and Bo Ørting. "Ahipa." In *Roots and Tubers: Ahipa, Arracacha, Maca, and Yacon*, edited by M. Hermann and J. Heller, pp. 13–74. Rome: International Plant Genetics Resources Institute, 1997.

Souden, David. *Stonehenge Revealed*. New York: Facts on File, 1997.

Sowokinos, Joe. "To Chip or Not to Chip." *Valley Potato Grower* (February, 1994): 36–37.

Sperling, Louise. "Farmer Participation and the Development of Bean Varieties in Rwanda." In *Diversity, Farmer Knowledge, and Sustainability*, edited by Joyce Lewinger Moock and Robert E. Rhoades, pp. 96–112. Ithaca: Cornell University Press, 1992.

Stahl, Peter, ed. *Archaeology in the Lowland American Tropics*. Cambridge: Cambridge University Press, 1995.

Stanley, Doris. "Potatoes Once Again under Attack." *Agricultural Research* 45 (May, 1997): 10–13.

Stone, Doris, ed. *Pre-Columbian Plant Migrations*. Cambridge, Massachusetts: Harvard University Press, 1984.

Stuart, William. "The Value of Good Seed Potatoes." *American Potato Journal* 3 (1926): 83–85.

Sutherland, J. A. "A Review of the Biology of the Sweetpotato Weevil *Cylas formicarius*." *Tropical Pest Management* 32 (1986): 304–15.

Talbot, Charles H. "America and the European Drug Trade. In *First Images of the Americas: The Impact of the New World on the Old*, edited by Fredi Chiappelli, 2 vols., pp. 833–44. Los Angeles: University of California Press, 1976.

Tapia, Mario, H. Gandarillas, S. Alandia, A. Cardozo, and A. Mujica. *Quinua y kañiwa: cultivos andinos*. Bogotá: Instituto Interamericano de Ciencias Agrícolas, 1979.

Tapia, Mario, and A. Rosas. "Seed Fairs in the Andes: A Strategy of Local Conservation of Plant Genetic Resources." In *Cultivating Knowledge: Genetic Diversity, Farmer Experimentation, and Crop Research*, edited by Walter de Boeuf, Kojo Amanor, and Kate Wellard, pp. 111–18. London: Intermediate Technology Publishers, 1993.

Thakur, K. C., M. D. Upadhya, S. N. Bhargava, and A. Bhargava. "Bulk pollen extraction procedures and the potency of the extracted pollen." *Potato Research* 37 (1994): 245–48.

Thiele, Graham. "Informal Seed Systems in the Andes: Why Are They Important and What Should We Do With Them?" *World Development* 27 (1999): 83–99.

Thiele, Graham, André Devaux, and Carlos Soria. "Innovación tecnológica en la papa: de la oferta de tecnología al impacto macro-económico." Cochabamba: PROINPA, 1995. Mimeographed.

Thiele, Graham, G. Watson, and R. Torrez. "Cómo y dónde involucrar agricultores en la selección de variedades." Cochabamba: PROINPA, 1996. Mimeographed.

Thomas, Lewis. *The Lives of a Cell: Notes of a Biology Watcher*. New York: Viking, 1974.

Thomas, Michael M. "The Greatest American Shambles." *New York Review*, January 31, 1991, pp. 30–35.

Thompson, Warren. "Population," *American Journal of Sociology* 34 (1929): 959–75.

Thurow, Lester. "Asia: the Collapse and the Cure." *New York Review*, February 5, 1998, pp. 22–26.

Time-Life, eds. *The Human Dawn*. New York: Time-Life, 1990.

Topping, Audrey R. "Ecological Roulette: Damming the Yangtze." *Foreign Affairs*, September/October, 1995, pp. 132–46.

Torrez, R., J. Tenorio, C. Valencia, R. Orrego, O. Ortiz, R. Nelson, and G. Thiele. "Implementing IPM for Late Blight in the Andes." In *Impact on a Changing World: International Potato Center Program Report 1997–1998*, pp. 91–99. Lima: CIP, 1999.

Toussaint-Samat, Maguelonne. *A History of Food*. Translated by Anthea Bell. Cambridge: Blackwell, 1992.

Trefil, James. "Architects of Time." *Astronomy*, September, 1999, pp. 48–53.

Trognitz, Bodo, G. Forbes, and B. Hardy. "Resistance to Late Blight of Potato from Wild Species." *CIP Circular* (April, 1995): 6–10.

Trognitz, Bodo, M. Ghislain, C. Crissman, and B. Hardy. "Breeding Potatoes with Durable Resistance to Late Blight." *CIP Circular* (June, 1996): 6–9.

Troll, Carl. "The Cordilleras of the Tropical Americas: Aspects of Climatic, Phytogeographical, and Agrarian Ecology. In *Geo-Ecology of the Mountainous Regions of the Tropical Americas*, edited by Carl Troll, pp. 15–56. Bonn: Ferd. Dümmlers Verlag, 1968.

Tucker, John. "The Value of Seed Potato Certification to the Potato Industry." *American Potato Journal* 14 (1937): 39–45.

Tunali, Odin. "A Billion Cars: The Road Ahead." *World Watch* (January/February, 1996): 24–33.

Ugent, Donald, Tom Dillehay, and Carlos Ramírez. "Potato Remains from a Late Pleistocene Settlement in Southcentral Chile." *Economic Botany* 41 (1987): 17–27.

Ugent, Donald, and Linda W. Peterson. "Archaeological Remains of Potato and Sweet Potato in Peru." *CIP Circular* (September, 1988): 1–10.

United Nations (UN). *1979 Yearbook of World Energy Statistics.* New York: UN, 1981.

———. *1996 Yearbook of World Energy Statistics.* New York: UN, 1994.

———. *Demographic Yearbook.* New York: United Nations, various years.

United Nations Development Program (UNDP). *Human Development Report 1998.* New York: Oxford University Press, 1998.

United States Agency for International Development (USAID), CARE, and the International Potato Center (CIP). "Así vive el gorgojo de los Andes." Pamphlet. Lima: CARE-USAID-CIP, 1994.

———. "Ocho reglas de oro para el uso seguro de los plaguicidas." Pamphlet. Lima: CARE, 1994.

United States Bureau of the Census. *Historical Statistics of the United States, Colonial Times to 1970.* Bicentennial edition, in two parts. Washington, D.C.: U.S. Government Printing Office, 1975.

———. *Statistical Abstract of the United States.* Washington D.C.: U.S. Government Printing Office, various years.

United States Department of Agriculture (USDA). *Agricultural Statistics.* Washington, D.C.: U.S. Government Printing Office, various years.

United States Department of Agriculture, National Agricultural Statistics Service (NASS), *Agricultural Chemical Usage 1999 Field Crops Summary.* Online database. Washington, D.C.: NASS, 2000. Available at http://usda.mannlib.cornell.edu/reports/nassr/other/pucbb/aghc0500.txt

University of Idaho. "Potato Late Blight." Available at www.uidaho.edu/ag/plantdisease/lbhome.htm.

Upadhya, M. D., ed. *True Potato Seed (TPS) in South and South East Asia: Proceedings of the Regional Workshop for Researchers on True Potato Seed, Jointly Organized by the Indian Council of Agricultural Research and the International Potato Center, New Delhi, India, 4–8 January 1989.* New Delhi: Indian Council of Agricultural Research, 1990.

Upadhya, M. D., B. Hardy, P. C. Gaur, and S. G. Ilangantileke, eds. *Production and Utilization of True Potato Seed in Asia.* Lima: CIP, 1996.

van de Fliert, Elske. *Integrated Pest Management: Farmer Field Schools Generate Sustainable Practices: A Case Study of Central Java.* Wageningen: Veenman Drukkers, 1993.

van de Fliert, Elske, Rini Asmunati, Wiyanto, Yudi Widodo, and Ann R. Braun. "From Basic Approach to Tailored Curriculum: Participatory Development of a Farmer Field School Model for Sweetpotato." UPWARD (User's Perspective with Agricultural Research and Development) Conference paper. Yogyakarta-Java, October 2–5,1995.

van de Fliert, Elske, and Rini Asmunati. "Identification of IPM and IPM Training Needs for Sweetpotato in East and Central Java." Progress Report I. Yogyakarta: CIP, 1995.

Vasey, Daniel E. *An Ecological History of Agriculture.* Ames: Iowa State University Press, 1992.

"Vegetable Vaccines." *Discover,* September, 1998, p. 27.

Verano, John W., and Douglas H. Ubelaker. "Health and Disease in the Pre-Columbian World." In *Seeds of Change,* edited by Herman J. Viola and Carolyn Margolis, pp. 209–23. Washington, D.C.: Smithsonian Institution Press, 1991.

Vietmeyer, Noel D. "Forgotten Roots of the Incas." In *Chilies to Chocolate: Food the Americas Gave the World,* edited by Nelson Foster and Linda S. Cordell, pp. 95–104. Tucson: University of Arizona Press, 1996.

————. *Lost Crops of the Inca: Little-Known Plants of the Andes with Promise for Worldwide Cultivation.* Washington, D.C.: National Academy Press, 1989.

Viola, Herman J. "Seeds of Change." In *Seeds of Change,* edited by Herman J. Viola and Carolyn Margolis, pp. 11–15. Washington, D.C.: Smithsonian Institution Press, 1991.

Viola, Herman J., and Carolyn Margolis, eds. *Seeds of Change.* Washington, D.C.: Smithsonian Institution Press, 1991.

Visaria, Leela, and Pavin Visaria. "India's Population in Transition." *Population Bulletin* 50 (October, 1995): 1–50.

von Hagen, Victor W. "America's Oldest Roads." *Scientific American,* July, 1952, pp. 17–21.

Wahid, Abu N. M. *The Grameen Bank: Poverty Relief in Bangladesh.* Boulder, Colorado: Westview Press, 1993.

Walden, Hilaire. *The Potato Cookbook.* New York: Smithmark, 1995.

Walker, Geoffrey J. *Spanish Politics and Imperial Trade, 1700–1739.* Bloomington: Indiana University Press, 1979.

Walker, Thomas S. "Patterns and Implications of Varietal Change in Potatoes." In *Social Science Department Working Paper Series.* Lima: CIP, 1994. Mimeographed.

————. "Trip Report to Vietnam and China: 18 October–2 November 1997." Internal document. Lima: CIP, 1997.

Walker, Thomas S., and Charles C. Crissman, eds. *Case Studies of the Economic Impact of CIP-Related Technology.* Lima: CIP, 1996.

Walsh, Jane MacLaren, and Yoka Sugiura. "The Demise of the Fifth Sun." In *Seeds of Change,* edited by Herman J. Viola and Carolyn Margolis, pp. 17–41. Washington. D.C.: Smithsonian Institution Press: 1991.

Weeks, John R. *Population: An Introduction to Concepts and Issues.* 7th ed., rev. Belmont, California: Wadsworth, 1999.

Wenke, Robert J. *Patterns in Prehistory.* 3rd ed., rev. New York: Oxford University Press, 1990.

Wesselman, Hank. *Spiritwalker: Messages from the Future.* New York: Bantam, 1995.

West Bengal Agricultural Research Service Association (WBARSA). "Farm Focus: Potato." *Technical Bulletin of the Agricultural Research Service Association.* Calcutta: Galaxy Printers, 1996.

Whalon, Mark E., Utami Rahardja, and Patchara Verakalasa. "Selection and Management of *Bacillus Thuringiensis*–Resistant Colorado Potato Beetle." In *Advances in Potato Pest Biology and Management,* edited by G. W. Zehnder, M. L. Powelson, R. K. Jansson, and K. V. Raman, pp. 309–21. St. Paul, Minnesota: American Phytopathological Society, 1994.

Wheatley, Christopher, Lin Liping, and Song Bofu. "Enhancing the Role of Small-Scale Sweetpotato Starch Enterprises in Sichuan, China." In *International Potato Center Program Report 1995–1996,* pp. 270–75. CIP: Lima, 1997.

Whitmore, Thomas M. *Disease and Death in Early Colonial Mexico: Simulating Amerindian Depopulation.* Boulder, Colorado: Westview, 1991.

Wilkie, James W., Eduardo Alemán, José Guadalupe Ortega, eds. *Statistical Abstract of Latin America.* Vol. 35. Los Angeles: UCLA Latin American Center Publications, 1999.

Willock, Colin. *Africa's Rift Valley.* New York: Time-Life Books, 1974.

Wilson, Jill E., Finau S. Pole, Nicole E. J. M. Smit, and Pita Taufatofua. *Agro-Facts: Sweet Potato Breeding.* Apia, Western Somoa: University of the South Pacific Institute for Research, Extension, and Training in Agriculture (IRETA), 1989.

Wolfe, G. William. "The Origin and Dispersal of the Pest Species *Cylas* with a Key to the Pest Species Groups of the World." In *Sweet Potato Pest Management: A Global Perspective,* edited by Richard K. Jansson and Kandukuri V. Raman, pp. 13–43. Boulder, Colorado: Westview, 1991.

Woolfe, Jennifer A. *The Potato in the Human Diet*. Cambridge: Cambridge University Press, 1987.

———. *Sweet Potato: An Untapped Food Source*. Cambridge: Cambridge University Press, 1992.

World Bank. *World Tables*. Vol. 2, *Social Data*. Washington, D.C.: World Bank, 1983.

———. *World Development Report 1984*. New York: Oxford University Press, 1984.

———. *World Development Report 1994*. Washington, D.C.: Oxford University Press, 1994.

———. *World Development Report 1998–1999: Knowledge for Development*. Washington, D.C.: Oxford University Press, 1999.

———. *Entering the 21st Century: World Development Report 1999/2000*. Washington, D.C.: Oxford University Press, 2000.

Wright, Gavin. *Old South New South: Revolutions in the Southern Economy Since the Civil War*. New York: Basic Books, 1986.

Yamamoto, Norio. "The Ecological Complementarity of Agro-Pastoralism: Some Comments." In *Andean Ecology and Civilization: An Interdisciplinary Perspective on Andean Ecological Complementarity*, edited by Shozo Masuda, Izumi Shimada, and Craig Morris, pp. 85–99. Tokyo: University of Tokyo Press, 1985.

Yang, Dali L. *Calamity and Reform in China: State, Rural Society, and Institutional Reform Since the Great Leap Forward*. Stanford: Stanford University Press, 1996.

Young, Allen M. *The Chocolate Tree: A Natural History*. Washington, D.C.: Smithsonian Institution Press, 1994.

Yunas, Muhammad, with Alan Jolis. *Banker to the Poor*. London: Arum Press, 1998.

Zalewski, James. "Russet Burbank: Is It Here to Stay?" *Valley Potato Grower* 112 (1992): 20–24.

Zeballos, H. Hernán. *Aspectos económicos de la producción de papa en Bolivia*. Lima: CIP, 1997.

Zehnder, G. W., M. L. Powelson, R. K. Jansson, and K. V. Ramanet, eds. *Advances in Pest Biology and Management*. St. Paul, Minnesota: American Phytopathological Society, 1994.

Zhang, Dapeng, Marc Ghislain, Zósimo Huamán, Ali Golmirzaie, and Robert Hijmans. "RADP Variation in Sweetpotato (*Ipomoea batatas* (L) Lam) Cultivars from South America and Papua New Guinea." *Genetic Resources and Crop Evolution* 45 (1998): 271–77.

Zhang, L., J. F. Guenthner, R. B. Dwelle, and J. C. Foltz. "U.S. Opportunities in China's Frozen French Fry Market." *American Journal of Potato Research* 76 (September–October 1999): 297–304.

Zimmerer, Karl S. *Changing Fortunes: Biodiversity and Peasant Livelihood in the Peruvian Andes*. Los Angeles: University of California Press, 1996.

Zimmerer, Karl S., and David S. Douches. "Geographical Approaches to Crop Conservation: Partitioning of Genetic Diversity in Andean Potatoes." *Economic Botany* 45 (1991): 176–89.

Zuckerman, Larry. *The Potato: How the Humble Spud Rescued the Western World*. Boston: Faber and Faber, 1998.

Index

Chata Rota variety of potato, 169
chaucha species of potato, *77*
Chauvet cave site, 9–10
Chejchi variety of potato, 95
chemical controls, bacterial wilt, 195. *See also* pesticides
chicha, 57
Chile, 41, 59
chili peppers, 20, 21, 22–23
Chimborazo Province, Ecuador, 119–31
Chimú culture of Peru, 55
China: agricultural reform in, 147–48; fertility rates in, 155; food supply in, 156; local agriculture in, 265–68; potato production in, 5, 31–32, 33; soybean domestication in, 20; sweet potato in, 251–58
chips, potato: consumption of, 45–46; in Egypt, 187, 188; Indian potatoes for, 215; manufacturers of, 48; varieties for, 47
chips, sweet potato, 238–39
Chola variety of potato, 104, 106
Christianity, and time cycles, 15
chuño, 58, 77
Chuquisaca Province, Bolivia, 85
Cieza León, Pedro de, 55
CIP (International Potato Center): in Bolivia, 84; breadth of research of, 230; in China, 252, 256; in East Africa, 196; in Ecuador, 105, 113; and late blight resistance, 167; local agricultural focus of, 277–78; methods of, 272–74; mission of, 61–62, 202; pest management projects of, 170–76; potato accession collection by, *77, 78–79;* and true potato seed, 205, 216, 221; and virus testing, 222
class, social: and land distribution in Ecuador, 103–104, 115; and potato status, 3. *See also* social structure
classification of potato, 39–41, 76–77
clean seed issue, 85, 163, 195–96, 198, 273
climate: and Andean zones, 55–59, 60, *89;* and corn in Europe, 23–24; of East Africa, 193, 194; of Ecuador, 104, 112, 120; for ideal potato, 37; of India, 211–12, 213; of Ireland, 25, 136; production projects based on, 89; and rice suitability in Asia, 31; for sweet potatoes, 229, 230. *See also* weather
coastal valleys, 59
Cobo, Bernabé, 27, 57

coffee, origins of, 19
cold storage in India, 212
collection vs. domestication of plants, 11
collective agriculture, 146–48
Collins, Wanda, 167
Communism and land reform, 130, 146–47
communities: and individualism in Ecuador, 115; seed production leadership in, 123, 125, 130; for seed production projects, 122. *See also* local agricultural considerations
CONDESAN (Consortium for the Sustainable Development of the Andean Ecoregion), 113
consciousness, human, and astronomy, 15
Consortium for the Sustainable Development of the Andean Ecoregion (CONDESAN), 113
Consultative Group on International Agricultural Research (CGIAR), 62
consumption: in Bolivia, 85; in Ecuador, 106; in Egypt, 178; in Ireland, 25; and lost crops, 69–70, 75; in Poland, 33; of sweet potato, 248–49; urban concentration of, 48–49; in U.S., 43, 45–46, *47*
Cook, Sherbourne, 26
cooking. *See* preparation
cooperatives, and Ecuadoran farming, 117
corn: domestication of, 12, 20, 22; and Incas, 56–57; Old World popularity of, 23–24; origins of domestication of, 20, 22; production of, 34; yields of, 34
Cortés, Hernán, 63
COTESU (Swiss Agency for Technical Cooperation), 84
cottagers, Irish, 25, 139, 144
Counterblaste to Tobacco (James I), 39–40
cowpeas, origins of, 19, 20
CPRI (Central Potato Research Institute), 211
Crissman, Charles, 106, 114–15, 116
crop-lien system, 145
cropping systems. *See* agriculture
Cruza variety of potato, 196–97, 200–201
Cuba, 130
cucumbers, origins of, 20
cultivation: by early Europeans, 25; ease of, 136–37, 139; in East Africa, 197–98; in Ecuador, 104–106; in Egypt, 183; of family garden, 37–38; in India, 219–20; of lost crop roots and tubers, 64–65, 70–73; of

cultivation (*cont.*)

    manioc (cassava), 235–36; minituber method of, 223; and Solanaceous family, 41; of sweet potato, 230–31, 239, 243, 247, 248; by TPS method, 224, 257; in Tunisia, 175

cultural issues: and family planning, 277; in food production and population growth, 150–51, 152–57; in Indonesia, 244, 245–47; Spanish-Indian relations as, 63–64; and sweet potato preparations, 239; and symbolism of potato, 42. *See also* social structure

Cuzco, 53

dairy products, 20

Davis, Kingsley, 149

day laborers, Ireland, 137

de Bary, Anton, 162

DECIS (deltamethrin), 171, 175

dehydrated potato flakes, consumption of, 45–46

delayed-return strategies, 10

deltamethrin (DECIS), 171, 175

demographic transitions and population dynamics, 149–57

Deng Xiaoping, 148

Desirée variety of potato, 99

Devaux, André, 86, 91

developing countries: fungicide usage in, 163; growth of potato in, 5; pest management vs. insecticides for, 170–76; population dynamics in, 155–56; and potato variety choices, 46. *See also individual countries*

Diamond variety of potato, 187

diffused-light storage silos, 126–27, *127*, 197, 198

Dirt (Logan), 3

diseases, human: and Irish potato famine, 140; and malaria, 197; in New vs. Old World, 26–27; and starvation, 150

diseases, plant: and Andean vulnerabilities, *89;* and aphid populations, 211; and bacterial wilt, 195–96, 257; diversity as weapon against, 161–62; fungicides for, 111, 163–64, 198; horizontal resistance to, 166–67; and mosaic virus, 236; natural enemies approach to, 177–84; of sweet potato, 231; and true potato seed, 207. *See also* late blight; viruses

diversity, ecological: in Andean roots and tubers, 76; of andigena and tuberosum species, 77; vs. consumption levels, 75; and disease resistance, 161–62; and endangerment of potato, 60–75; and minituber success, 217; and preservation of Andean plants, 113–14; and preserving potato, 78–80; of sweet potato, 233; and vertical climate compression in Andes, 55–56, 60; working with, 262

Dokolo, Uganda, 238–39

domestication of plants, 11. *See also* agriculture

draft animals, Americas' lack of, 23

drought, and sweet potato, 238–39

drought tolerance, in Bolivia projects, 88

drug substances from plants, 10–11, 22, 64–65

Early Rose variety of potato, 44–45

East Africa. *See* Africa

ecotourism, 117

Ecuador: achira in, 74; and Chimborazo project, 119–31; and climatic tiers, 59; fungicide usage in, 163–64; geographical overview of, 102–103; history of, 103–104; late blight in, 162; local agricultural considerations in, 263; lost crop consumption in, 70; pesticide use in, 106–108; potato cultivation in, 104–106; and seed multiplication project, 119–31; and watershed project, 113–18

Ecudal, 106

education and training: CIP focus on, 273; in East Africa, 198–99; in Ecuador, 117, 119–20, 121, 123; in Egypt, 182–83; importance of, 264–65; in India, 213, 220–21; in Indonesia, 241–50; in late blight control, 168–70; and population dynamics, 155; and sweet potato, 238, 239, 257; in Vietnam, 224

eggplants, origins of, 19, 20

Egypt: and ancient astronomy, 16; and biological controls project, 177–84; french fries in, 186–89; and true potato seed in, 209

El Angel, Ecuador, 115

El-Bedewy, Ramzy, 178–79, 182, 184, 188, 209

electrification, rural, 126

ELISA (Enzyme-Linked Immunosorbent Assay), 222, 255

Elly, Samuel, 139
El Niño weather effects, and late blight, 167
emigration, potato famine, 142, 144
endangered crops. *See* lost crops
engineering, Incan, 55
England. *See* Britain
environmental issues in pesticide usage, 170–71, 179
environmental plasticity, 78
Enzyme-Linked Immunosorbent Assay (ELISA), 222, 255
Epstein, T. Scarlett, 276
erosion problems, Ecuador, 113–14, 115, 125
Esperanza variety of potato, 121, 125
Espinosa, Patricio, 69, 70
Essay on the Principle of Population (Malthus), 135, 151
Estenssoro, Victor Paz, 83, 84
Estrada, Nelson, 87, 93
Europe: vs. Americas, 18–28; lack of true potato seed in, 208; pesticide usage in, 107; population dynamics in, 135, 151, 152–53; potato production in, *32*, 33; yields in, 38

Fahem, Mohammed, 174–75
family farms. *See* small farms
family size and population dynamics, 152, 280
farmer field schools, 241–50
farmers: and individualism in Ecuador, 116; and large vs. small in Ecuador, 117; and pesticide safety rules, 108; and reaction to minitubers, 220; and small vs. large, 145–46, 165, 166. *See also* education and training
Farmers Alliance, 146
Farm Fry Corporation, Egypt, 186–89
fast-food market, 45–46
fava bean, 20
feed, animal. *See* animal feed
female literacy and fertility rates, 155
Fermoy workhouse, 140
fertility rates: China, 155; and food production, 149–50, 151–52; in France, 143; in Ireland, 136; in Japan, 157; and property subdividing, 138; trends of, 156; in Tunisia, 173
fertilizer, 220, 255
fibers, from Americas, *21*
flour, sweet potato, 231, 240, 249

flowers, solanaceous family, 40–41
flush-out system, 193–201
foliage, potato, toxicity of, 40
food production and population, 4–5, 135, 149–57
food security, 155, 235–40
foot plow *(taklya)*, 60
Forbes, Greg, 162, 164
forests, *57*
FORTIPAPA potato project, 105, 106, 108, 119–31, 263
France, 47–48, 143, 153
Franco, Javier, 98, 99, 101, 263
free-trade ideology and potato failure, 136, 140, 142
French, Edward, 162
french fries: in China, 254; in Ecuador, 105; in Egypt, 186–89; in Europe, 47–48; in U.S., 43, 44, 45
fresh vs. frozen potatoes, and french fries, 47–48
fresh vs. processed potatoes, 33–34
Fripapa variety of potato, 105, 106, 111, 128
Frito-Lay, 106
frost: and Bolivian potato projects, 87–88, 89; and late blight, 163; search for tolerance to, 92–94
frozen-potato industry, 45, 47–48. *See also* processed potatoes
fruits, from Americas, *21*
Fry, William, 165
Fund for Agricultural Development (FUNDAGRO), 113
fungicides, 111, 163–64, 198

Gabriela variety of potato, 106, 121, 125
García, Willman, 92, 93, 94, 95–96
Garhbeta Block Office, India, 220–21
Garnet Chili variety of potato, 44–45
GDP, United States, 278–79
Gendarme variety of potato, 95, 99, 101
gendered labor division in agriculture. *See* women
genetic engineering: and early plant domestication, 11–12; and potato, 269–76. *See also* breeding; diversity, ecological
geographical range, 41
germ plasm bank, 68–69, 273–75
gin, 142
ginger, origins of, 20

lowland tropics and root/tuber cultivation, 12–13. *See also* yungas
Lozada, Sánchez de, 92
lunar calendars, 15

maca roots, 59, 64, 65
Machu Picchu, 54
Madkour, Magby, 181
Mahendra Farm, 215–16
Maine potatoes, 44, 46, 47
Majín, Luis, 129–30
malaria, 197
Malaysia, infant mortality in, 153
Malthus, Thomas, 4, 135, 138, 148, 149, 151
manioc (cassava), 12–13, 235, 238
Mann, Jagbir Singh, 221
Mann, Sarla, 221
Mantaro valley, Peru, 277–78
Mao Zedong, 147, 148
marketing of crops, 70, 218–19. *See also* consumption
marriage patterns, 152, 157
mashua tubers: collection of, 62; consumption of, 70, 75; cultivation and preparation of, 64, *66*, 68–69, *69*
maturation times: of andigena species, 104–105; average of, 37; in East Africa, 198–99; and factors affecting, 120–21; in India, 211, 213–14; and minituber method, 217–18; and sweet potato, 237, 247, 256–57; and tuber moth, 172–73
mauka roots, 57, 62, *66*, 71, 75
Mayans, 15, 16
medicinal uses for plants, 10–11, 22, 64–65, 68
men: and division of labor, 101; traditional farming work of, 60
Merino, Fausto, 119, 121, 126, 130
Mesoamerica: main crops of, 22; tomato and, 24, 40
mestizo farmers, 104, 116–17
metalaxyl, 165
Mexico, trade capabilities vs. Peru, 63–64
Michigan potatoes, 46
microclimates in Andes, 59
microwave ovens, 43, 45
mid-altitude mountain valleys, 55, 56, *57*, 89, 94
Midnapore district, India, 213, 218–19
military regimes, Bolivia, 84
Millardet, Alexandre, 162

millet, origins of, 19
mining, Bolivia, 83, 84
minitubers, 216–21, 223
Minnesota potatoes, 47
Mira River valley, Ecuador, 114
Mitchell, John, 141
Moawad, Galal M., 179
modernization of agriculture, and small farmers, 145–46
Mojkerto district, Indonesia, 242–43
Mondial variety of potato, 174
Monte Verde, Chile, 18
Moore, David, 162
moral issues: and Irish potato famine, 136, 138–39; and potato reputation, 142
mortality rates and population dynamics, 151–52, 153–54, 156
Mortimore, John, 208
mosaic virus, 236
mother plants for seed production, 199–200
multiplication of seed: in East Africa seed program, 198–200, *200*, 202–204, *203*; in Ecuadoran project, 119–31; and minituber advantages, 218; in sprout cuttings method, 126, *127*, 128–29; and sweet potato vines, 255

Namulonge Research Station, Uganda, 235, 237
naranjilla, 40
narcotic plants and potato, 39–40
NARO (National Agrarian Research Organization), 195, 235
National Agrarian Research Institute (INIAP), 104, 105, 106
National Agrarian Research Organization (NARO), 195, 235
National Institute for Agronomic Research (INRAT), 170–76
National Performance Trials (NPT), 233
National Revolutionary Movement (MRN), Bolivia, 83
Native Americans. *See* indigenous peoples of Americas
natural enemies approach to pest control, 177–84, 241, 242
Navia, Oscar, 163–64
nawalla, 183–84
Ndolo, Philip, 239
Near East and origins of agriculture, 19, 20

Santa Isabel variety of potato, 128
savannas, *57*
savings and loan crisis, U.S., 285
Schmiediche, Peter, 224
scientific nature of agriculture, 11, 138
Scotland, potato failure in, 143
seasonal agricultural cycles, 14–17, 53
security, food, 155, 235–43
seedbed training programs, 101–102
Seed Growers Association, 197
seed multiplication. *See* multiplication of
  seed
seed production: in Africa, 193–204; in Bolivian projects, 85–86, 86–90; in East Africa programs, 194; in India, 211–25; local
  control over, 96–97; and sweet potato,
  233, 256–57; of true potato seed, 205–25;
  in Tunisia, 174
seed quality, importance of, 261–67
sharecroppers and land distribution, 146
Shepard, Merle, 253
Shepody variety of potato, 47
Shobol Association, 129
Sichuan province, China, 252–55
sierra, Ecuadoran, 103, 105
small farms: Andean structure of, 60, 117; decline of, in U.S., 44; in Ecuador, 106, 120;
  late blight costs for, 166; as major potato
  producers, 165; and modernization of agriculture, 145–46; perspective of, 5–6. *See
  also* land distribution
smallpox, 26–27
Smit, Nicole, 237, 239
smoking, 22, 39
social structure: and cultural symbolism of
  potato, 42; and disease effects on native
  populations, 27; and ease of potato cultivation, 136–37; in Ecuador, 105, 114, 115,
  120, 123; and food production and population, 150, 152–57; in India, 218; in Ireland, 25, 144; and land reform, 145–48;
  and potato in Asia, 33; and potato in
  U.S., 43; and sweet potato preparations,
  239; and urbanization in U.S., 44. *See also*
  cultural issues
soil building techniques, 215, 257. *See also*
  erosion problems; rotation, crop
Solanaceae family, 39–41
Solanum genus, 41, 76–77
solar calendars, 14–15

sorghum, origins of, 19, 20
Soroti district, Uganda, 237–40
South, the American, land distribution in,
  145–46
South America, as origin of sweet potato,
  229–30. *See also individual countries*
South Korea, land reform in, 131
South Pacific, 19, 230
Soviet Union, land reform in, 146–47
soybeans, 19, 34
Spain, 24, 26–27, 62–64
specialization in cultivation
  andmanufacturing, 44, 48–49
species, potato, 41
spiders, as natural enemies, 241, 242
Spiritwalker (Wesselman), 17
SPK 004 sweet potato vines, 233, 239
sprout cuttings method, 126, 127–28, *128*
Spunta variety of potato, 174
squashes, 20, *21*, 22
Sri Lanka, infant mortality rates in, 154
Sriyono, 247–48
Stalin, Joseph, 146
starch: achira as source of, 74, 75; and potato
  nutrition, 36; and sweet potato, 229–34,
  251, 253, *254;* and tuber moth devastation,
  188
starvation: as population control, 150; in Soviet Union, 146. *See also* Ireland
stimulants, *21*, 22, 39
stone age culture, 9–11, 14–15
Stonehenge, 14–15
storage: in Bolivia, 86; in diffused-light silos,
  126–27; in East Africa, 197; in Ecuador,
  128; in Egypt, 183–84; and food security,
  237; and grain advantage, 13; in India,
  211–12, 219; and lost crop capabilities, 64;
  of sweet potato, 238; and tuber moth
  problems, 94; in Tunisia, 174–75; in Vietnam, 224; virus issues of, 85, 222
The Story of B (Quinn), 15
striga weed, 236
subdividing of property, 137–38, 143
sugarcane, origins of, 19
sugar substitute, yacón as, 73
Sumerians, 16
super seed, 222–25
sweet potato: in Africa, 229–34; in China,
  251–58; ease of cultivation of, 12–13; vs.
  potato, 41–42, 230–31, 232; preservation

Vietnam, 223–24, 238, 243
village life, perspective of, 5–6
vine cutting, virus-free sweet potato, 255
viruses: and aphids, 211; in Bolivia, 85, 95; and lost crops, 68; and mosaic, 236; as pest control, 175, 177, 178–80, 182, 184; and TPS advantages, 222; types of, 222
virus-free sweet potato cuttings, 255
Vitamin A deficiency, and sweet potato, 233
vodka, 25

Wallace, Alfred Russel, 15
waraji beverage, 239
warm valleys. *See* yungas
Washington state, and Russet-Burbank, 46
watershed project, Ecuador, 113–18
Waych'a variety of potato, 93
weather: El Niño effects on, and late blight, 167; and potato disease propagation, 162. *See also* climate
websites, 285–86
weevil: in potato, 108–12, 122, 125; in sweet potato, 231, 237, 239, 247
Wesselman, Hank, 17
West Bengal, India, 210–21
West Java, Indonesia, 242–50
wheat, 19–20, 34, 43, 211
Wheatley, Chris, 251, 253
Whole Art of Husbandry (Mortimore), 208
Wiggins, John, 138
wild species of potato, 20, 76–80, 88

Willis, N. P., 137
Wiyanto, 244
women: and division of labor, 101; in East Africa, 198–99; in Ecuador, 127–28; and literacy and fertility rates, 155; Quechua-speaking, 124; in sweet potato production, role of, 235, 238, 239; work of, 60
workhouses, Irish, 140
Wumeng Agricultural Institute, 257

yacón roots, 57, 62, *67, 73, 74*
yams, 10, 12, 19, 230
yautia (cocoyam), 12–13
yields: in Bolivia, 85; in East Africa, 195; in Ecuador, 104, 105; of general potato, 34; in India, 213; of Maine vs. Idaho potatoes, 44; by minituber method, 216–17, 223; of potatoes vs. grains, 38; and seed certification, 85–86; of sweet potato, 231, 243, 255–56; and true potato seed, 205
Young, Arthur, 25
Yukon Gold variety of potato, 47
yungas: and late blight problems, 94; and lost crop cultivation, 73; as tropical produce source, 59; and tuber moth problems, 89; and vertical climate compression, 55, 56, *57*
Yungay variety of potato, 167
Yusuf, Mohammed, 195

Zhang Dapeng, 256

JAMES LANG is associate professor of sociology and former director of the Center for Latin American and Iberian Studies at Vanderbilt University. He is the author of four other books, including *Feeding a Hungry Planet,* which is about rice production, and *Inside Development in Latin America,* which is about community-focused, self-help projects.

ISBN 1-58544-138-4